New Frontiers in Nuclear Physics

New Frontiers in Nuclear Physics

Lecture Notes of
JSPS-INS International Spring School

Shimoda and Tokyo, Japan 1993

Editors:

S. Homma
Y. Akaishi
M. Wada

Institute for Nuclear Study, University of Tokyo
3-2-1 Midori-cho, Tanashi, Tokyo, 188 Japan

World Scientific
Singapore • New Jersey • London • Hong Kong

Published by

World Scientific Publishing Co. Pte. Ltd.

P O Box 128, Farrer Road, Singapore 9128

USA office: Suite 1B, 1060 Main Street, River Edge, NJ 07661

UK office: 73 Lynton Mead, Totteridge, London N20 8DH

NEW FRONTIERS IN NUCLEAR PHYSICS
Lecture Notes of JSPS–INS International Spring School

ISBN 981-02-1617-3

Cover design by courtesy of *S. Keiko*

Printed in Singapore by Utopia Press.

Preface

"The JSPS-INS International Spring School on New Frontiers in Nuclear Physics" was held at Shimoda and Tokyo from March 8 through 19, 1993. The spring school, the first of its series, was sponsored by the Japan Society for the Promotion of Science and the Institute for Nuclear Study, University of Tokyo. The aim of the school was to stimulate young Asian physicists toward frontiers of today's nuclear physics under free and informal atmosphere.

The lectures at the school covered a wide range of subjects in nuclear physics; nucleon structure, heavy-ion physics, high-spin states, symmetries, astrophysics with nuclei, and exotic atoms and nuclei. The school under these subjects was quite timely since the Institute for Nuclear Study has been strongly promoting "Japanese Hadron Facilities", in which nuclear physics is expected to play an important role. The lectures were very exciting and lively, thanks to the efforts of all the lecturers.

The Lecture Notes are the compilation of the papers based on these excellent lectures given at the school. We would like to express our hearty thanks to the lecturers for kindly accepting our requests to give the lectures and also for preparing the manuscript for publication.

We sincerely hope that the outcome of the school would significantly contribute to the future development of this field of physics in Asia.

Finally, we would like to express our thanks to Ms. K. Hata for her excellent assistance in editing these Lecture Notes.

T. Yamazaki
President of the School

S. Homma
Chairman of the Organizing Committee

S. Homma, Y. Akaishi, and M. Wada
Editors

Contents

Trends and Perspectives on Heavy Ion Physics

Paul Kienle

Physik-Department E12,
Technische Universität of München,
D-8046 Garching, Germany

Lecture 1

Heavy ion research has developed during the last decennium to a frontier science of nuclear physics. What is it all about? It focuses on investigations of processes induced by fast heavy ions passing through matter, such as nucleus-nucleus, ion-atom and ion-molecule collisions. Using fast heavy ions as a tool, heavy ion research is an interdisciplinary field of science with novel applications in physics, chemistry, biology and medicine. It is fast growing in its fundamental as well as applied aspects, and it contributes to the renaissance of nuclear physics.

1 A Historic Introduction to Heavy Ion Physics

The development of heavy ion physics is intimately related with the art of acceleration of nuclei of all elements to energies above the Coulomb barrier (6 MeV/nucleon for all ions except the lightest ones bombarding uranium as target). The birthplace of heavy ion physics is Berkeley, California. Luis W. Alvarez accelerated in 1940 with the cyclotron of E.O. Lawrence completely ionized carbon ions to 50 MeV energy. The cyclotron frequency ω_c of $^{12}C^{6+}$, $\omega_c = (q/A)B$, is similar to that of deuterons in a field B, because their specific charges $q/A = 0.5$ are about the same. Since then cyclotrons are in use for acceleration of heavy ions. The maximum specific energy T/A, which can be achieved with a cyclotron with maximum orbit radius R_m and magnetic field B_m is given by the simple formula

$$T/A = K(q/A)^2 \text{ with } K = 48.24(B_m R_m)^2 \tag{1}$$

B_m in Tesla, R_m in m.

The specific charge (q/A) should be as high as possible to reach high energies. Thus cyclotron acceleration has profited most from the development of high charge state ion sources, such as the electron cyclotron resonance (ECR) sources. One main avenue of heavy ion research was directed towards the synthesis of heavy elements. In 1958, Ghiorso et al, reported the synthesis of element 102, using a heavy ion reaction: ^{246}Cm(^{12}C, 4n) $^{254}_{102}$No, which caused a big, not yet completely resolved controversary with Flerov et al, who later synthesized Nobelium using also heavy ion reactions.

In 1974 Oganessian et al pioneered a different type of heavy ion reaction, termed "cold fusion" to synthesize heavy elements. Instead of bombarding the heaviest elements such as Cm with light projectiles, targets around the double closed shell nucleus ^{208}Pb are bombarded with medium heavy projectiles, such as ^{48}Ca, yielding a relatively "cold" compound nucleus at bombarding energies close to the Coulomb barrier. Thus only one or two neutrons have to be evaporated by the compound nucleus on its way to a particle stable residue. This reduces the strong fission competition, effective in each evaporation step compared with the usual "hot" 4n-evaporation channels.

On August 21, 1982, Armbruster and his group at GSI Darmstadt succeed using the cold fusion reaction, ^{209}Bi (^{58}Fe, n) ^{266}Mt to synthesize the heaviest element known up till now which was named recently Meitnerium, Mt, in honor of Lise Meitner.

A second important direction of heavy ion physics was started by Zucker and Wyly in 1953. They answered the question whether a hydrogen bomb explosion may burn the atmosphere and the water of the earth by systematic measurements of relevant ^{14}N-induced reaction cross sections, with a clear no. This was the beginning of ongoing investigations of low energy nucleus-nucleus collisions.

A third early arena of heavy ion studies has lead to a lively field of nuclear structure investigations. In 1959 Stephens and Diamond succeeded in the first observation of heavy ion induced multiple Coulomb excitation of nuclei and in 1962 Gugelot and Morinaga founded the "in beam" γ-spectroscopy following fusion reactions as a powerful tool to study high spin states in nuclei. This field is blooming in the US and in Europe, where very powerful high-resolution 4π-γ-detectors have been used and are further developed for interesting high spin studies. Where is the J-ball, somebody may ask looking at the Japanese scene. The "fifties" brought a new approach to heavy ion acceleration. The AEC contracted the construction of two drift tube linacs of the Alvarez type for acceleration of heavy ions, the HILAC in Berkeley and the HIA in Yale. Both accelerated heavy ions up to argon to energies of 10 MeV/nucleon. The HILAC was mainly used for nuclear synthesis and excitation of high spin states in nuclei, as already mentioned. The HIA was devoted mainly to the investigation of nuclear reaction mechanisms. In course of these experiments Kaufmann and Wolfgang discovered a new type of heavy ion reaction, called "Deep Inelastic Scattering". In semiperipheral collisions the two colliding nuclei form a dinuclear system for a short time, while rotating and exchanging all their kinetic energy above the Coulomb barrier and many nucleons, to separate again after a contact time of about 10^{-21} s. These dissipative reactions were investigated in detail during the seventies when heavier beams became available, first at ALICE a linac-cyclotron combination at Orsay, France and about at the same time at the Super HILAC in Berkeley, with which ions up to Xenon could be accelerated in the first construction phase.

Finally in 1975 the UNILAC at GSI Darmstadt, constructed by Schmelzer and his team, succeeded in acceleration of all ions up to uranium to energies of about 8.5 MeV/u. A broad research program started around the UNILAC including heavy element synthesis, structure of nuclei far off stability with studies of new decay modes, such as proton decay of nuclei at the proton drip line and high spin physics using multiple Coulomb excitation and inverse reaction kinematics for life time measurements. Dissipative reaction have been investigated intensively with the heaviest projectiles, such as uranium and lead, by making use of large area position sensitive detectors and kinematic coincidences. Interdisciplinary studies in the fields of atomic and biophysics were successfully started. Especially the investigations of high Z-hydrogen like ions, and quasi atoms with Z ranging up to the critical $Z(172)$, for which the binding energy becomes larger than 2 mc^2, the rest-mass of an electron-positron pair, were pushed forward. In the high Coulomb field of high Z-quasi atoms electron positron pair creation was observed and anomalous monoenergetic pairs were discovered.

During the "sixties and seventies" heavy ion research profited from the large Tandem accelerators with terminal voltages V up to 20 MV, which provided energy variable, high resolution, low emittance heavy ion beams with energies $T = V(1+q)$, in which q denotes the accelerated charge state after the terminal stripper. At the moment a laboratory in Strasbourg, France, works on a 35 MV Tandem accelerator. In order to increase the energy of Tandem beams, special linacs were constructed for post acceleration. For the linacs either superconducting cavities (Argonne, Stony Brook, Legnaro) high shunt impedance interdigital line cavities (Munich) or single resonators (Heidelberg) were developed and used at various installations. This led to a proliferation of low energy heavy ion beams with many nice nuclear reaction studies, such as nuclear molecular resonances, transfer reactions, inelastic scattering, fusion below the barrier, fission experiments and many more. Tandems and Tandem post accelerators were also widely used for in beam γ-spectroscopy.

Severe competitors for tandem generators to accelerate heavy ions, became various kinds of variable frequency cyclotrons with K-numbers up to 130. Especially in connection with the development of high charge state ion sources, these cyclotrons are even now a-days much used sources for low energy heavy ion beams.

During the eighties a rush to higher and higher energies of heavy ion beams started. Large sector focussing cyclotrons were built in Berlin (HMI), CAEN (GANIL) and Tokio (RIKEN), with K-numbers up to 400. About at the same time superconducting compact cyclotrons were developed at various institutions (Michigan State University, Chalk River, Milano, Orsay, Groningen) with K-numbers up to 800. All these cyclotrons profit also from high charge state ion sources.

These large cyclotrons were especially used for the production and study of the properties of hot nuclear matter and for nuclear fragmentation experiments to produce radioactive beams such as ^{11}Li.

During the midseventies projects began to make use of illustrious, old p-synchrotrons for acceleration of heavy ions, such as the Bevatron in Berkeley with the Super-HILAC as an injector and the synchrophasotron in Dubna. Ions with $q/A = 0.5$ could be accelerated to 2.1 GeV/u and 4.1 GeV/u respectively, though with rather limited beam qualities. After an upgrade the Bevalac was able to accelerate all elements up to uranium to energies of about 1 GeV/u. With the heavy beams first collective effects in nucleus-nucleus collisions were found, which gave some hope that the equation of state of nuclear matter at elevated temperatures and densities may be studied. Fragmentation studies, particle production below threshold and Dilepton spectroscopy complemented this pioneering Bevalac program until its termination in 1993. It must be also mentioned that a very successful cancer therapy program has been initiated in Berkeley using light ion beams and many first atomic physics experiments, in the field of atomic processes and spectroscopy has been carried out using fast heavy ion beams.

The BEVALAC program is now taken up and developed further by GSI Darmstadt using high quality heavy ion beams from SIS/ESR, a heavy ion synchrotron in combination with a cooler ring, taken into operation in 1990. SIS-beams with energies up to 2 GeV/u and ions up to uranium are used for programs to study dense, heated and to nucleon resonances excited hadronic matter using global reaction investigations as well as particle production experiments. Using the beam-fragmentation method radioactive beams for various nuclear structure studies far off stability are produced. Multifragmentation of nuclei was studied under various conditions, including explosion type of reaction channels in the most central collisions. Relativistic heavy ions are used for Coulomb excitation of high lying excited states, such as giant resonances including multiphonon excitation. In the ESR heavy ion beams of stable as well as unstable nuclei are cooled to the highest possible phase space densities. Using such novel methods boundstate β-decay was observed for the first time, laser transitions between the hyperfine states of hydrogen like Bi, and X-ray transitions of hydrogen like uranium were precisely measured.

Still higher energy heavy ion beams were produced using the AGS at BNL and the CERN accelerator complex including the SPS. In 1986/87 ^{16}O and ^{32}S beams were accelerated at CERN to energies up to 200 GeV/u. In 1992 Au-beams were brought up to 10 GeV/u in Brookhaven. Starting from 1994 CERN will have 160 GeV/u Pb-beams, and around the turn of the century RHIC a Relativistic Heavy Ion Collider at BNL will provide nucleus-nucleus collisions at $\sqrt{s} = 200$ GeV in the nucleon-nucleon system, and if the LHC at CERN will be built heavy ion collisions at $\sqrt{s} = 7$ TeV may be studied.

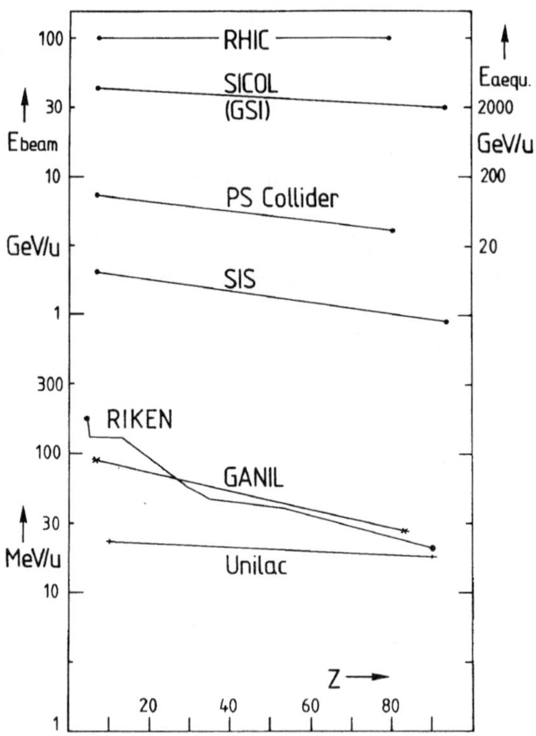

Fig. 1. Beam energy per nucleon versus atomic number for some selected heavy ion acceleration facilities. For collider projects like RHIC the equivalent fixed target beam energies are given. SICOL and PS-Collider are concepts which are not realized. The LHC collider project with beam energies of 3.5 TeV/u should be added. In the meantime Au beams were accelerated in the AGS to 10 GeV/u. CERN expects to have Pb-beams from the SPS with energies of 160 GeV/u by end of 1994.

The main goal of studying very high energy collisions is to produce hadronic matter at a very high energy density for which a transition to the quark gluon plasma is predicted.

Fig. 1 shows a presentation of the beam-energy per nucleon versus atomic number characteristic of some typical heavy ion accelerator facilities. Linacs and cyclotrons dominate the low energy regime up to 100 MeV/u, whereas above this energy synchrotrons provide beams up to 200 GeV/u, such as the CERN-SPS, which will accelerate Pb-beams to energies of 160 GeV/u in late 1994.

For the study of nucleus-nucleus collisions at still higher energies heavy ion colliders are needed. At KEK a PS heavy ion collider with beams of 4-7 GeV/u has been discussed. GSI started a conceptual study of a collider with energies up to 50 GeV/u. None of these projects may be realized. But as mentioned before, BNL constructs a double ring collider using superconducting magnets (B_m = 3.8 T) in an existing 3.8 km circumference tunnel to reach maximum energies of 100 GeV/u per beam (RHIC). CERN's intention to accelerate also heavy ions in their proposed LHC to $\sqrt{s}$ = 7 TeV has already been mentioned. The most difficult problem with colliders is concerned with reaching adequate reaction rates. We will mention this luminosity question in the next paragraph.

2 The Main Tools of Heavy Ion Research: Universal Heavy Ion Acceleration Systems.

In the following we will discuss in detail the universal heavy ion acceleration complex at GSI-Darmstadt; UNILAC/SIS/ESR, which provide heavy ion beams up to 2 GeV/u with high phase space densities and luminosities for certain experiments. Before considering in more detail the applied acceleration and cooling methods, we like to make a remark concerning high luminosity heavy ion colliders which offer the only cost effective way to study very high energy nucleus-nucleus collisions. The c.m. energy of a collider is given by $\sqrt{s} = 2E_B$ with E_B denoting the beam energy. The luminosity L is given by the expression

$$L = \frac{N_1 \cdot N_2 \cdot f \cdot \gamma}{4\pi \langle\sigma_1\rangle\langle\sigma_2\rangle} \tag{2}$$

$N_{1,2}$ are the number of ions per colliding bunch, f is the frequency of the colliding bunches $4\pi\langle\sigma_1\rangle\langle\sigma_2\rangle$ denotes the collision area and γ the relativistic factor. The luminosity can be increased by optimalisation of the number of ions per bunch, the bunch collision frequency and by minimalisation of the collision area. This has to be done against the large space charge forces which drive the highly charged beams into instabilities. The only way to achieve this is by applying phase space cooling. We will discuss one method, electron cooling, which is very efficient for not too high energies. For high energies the only method developed until now is stochastic cooling, for coasting beams. What one would like to have for high energy colliders would be the development of cooling for bunched beams also, which means an efficient suppression of the coherent beam signals from the revolving bunches.

In following we like to present the UNILAC/SIS/ESR-complex of GSI, which is an optimized heavy ion acceleration facility for energies up to 2 GeV/u, producing beams

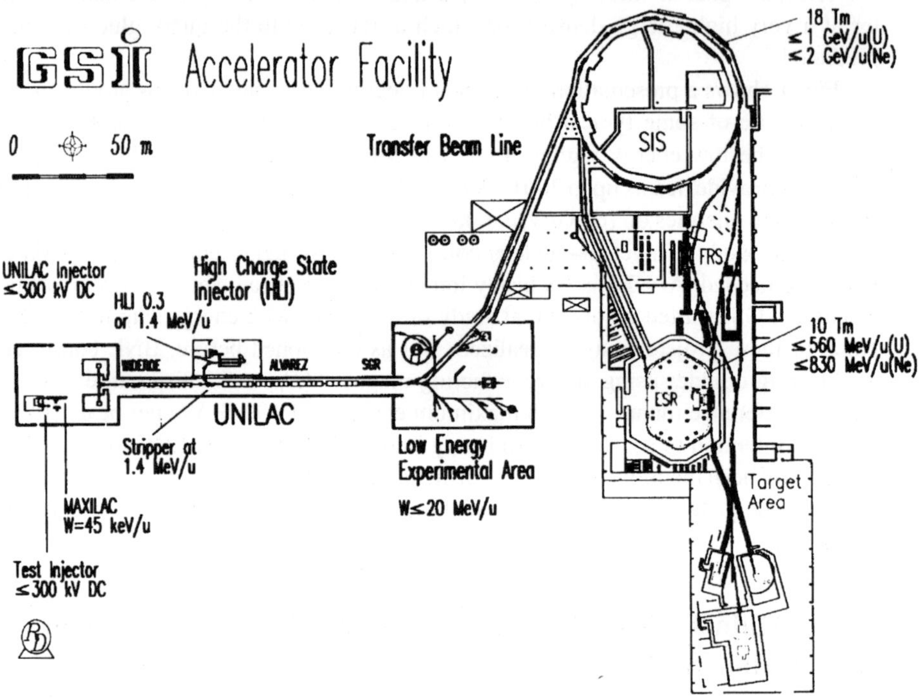

Fig. 2. GSI heavy ion accelerator facility, a combination of the UNILAC, a heavy ion synchrotron SIS and the experimental storage ring ESR. The post stripper section of the UNILAC has two injectors, a Wideroe type linac with stripping at 1.4 MeV/u and a recently built high charge state injector (HLI) with an ECR source, RFQ and IH-linac, The low energy experimental area is served with energy variable heavy ion beams up to 20 MeV/u. The high energy target area with 3 caves is served with beams from SIS (U<1 GeV/u) and cooled beams from the ESR (U< 560 MeV/u). The cooled beams of the ESR may be reinjected in SIS and accelerated to higher energies (U = 1.3 GeV/u).

of completely stripped ions, and radioactive beams of high phase space density. Fig. 2 shows the general lay out of this facility, which consists of an universal linear accelerator the UNILAC, capable of acceleration of all elements to energies up to 20 MeV/u, a medium energy heavy ion synchrotron with an energy up to 2 GeV/u, depending on the specific charge of the injected ions, and the experimental storage ring ESR, which is designed to store and cool heavy ion beams up to energies of about 560 MeV/u.

These three accelerators are connected in various ways. The UNILAC may serve as an injector into SIS, accelerated and bunched beams from SIS, may be transferred to the ESR and after cooling reinjected into SIS for further acceleration.

The UNILAC is a universal linear accelerator, with three different accelerator sections. A Widerroe type low β-structure, which accelerates low specific charge state ions ($q/A \approx 0.04$) out of Penning ion sources from 12 keV/u up to 1.4 MeV/u. For more effective acceleration in 4 Alvarez type linacs, the specific charge state is raised to $q/A \sim 0.1$ by passing the ions through a gas stripper target and is then accelerated to 11.4 MeV/u. This is also the standard injection energy into SIS. For energy variation several separately phased single resonators can be used for further acceleration or deceleration with any combination of Alvarez tanks. With an overall length of 120 m a maximum energy of about 20 MeV/u may be reached.

In order to run a low energy program parallel to a low duty factor ($< 1°/_{\infty}$) SIS- injection with independent choice of ion species and energy, a second prestripper injector into the Alvarez sections of the UNILAC was built. This high charge state injector makes use of recent developments of high charge state ion sources and advanced acceleration techniques. It consists of a 14.5 GHz ECR-ion source, a radiofrequency quadrupole (RFQ) linac for energies up to 300 keV/u, followed by a high shunt impedance interdigital line linac up to 1.4 MeV/u. The ECR-ion source delivers up to 5 μA $^{28+}$U ions, the specific charge of which is high enough to be accelerated by the Alvarez sections without further stripping. The RFQ and interdigital line acceleration structures operate at the same frequency as the Alvarez linac (108 MHz), so that each oscillation can be loaded with particles (9 ns microstructure). The macro structure duty factor can be up to 50%. For the heaviest ions, the Alvarez tanks limit it to 20%. For SIS injection several methods are studied to raise injection currents by a factor of 100-1000 relative to presently available currents from Penning-sources operated at low duty factor. The most elegant solution seems to be based on magnetically pulsed ECR-sources, with which the necessary pulse current amplification was observed in a first test.

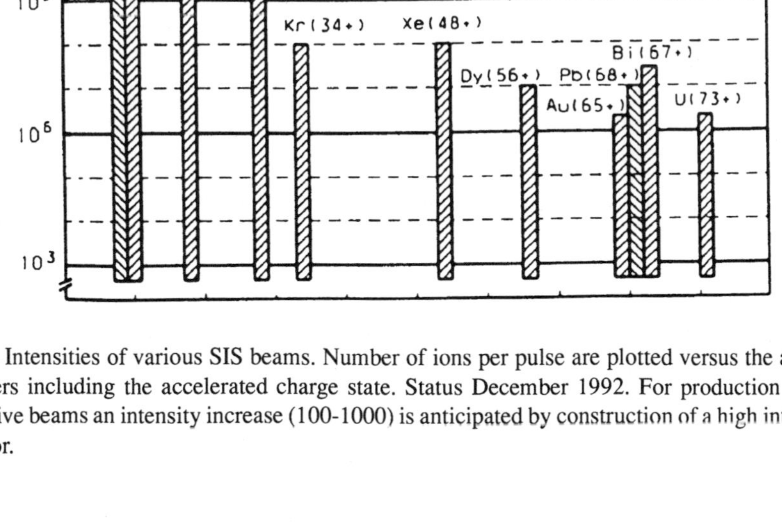

Fig. 3. Intensities of various SIS beams. Number of ions per pulse are plotted versus the atomic numbers including the accelerated charge state. Status December 1992. For production of radioactive beams an intensity increase (100-1000) is anticipated by construction of a high intensity injector.

In order to achieve high currents from SIS, multiturn injection is applied with about 20 turns effectively. Four bunches are formed with the r.f. followed by a synchronous acceleration up to a maximum bending power of 18 Tm with a repetition rate of about 1 Hz. Following acceleration, the bunches can either be directly transferred to the ESR, with stripping to high charge states in between, or slowly extracted within a few seconds and transferred to the experiments in the target hall. Fig. 3 shows the maximum intensities obtained so far with various ions accelerated in SIS including the limits set by incoherent space charge effects at 11.4 MeV/u injection energy.

The synchrotron is connected with the ESR a storage ring with a maximum bending power of 10 Tm, half the circumference of SIS, a large aperture magnetic lattice, r.f. cavities for acceleration, deceleration and bunching, and a powerful electron cooling device. The lattice of the ESR is shown in Fig. 2. It has two straight sections one of which contains the cooler and the other a gas jet target for experiments with the circulating cooled ion beam. The ESR is operated at a vacuum of 10^{-11} mb in order to allow long storage times.

3 Electron Cooling of Heavy Ions.

Phasespace cooling of protons, stored in a ring and brought in interaction with a cold electron beam of a certain length through which they pass repeatedly was first proposed by Budger in Novosibirsk and successfully demonstrated with a small test ring. A detailed treatment of electron cooling and first experimental experiences with various cooler rings is given by Poth. Here we like to give only a very elementary treatment of electron cooling to illustrate some important basic features of this novel method.

Let us consider a stored ion beam, which interacts in a straight section of a storage ring with a "cold" electron beam of well defined velocity $\vec{v_e}$, electron density n_e over a length L. The electron beam is always renewed, while the ion beam with a velocity distribution $f(\vec{v_i})E$, with $\vec{v_i} \approx \vec{v_e}$ passes repeatedly through the electron target.

In the reference frame of the cold electron beam $\vec{v_e} = 0$, the ions have a statistical distribution of their velocities $f(\vec{v_i})$ which may be associated with a temperature. Due to Mott-scattering of the ions on the cold electrons, the electrons gain energy just as the ions loose it. As the electrons are always renewed, the ions "cool" down due to the energy loss, in the electron target with $\vec{v_e} = 0$ density n_{e} and length L_e which we can easily estimate using Bethe-Bloch's formula:

$$-\frac{dE}{dx} = -\frac{\vec{F}d\vec{x}}{dx} = \frac{4\pi z_i^2 e^4}{m_e v_i^2} \cdot n_e L_e \tag{3}$$

From this we can derive an expression for the cooling force $\vec{F}$:

$$\vec{F} = \frac{4\pi Z_i^2 e^4}{m_e} n_e L_e \cdot \frac{\vec{v_i}}{v_i^3} \tag{4}$$

Assuming an electron velocity distribution $f(\vec{v_e})$, we can derive a more realistic expression for the cooling force

$$\vec{F} = -\frac{4\pi Z_i^2 e^4}{m_e} n_e L_e \int \frac{(\vec{v_i} - \vec{v_e})}{|\vec{v_i} - \vec{v_e}|^3} f(v_e) d\vec{v_e} \tag{5}$$

The result its shown schematically in Fig. 4 in which F is plotted versus v_i. For zero ion velocity in the cold electron reference frame the cooling force vanishes to increase proportional to v_i like a friction force. For $v_i \gg v_e$ the cooling force drops with $1/v_i^2$.

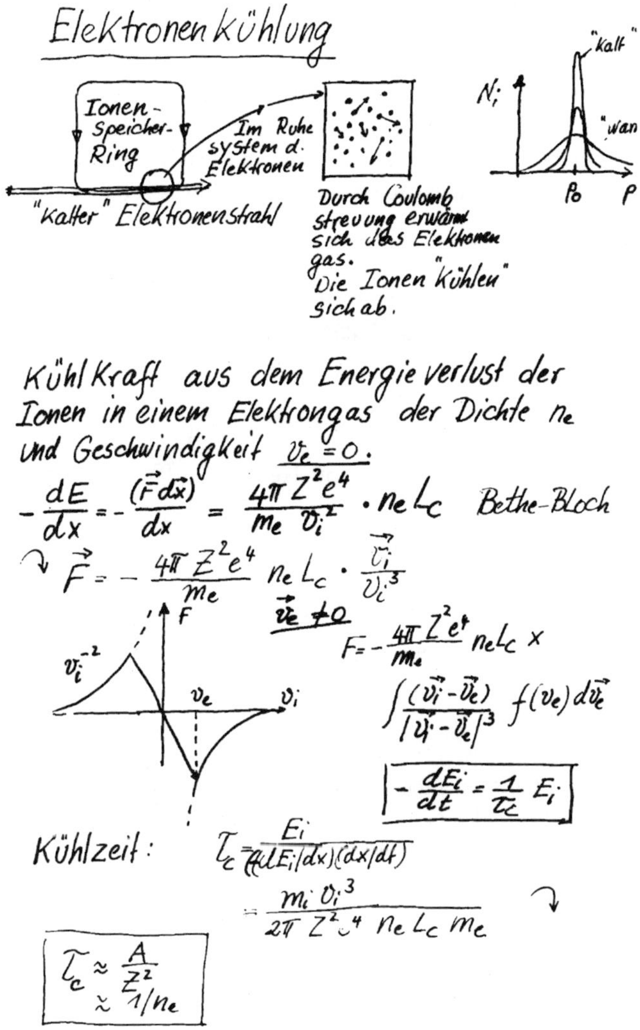

Fig. 4. A simple explanation of electron cooling and derivation of the cooling force taken from an original german lecture of the author (english version see text).

The rate of change of the ion energy E_i can be related to the cooling time τ_c using the expression

$$-\frac{dE_i}{dt} = \frac{1}{\tau_c} \cdot E_i \tag{6}$$

from which τ_c is derived

$$\tau_c = \frac{E_i}{-(dE_i/dx)(dx/dt)} = \frac{m_i v_i^3}{2\pi Z_i^2 e^4 n_e L_e m_e} \tag{7}$$

The cooling time scales with A/Z^2 and is inversely proportional to the area density of the electron beam n_e L_e .

Thus completely stripped high Z ions are cooled faster than low Z ions. In the following some results of heavy ion cooling experiments from the ESR will be presented.

Fig. 5 shows some cooling force measurements, of electron cooling experiments at the ESR with an electron density n_e = 10^6 cm^3 and 150 MeV/u Ne^{10+} and 230 MeV/u Bi^{82+} beams. Qualitatively the expected behavior of Fig. 4 seems to be reproduced.

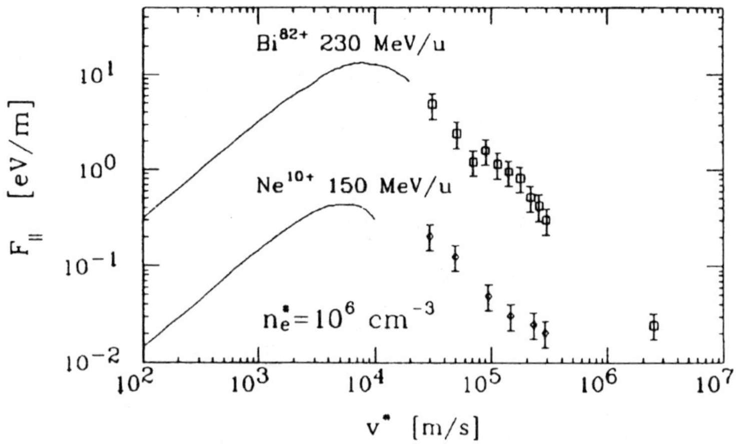

Fig. 5. Cooling force as function of the longitudinal relative velocity of Ne^{10+} at 150 MeV/u and Bi^{83+} at 230 MeV/u normalized to $n_e = 10^6$ cm^3. Straight lines are smoothed data, from measurements of the particle distribution with heating by rf noise, data points were determined by high voltage stepping.

Information concerning the velocity distribution of cooled coasting beams can be obtained from the Schottky-noise frequency spectrum of the current fluctuations induced in capacitive pick up plates, given as

$$s(\omega)_n = \int \langle \gamma I_n(0) \gamma I_n(\tau) \rangle \exp(i\omega\tau) d\tau \tag{8}$$

The expression in brackets is the autocorrelation function of the current fluctuations. A maximum of the noise power is found at the nth harmonic of the revolution frequency, which can be measured precisely. For small particle numbers one can readily deduce also the momentum spread $\Delta p/p$ of the beam which is simply related to the frequency spread $\Delta f/f$ in the Schottky noise spectrum using $\Delta f/f = \eta\gamma(\Delta p/p)$ where the parameter

$\eta = 1/\gamma^2 - 1/\gamma_{tr}^2$ contains the relativity γ-factor and the transition γ-factor of the ESR lattice (γ_{tr} = 2.67). Fig. 6 shows the results from cooling 150 MeV/u Kr^{6+} using a 80 keV electron beam of 1 A. The Schottky noise frequency signal of the 40th harmonic of the particle revolution frequency reveals a width of about 35 Hz at 56 MHz, which corresponds to a momentum spread $\Delta p/p \sim 10^{-6}$. This is by more than three orders of magnitude smaller than the width of the initially injected beam of $\Delta p/p \sim 2 \times 10^{-3}$. In addition small emittances of the order of 0.1π mm mrad have been observed for the cold beams.

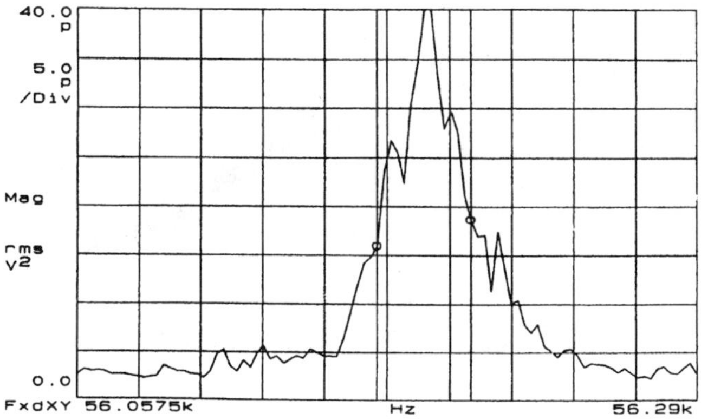

Fig. 6. Schottky noise frequency spectrum of a coasting $^{86}Kr^{36+}$ beam at 150 MeV/u with about 10^5 ions in the ring cooled with 1 A electrons. $\Delta f/f$ = 35 Hz/56 MHz measured in the 40th harmonic.

Another important characteristic of a heavy ion cooler ring is the recombination rate of the highly charged ions in the ring, because it limits the storage time of the ions. With ultra high vacuum in the ring, the storage time is only limited by the recombination rate of the ions in the electron cooler. For completely stripped ions two processes may contribute to the recombination. Radiative electron capture, with a cross section, which is given approximately by the expression

$$\sigma_R \cong 2 \times 10^{-22} \cdot \frac{E_k}{nkTe} \text{ cm}^2 \tag{9}$$

E_k = K-binding energy, n = main quantum number, Te = electron temperature
It scales with Z^2 of the ions. A second process is a kind of dielectronic recombination of the cooling electrons in the Coulomb field of the ion. One electron interacts with a second one, such that one electron is scattered to a higher energy continuum state and the other is captured by the ion. It is the inverse process to electron impact ionization and can be calculated this way. It becomes important for high electron densities in the cooler, high Z-ions and very low velocities. Although there is some indication of the contribution of this process from recent observation of characteristic recombination X-

rays from the cooler, the main recombination process seems to be radiative electron capture.

Fig. 7 shows measurements of total recombination rates, by detection of the charge changing process in the cooler target of the ESR using charge state analysis with the bending magnet following the cooler. The cooler current was 1 A for the Ar^{18+} and 0.5 A for the Au^{79+} beams. The measured recombination rate scales linearly with the electron current density and with the square of the ion charge Z as expected for radiative recombination. This is shown in Fig. 8 which displays the recombination rate normalized to I_e = 0.1 A as function of Z. While the recombination rate is linearly increasing with

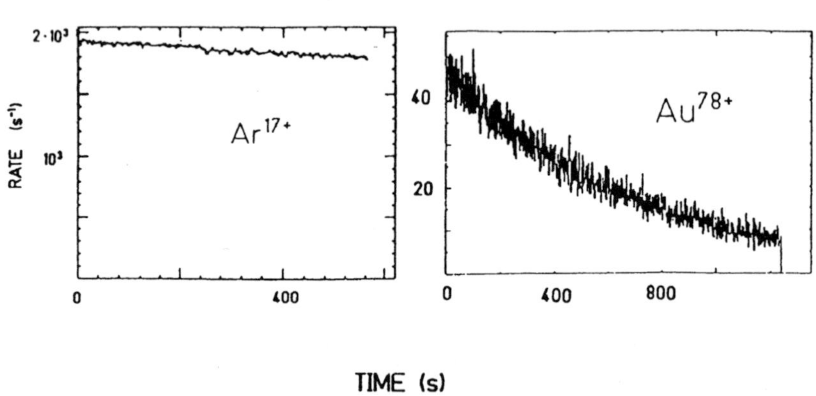

Fig. 7. Lifetime of Ar^{18+}(a) and Au^{79+}(b) beams due to radiative capture in the cooler section as measured by the rate of Ar^{17+} and Au^{78+} respectively, impinging on a particle detector in the deflection section behind the cooler. The cooler current was 1 a for Ar^{18+} and 0.5 A for Au^{79+} beams.

I_e, only a weak influence of I_e on the momentum spread and the transverse emittance has been observed. This observation allows already an efficient cooling with electron currents as small as 0.1 A. From an extrapolation of the results displayed in Fig. 8 one expects storage times for bare U^{92+} ions of the order of one hour, with effective electron cooling, This was recently indeed observed. Thus we conclude electron cooling works well for medium energy bare ions up to uranium.

The large aperture of the ESR magnet lattice allows the simultaneous storage and cooling of several charge states. This unique feature is manifested in Fig. 9 which shows a Schottky noise spectrum of cooled Bi^{82+}, Bi^{81+} and Bi^{80+} ions of 240 MeV/u circulating in the ESR. Bi^{82+} has the highest circulation frequency, because it revolves on the most inner orbit. The lower charge states of Bi are populated by successive electron capture processes in the cooler device. They revolve on outer orbits which is reflected by lower revolution frequencies. Note, that all ions independent on their charge states are cooled to the same velocities. Therefore the relative frequency differences $\Delta f/f_q$ are determined only by orbit length differences ΔL due charge change differences Δg:

$$\Delta f/f_q = \Delta L/L_q = \gamma_{tr}^{-2}\Delta q/q. \tag{10}$$

As $\Delta q/q$ is precisely known γ_{tr} can be measured with precision. This information is also needed for calibrating mass measurements, which can be carried out with cooled radioactive nuclei stored in the ring.

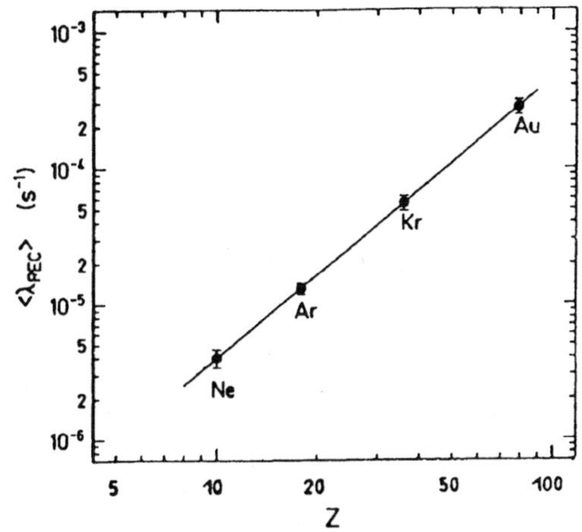

Fig. 8. Experimentally determined recombination constants λ_{REC} for various completely stripped ions stored and cooled in the ESR. The recombination constants are normalized on $I_e = 0.1$ A.

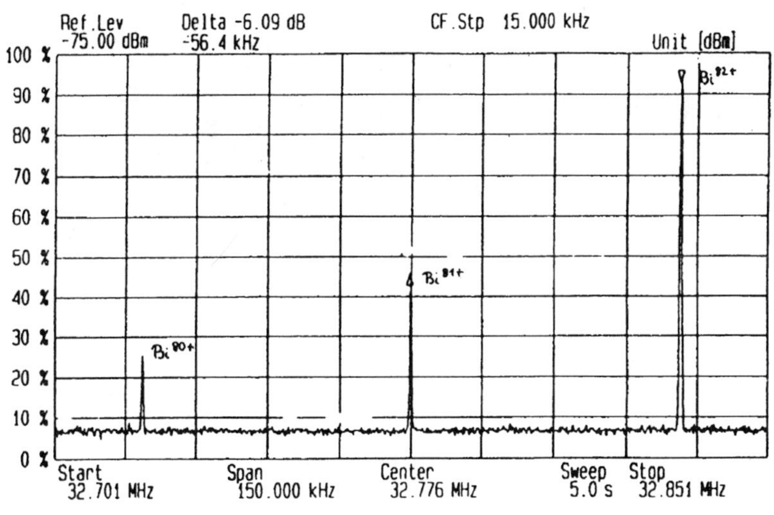

Fig. 9. Schottky noise frequency spectrum of three charge states of Bi-ions of 240 MeV/u being cooled in the ESR-ring.

Lecture 2

A chinese student gave after lecture 1, the following comment: "What are all these cyclotrons for? Where is the physics?" I shall try to answer his questions in the second lecture. But before doing so I take the privilege to give the following comment: "Van der Meer had to invent the stochastic cooling method, and CERN had to build coolerrings for antiprotons and convert the SPS into a Sp-S$\bar{p}$-collider, before Carlo Rubbia with UA1, a huge detector could find the Z_0 and the W-boson".

In the following we will talk about some selected topics of heavy ion physics, such as; synthesis of heavy elements, nuclear dynamics up to 2 GeV, and bound β-decay.

4 Synthesis of Heavy Elements

There is strong interest in studying properties of nuclei far off stability. One of these research efforts is focussed on studying the properties of nuclei of the heaviest elements another one is to study nuclei at the proton drip line. We shall talk first a bit about heavy element research.

Fig. 10 shows schematically the various approaches used for the synthesis of the heaviest elements. The classical one is based on multiple neutron capture in an as high as possible neutron flux followed by β-decay. The heaviest nucleus synthesized in this way is ^{257}Fm. The standard heavy ion fusion reaction for heavy element synthesis is the bombardment of heavy, mostly radioactive, targets with relatively light projectiles.

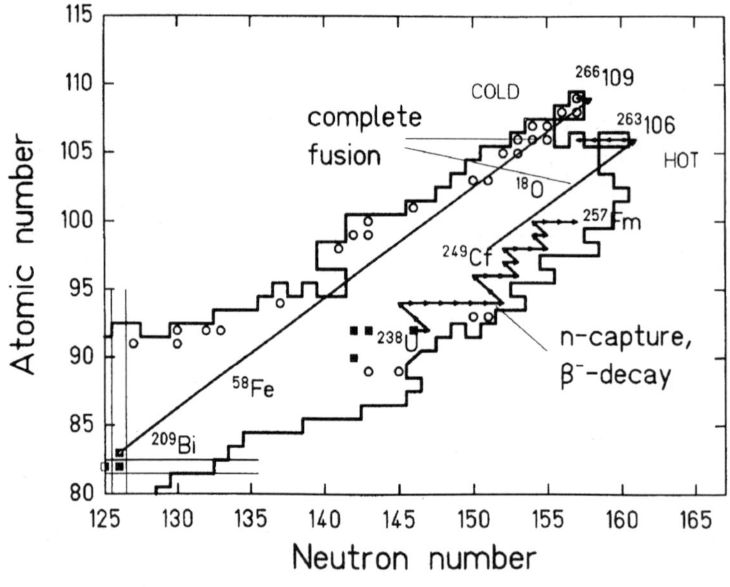

Fig. 10. Illustration of three classical ways for the synthesis of heavy elements: Cold and hot complete fusion and multiple neutron capture.

The heaviest element synthesized in this way is ^{263}106, using the ^{249}Cf (^{18}O, 4n) ^{263}106 reaction. The disadvantage of this type of reaction is the high excitation energy brought into the nucleus at bombarding energies close to Coulomb barrier, which leads to a 4n evaporation channel. The residue production cross section is limited by the fission competition in each neutron evaporation step. These reactions are termed "hot" fusion reactions.

In contrast to this Oganessian et al introduced the cold fusion reaction, which allowed the synthesis of the heaviest elements. In the cold fusion reaction, targets around the doubly magic ^{208}Pb-nucleus are bombarded with medium heavy projectiles. The excitation energy of the compound nucleus at the Coulomb barrier is so low, that only one or two neutrons may be evaporated to generate a particle stable residual nucleus. The heaviest nucleus known until now was synthesized at GSI with the cold fusion process, ^{209}Bi(^{58}Fe,n) ^{266}109.

Fig. 11 shows the measured heavy element production cross sections for cold fusion reactions as function of the entrance fissility parameter. Note the strong reduction of the fusionability with increasing Z. It is the hindrance for fusion at the barrier which limits the usefulness of this reaction type for the synthesis of still heavier nuclei.

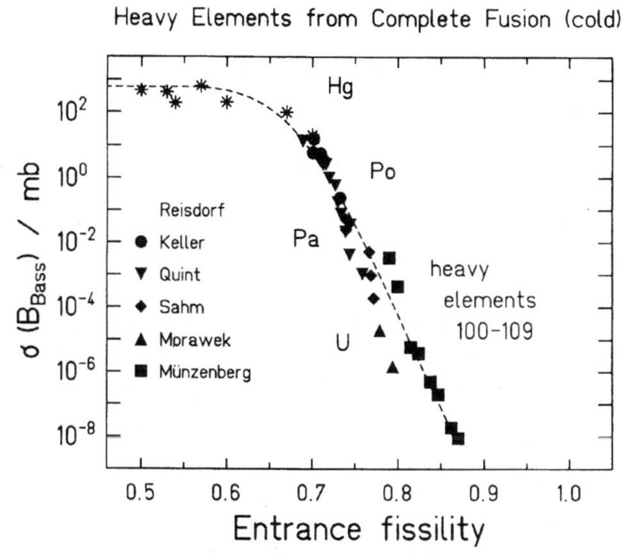

Fig. 11. Fusion cross sections as function of the entrance fissility parameter.

In order to identify nuclei with production cross sections in the order of 10 pb a combination of methods, which permits the identification of one nucleus, is useful. Fig. 12 shows the scheme of a Seperator for Heavy Ion fusion Products, SHIP, used at GSI for the heavy element synthesis work. An intense beam of heavy ions, with energies chosen to optimize the synthetisation yield, is focussed on a thin target, (0.5 mg/cm^2) which is rotated in order to decrease the beampower deposited into the target area. The evaporation residues (ER) emitted in beam direction have velocities $V_{ER} = v_p(Ap/A_{ER})$ which

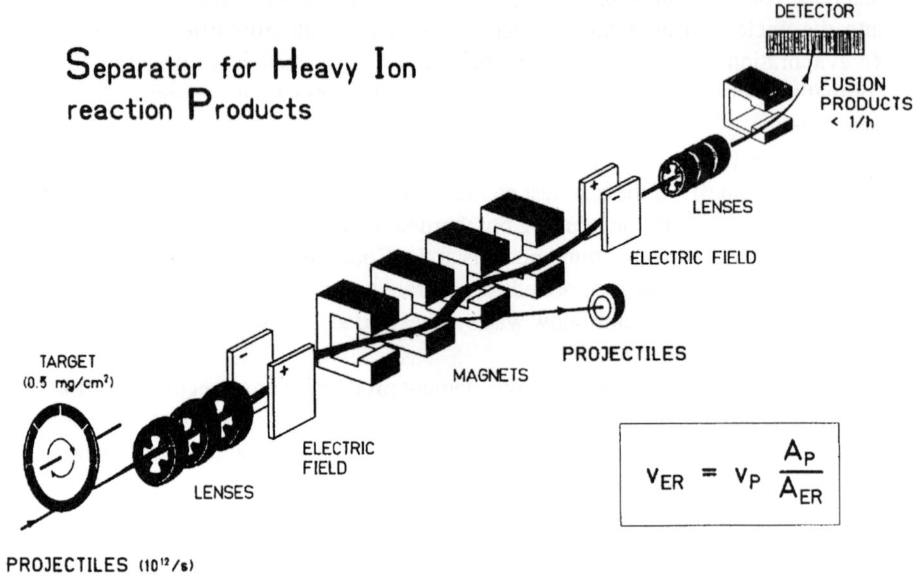

Fig. 12. Scheme of SHIP a separator for heavy ion reaction products.

are smaller than the projectile velocity v_p by the ratio of the projectile to the evaporation residue mass (Ap/A_{ER}). This velocity difference is the basis of the separator which is a kind of double Wien filter composed of crossed electric and magnetic fields. The fields are set such that the slow evaporation residues pass the filter, whereas the faster projectiles become deflected. Recently a magnetic deflection was added which makes the instrument also a mass separator. At the focal plane a position sensitive silicon detector may be places, which records the time and position of the implanted ER and the consecutive decay chains. For each decay the position, time and decay energy, in case of α-decay or fission, is marked. If the decay chains of the daughter nuclei are known, the mother may be identified with a single event.

Fig. 13 shows the end of the chart of nuclei with the elements Ns (Nielsbohrium) Hs(Hassium) and Mt (Meitnerium) for which the indicated isotopes were synthesized using cold fusion reactions and their decay characteristics were determined. From the α-decay energies the masses were determined from which the fission barriers B_f of the heaviest elements could be reliably deduced and plotted in Fig. 14 for nuclei with $N - Z$ = 48 as function of the element numbers. The dashed curve is what one expects for B_f from a liquid drop model. As B_f decreases rapidly with Z, heavy elements would be unstable against fission. Yet the observed fission barriers are not decreasing, but range between 6 and 7 MeV. As a consequence no spontaneous fission has been observed from the heaviest nuclei synthesized at GSI. Alpha emission is the main decay mode. The half life of ^{266}Mt is about 3.4 ms with a decay energy of 11.10 MeV.

Theoretically fission barriers of the order of 6 MeV are predicted for nuclei in this region, caused by shell effects (dotted line). In fact even higher fission barriers are pre-

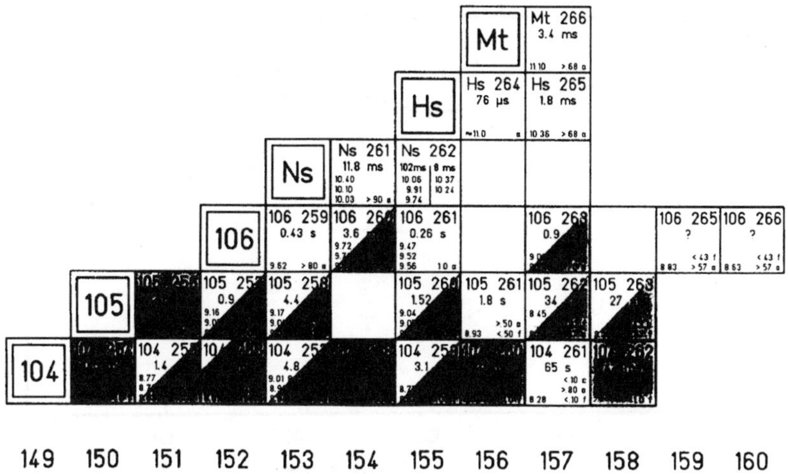

Fig. 13. Chart of nuclei for the heaviest elements known: Ns(Nielsbohrium) Hs(Hassium) Mt(Meitnerium).

dicted ($B_f \sim 9$ MeV) for nuclei with $Z \sim$112 and $N \sim$(174-6), forming a island of spherical shell stabilized nuclei, Until now only the sandbank of this island has been reached. Is it possible to synthesize these heavier more stable nuclei? In principle yes, but the cross section for the cold fusion reaction drops exponentially with Z, to values of around several pb for $Z = 110$. At GSI an experiment with an improved solid angle of SHIP and better detection efficiency is in preparation for synthesis of element 110 using the ^{62}Ni + ^{208}Pb $\rightarrow$ ^{269}110 +n reaction. Should this be successful one may try to produce ^{270}111 using a ^{209}Bi target and the same beam. As the α-emitting daughter decay chains are known, identification should present no new problem. How can one synthesize more neutron rich nuclei and heavier ones? There are several ideas being discussed, with some unsuccessful attempts to carry them out. One is to return to the hot fusion reactions using heavy targets, with better identification techniques. Attempts for synthesis element 110 were not successful until now. Another older idea is to “overshoot” the nucleus to be synthesized and use (HI, xn yn) type of reactions for the production. Multinucleontransfer reactions may still be another approach yet the basis of it is not well understood.

Finally we turn briefly our attention to the study of nuclei at the proton drip line. After the discovery of some proton emitters in the $Z > 50$ region at GSI and Munich during the early eighties, the field did not progress until recently when fusion product mass separators in combination with double sided Si-microstrip detectors were used at Daresbury and later at ANL to tackle these difficult experiments again. Fig. 15 shows energy spectra measured in an experiment using the Daresbury mass separator to separate A = 160 nuclei produced by bombarding a ^{106}Cd target with 300 MeV ^{58}Ni-ions. Fig. 15a displays all decays observed in the A = 160 focal plane region and Fig. 15b those occurring within 3 ms of the implantation of the ion into the same area of the

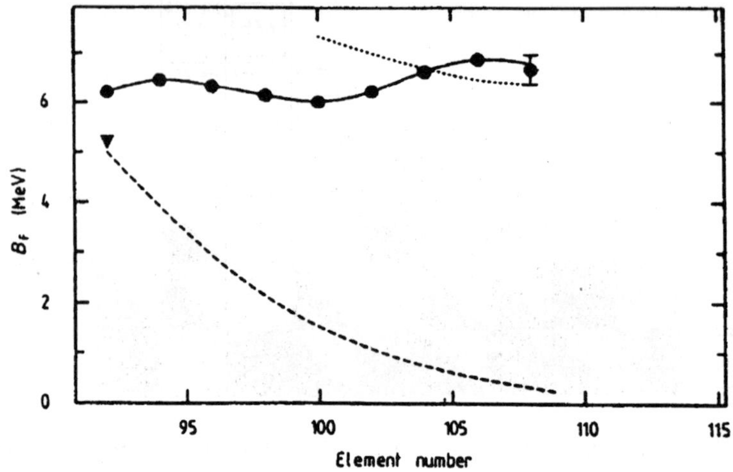

Fig. 14. Fission barriers B_f in MeV for heavy nuclei. Data points are measured values for $N-Z$ = 48 nuclei determined from α decay energies. Dashed line is the prediction of the liquid drop model, the dotted line shell corrections are included.

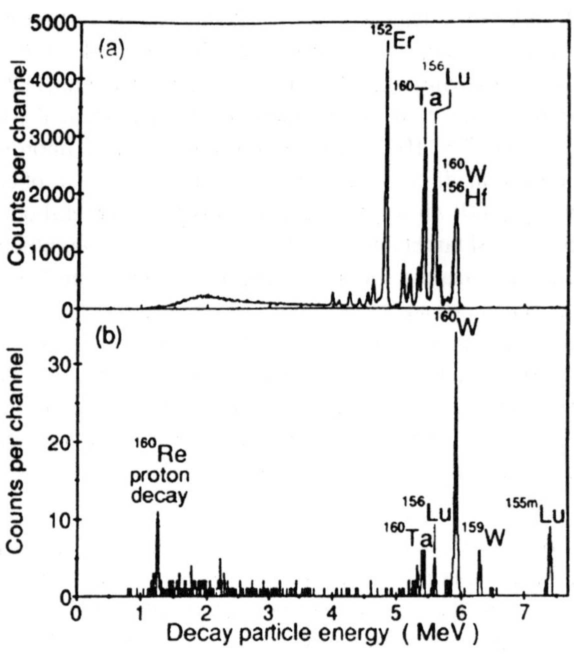

Fig. 15. Particle spectra recorded from A =160 mass nuclei produced by bombarding ^{106}Cd with 300 MeV ^{58}Ni-beams, separated by a mass spectrometer and implanted in a double sided microstrip detector. Fig. 15a displays all the decay particles observed while Fig. 15b only those within 3 ms of the implantation.

double sided Si-micro strip detector. The assignment of a 1.26 MeV proton decay line to ^{160}Re is based also on the observation of the ^{159}W daughter α-decay line with equal intensity in the mass region A = 160. The ^{155m}Lu α-decay peak arises from A = 155 ions implanted in the A = 160 focal point region in a lower charge state.

5 Nuclear Dynamics up to 2 GeV/u

In this chapter we shall discuss some selected topics of nuclear dynamics in the energy regime between 100 MeV/u and 2 GeV/u of current interest. First we will focus on peripheral collisions and describe some projectile fragmentation experiments with the final aim to produce cooled radioactive beams, measure their masses and study their reactions. High energy Coulomb fission will be presented shortly as a new powerful tool to produce and investigate heavy neutron rich nuclei. Exclusive Coulomb dissociation experiments with neutron rich radioactive beams, such as ^{11}Li will be presented as techniques to explore n-correlations in low density neutron halo nuclei. Finally the observation of multiple Coulomb excitation of giant dipole resonances using relativistic ion beams will be presented.

Then we turn to the phenomena on of multifragmentation and its observation using exclusive inverse reaction kinematic experiments. Finally we focus on central collisions with two subjects singled out: the observation of nuclear explosion producing fragments with high radial velocities, and the observation and study of "resonance matter" in subthreshold particle production experiments.

5.1 Nuclear fragmentation and electromagnetic dissociation

Many laboratories are engaged in the study of nuclear fragmentation and electromagnetic dissociation of fast projectiles pioneered by the LBL-Osaka-Tokyo group. The motivations are the production and study of neutron rich and deficient nuclei, the production of fast and cooled radioactive beams, the study of the structure of neutron halo nuclei and of nuclei far off stability using precision reaction spectroscopy with cooled beams and inverse quasielastic reactions.

Fig. 16 shows pictures of the nuclear fragmentation and electromagnetic dissociation reactions. In a nuclear fragmentation reaction light target nuclei such as ^{12}C are bombarded with fast heavy projectiles like 700 MeV/u ^{136}Xe. In a peripheral collision with ^{12}C some nucleons are chipped of ^{136}Xe and "Any" large fragment keeps on moving in the beam direction with about the projectile velocity.

We will show how one can separate these fragments from the many beam particles. For electromagnetic dissociation the target is a high Z-nucleus, such as ^{208}Pb. At close distance an intense flux of high energy virtual photons can excite collective modes in the projectile. This phenomenon may be utilized in various ways, such as to induce photofission in heavy nuclei, to excite multiphonon states or to investigate collective modes in neutron rich halo nuclei.

For separation of projectile fragments from beam particles a separation principle is applied as shown in Fig. 12, which was introduced with the fragment separator LISE at GANIL and which is used in a somewhat modified version at higher energies at the GSI

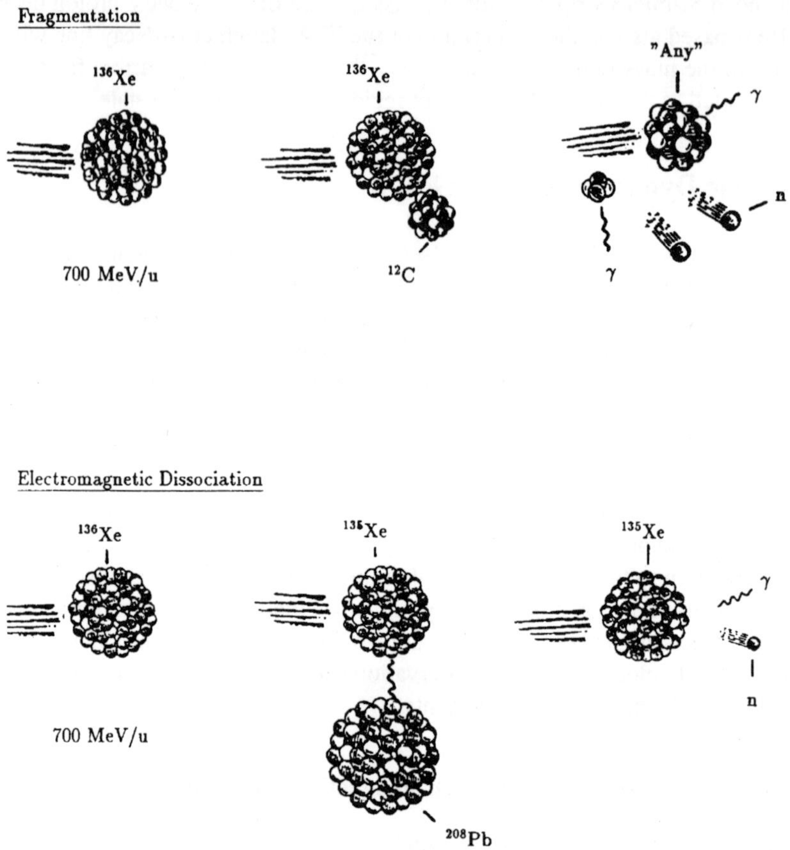

Fig. 16. Sketches of fragmentation and electromagnetic dissociation reactions induced by fast ^{136}Xe-nuclei (700 MeV/u) on ^{12}C- and ^{208}Pb-targets respectively.

with the fragment separator FRS. Fig. 17 shows schematically how a beam of ^{18}F-nuclei is separated from projectile fragmentation of a 500 MeV/u ^{40}Ar beam hitting a light production target. The first two dipole magnets focus all $A/Z = 2$ fragments through the second slit of the spectrometer achromatically. A time of flight TOF versus AE spectrum at this point is shown by the first insert. The beam contains various $A/Z = 2$ fragments including a small fraction of ^{18}F. This beam is passed through a specially formed degrader (achromatic) which introduces energy losses proportional to Z^2 for the various ions. The following spectrometer focuses then only one species with a certain Z in our case ^{18}F through the last slit of the spectrometer. ^{18}F is a radioactive beam which may be stored and cooled in the ESR to high phase space densities. This is illustrated in Fig. 18. In this case a 250 MeV/u ^{20}Ne beam was fragmented and all $A/Z = 2$ products including the original beam were transported through the FRS, which has no achromatic absorber introduced. This mixed beam was accumulated and cooled in the ESR. Fig. 18 shows Schottky noise frequency spectra for these mixed cooled beams in the ESR, taken

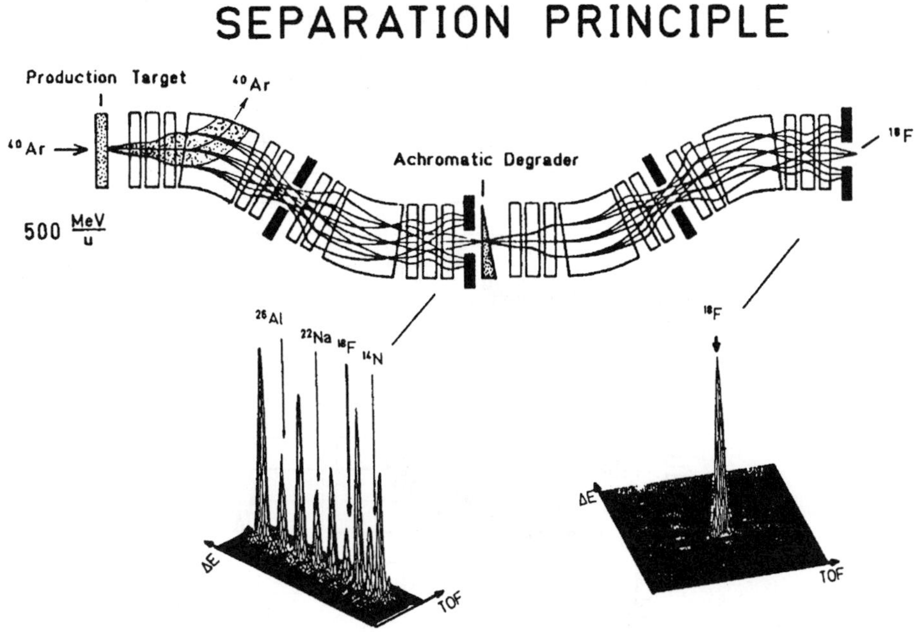

Fig. 17. Separation principle used at the FRS to separate projectile fragments from beam particles. The first stage separates in an achromatic mode according to Z/A. This is shown in the first insert in which the energy loss versus time of flight of the products which pass the first analyzer shits is shown. The spectrometer is set on $Z/A = 0.5$. The products pass then through an achromatic degrader in which they loose energy proportional to Z^2 . One isotope (^{18}F) is then separated using the second stage.

directly after injection and 12 hr later. Below the huge ^{20}Ne peak at slightly lower revolution frequency (lower value of Z/A) one recognized a narrow line ($\Delta f/f \sim 10^{-5}$) which corresponds to ^{18}F. After 12 hr this line has vanished because most of the ^{18}F has decayed. Note the very sharp line at the lowest frequencies; it most probably corresponds to α-particles. For low Z and low number of particles the Schottky signals are very narrow and can be therefore easily picked out of the noise. Plans exist at GSI to produce as a next step cooled ^{56}Ni-beams in the ESR to study collective excitations by inelastic scattering, such as the M1 spin flip excitations via (p p$\prime$) scattering in inverse reaction kinematics. A beam of medium energy cooled ^{56}Ni-ions will hit a hydrogen jet target and the proton recoils will be detected with Si-microstrip counters.

Next we like to discuss the results of an experiment in which a ^{238}U beam is fragmented and after selecting secondary beams of ^{237}U, ^{236}U and ^{235}U, their photofission was studied. Fig. 19 shows intensity distribution of the secondary beams in the final focal plane of the FRS including the corresponding photofission mass spectra. Such type of experiments can also be used to measure fragmentation cross sections of radioactivities, relevant for waste disposal on light targets in a very elegant way.

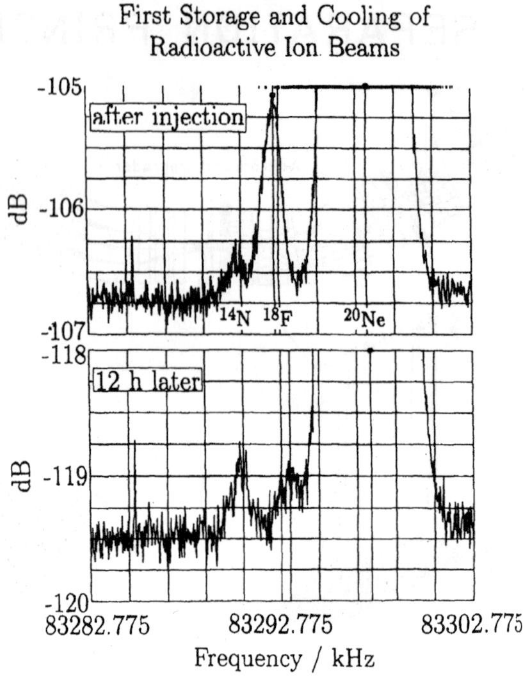

Fig. 18. Schottky noise frequency spectra of cooled ^{18}Ne beams (250 MeV/u) and fragments with $A/Z \approx 2$. Note the cooled radioactive beam ^{18}F after injection and its disappearance 12h later.

A result of a high resolution study of photofission of a 750 MeV/u beam of ^{238}U using a 1.2 g/cm^2 lead target is shown in Fig. 20. The fragment separator FRS was set to filter out fission products emitted in forward direction to pass through a multisampling ionization chamber. As one notes in the ΔE spectrum of the ionization chamber all elements in the light but especially also in the heavy fission group are easily resolved. This new technique can be applied to study photofission and fragmentation fission of heavy elements and to produce, identify and study neutron rich nuclei.

Next we like to present first results of Coulomb excitation experiments using fast heavy ions. The characteristic frequency of the exchanged virtual photons can be easily estimated using the expression

$$\hbar w \sim \hbar/\tau_{coll} = \hbar c \cdot \gamma \cdot \beta/R \tag{11}$$

with τ_{coll} = collision time, $\gamma = E/m_o c^2$ ~2-3 for SIS $\beta = v/c = \sqrt{\gamma^2 - 1}/\gamma$, R = nuclear radius.

It follows that for SIS energies the virtual photon spectrum with characteristic frequencies $\hbar\omega$ of about 20 MeV is ideally suited to excite giant resonances in nuclei. The virtual photon flux is so high, that one expects not only single excitation, but also measurable fractions of multiple excitations. Fig. 21 shows a schematic lay out of the set up used to study Coulomb excitation using fast heavy ion beams from SIS at GSI and exclusive detection of the energy and momenta of all outgoing particles.

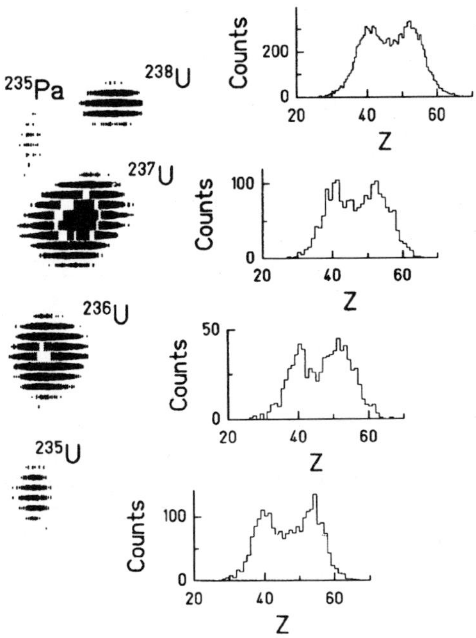

Fig. 19. Intensity distribution of ^{238}U fragmentation products in the focal plane of the FRS (left) and their corresponding fission mass distribution measured after several targets.

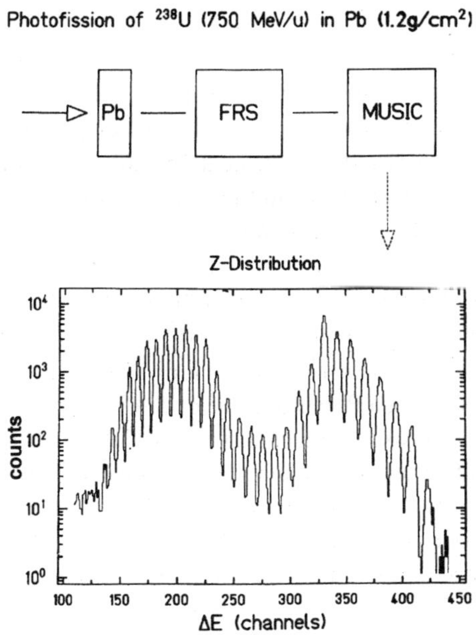

Fig. 20. Z-distribution of products originating from photofission of 750 MeV/u ^{238}U, measured with the FRS (Z/A) cut) and a Δ E determination using a multiple sampling ionization chamber (MUSIC).

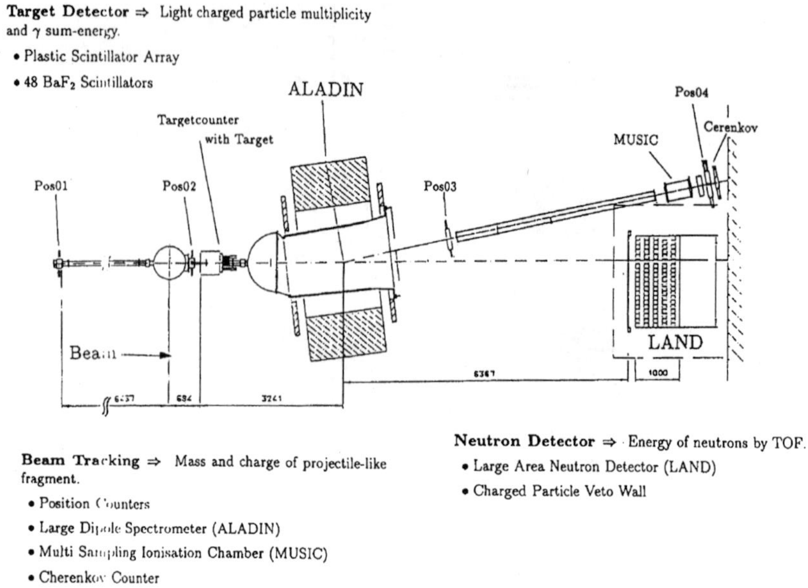

Fig. 21. Lay out of a detector to study Coulomb excitation using relativistic heavy ions. ALADIN is a large acceptance spectrometer for momentum analysis of the projectile like fragment. LAND is a large area neutron detector for time of flight analysis of fast neutrons evaporated from the moving fragment.

The outgoing projectile like particles are identified via momentum measurements in a large magnetic spectrometer, complemented by velocity measurements using time of flight and Čerenkov detectors and Z identification using a multiple sampling ionization chamber. The evaporation neutrons from the excited projectiles are emitted in forward direction and are detected by a large area position sensitive neutron detector (LAND). The neutron momenta are inferred from time of flight and direction measurements. The deexcitation γ-rays following neutron evaporation are detected using an array of 48 BaF_2-scintillation counters placed around the target.

Fig. 22 shows an excitation energy spectrum of ^{136}Xe from Coulomb excitation of a 700 MeV/u ^{136}Xe beam using a Pb-target. The excitation cross section per MeV energy is plotted as function of the excitation energy up to 40 MeV. The solid curve represents an analysis including the cross section of the GDR isovector excitation centered around 15.2 MeV (dashed curve) and smaller contributions from the GQR isoscalar excitation at 12.3 MeV and GQR isovector excitation at (22.1$\pm$0.7) MeV (dotted curve). The widths, excitation cross sections σ_{cx} and the energy weighted sum rule depletion in % are given in the table.

One notes in addition to the mentioned one phonon excitation strength additional strength located around 30 MeV, which is roughly twice the energy of the one phonon single GDR excitation. It is tentatively assigned to two phonon excitations. A verificating of the assignments given could be tested by measuring the excitation strength

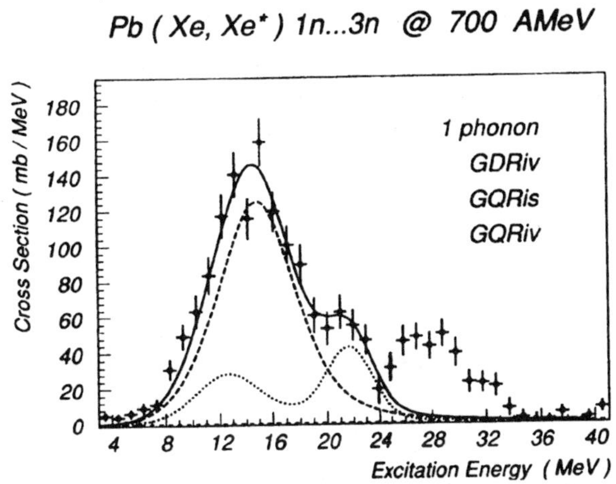

	E (MeV)	Γ (MeV)	σ_{CX} (mb)	EWSR depletion %
GDRiv	15.2	4.8	1050 ± 50	64 ± 3
GQRis	12.3	4.0	190	100
GQRiv	22.1 ± 0.7	≤ 5.4	185 ± 60	135 ± 44
$\int_{2\,MeV}^{36\,MeV} \frac{d\sigma_{tot}}{dE} dE = 1721 \pm$ mb			$\sigma_{nuc} \approx 100$ mb	

Fig. 22. Excitation energy spectrum of ^{136}Xe following Coulomb excitation. The analysis and the result of the analysis presented in the table is described in the text.

distribution as function of the bombarding energy.

As a last example of Coulomb excitation studies using relativistic heavy ion beams we show first results of Coulomb dissociation studies of 280 MeV/u ^{11}Li and 460 MeV/u ^{11}Be beams using the set up shown and described in Fig. 21.

In Fig. 23 we show as partial results neutron angular distributions from the reactions indicated. The ^{11}Li neutron angular distribution from a carbon target shows a width corresponding to an average momentum transfer of 28 MeV/c which can be accounted for by diffractive dissociation (solid line). The neutron angular distribution from Coulomb dissociation exhibits an angular width corresponding to a momentum spread of only 23 MeV/c. This reflects the weak binding of the two dissociated neutrons on the ^{9}Li-core. It may be interpreted as evidence for a neutron halo, with exponentially decreasing density, describe by a decay length of about 12 fm.

The ^{11}Be-neutron angular distribution from ^{10}Be +n break up reactions with C and Pb targets show much broader angular distributions with maxima at angles around 1.5° and FWHM of 78 MeV/c and 61 MeV/c for C and Pb targets respectively. The ^{11}Be+Pb $\rightarrow$ ^{10}Be+n+X neutron angular distribution may be reproduced by a direct Coulomb break

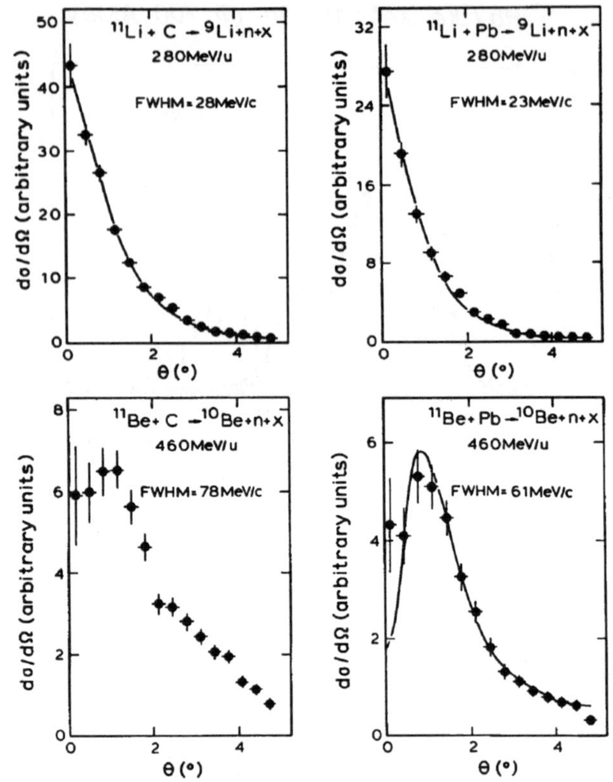

Fig. 23. Neutron angular distributions from the ^{11}Li and ^{11}B fragmentation and Coulomb dissociation processes respectively.

up calculation.

The data discussed originate from an experiment which allows to study the Coulomb dissociation of $^{11}\text{Li} \rightarrow {}^{9}\text{Li} + 2\text{n}$ exclusively by a measurement of the outgoing moment correlation of both neutrons and of ^{9}Li. Aim of this measurement is to gain information on the correlations of the two loosely bound neutrons in ^{11}Li.

5.2 Multifragmentation

Multifragmentation of nuclei is expected to occur as a reaction process at elevated nuclear temperatures. It was proposed that a nucleus heated close to the evaporation temperature of neutrons may instead dissociate in small pieces and decay in many fragments with the long range Coulomb interaction playing a special role in such a percolation process. For a long time experimental evidence of the actual occurrence of such a phenomenon was very scarce. In the following we describe an exclusive fragmentation experiment using Au-beams from SIS at 600 MeV/u energy which gives convincing evidence of the existence of the multifragmentation phenomenon.

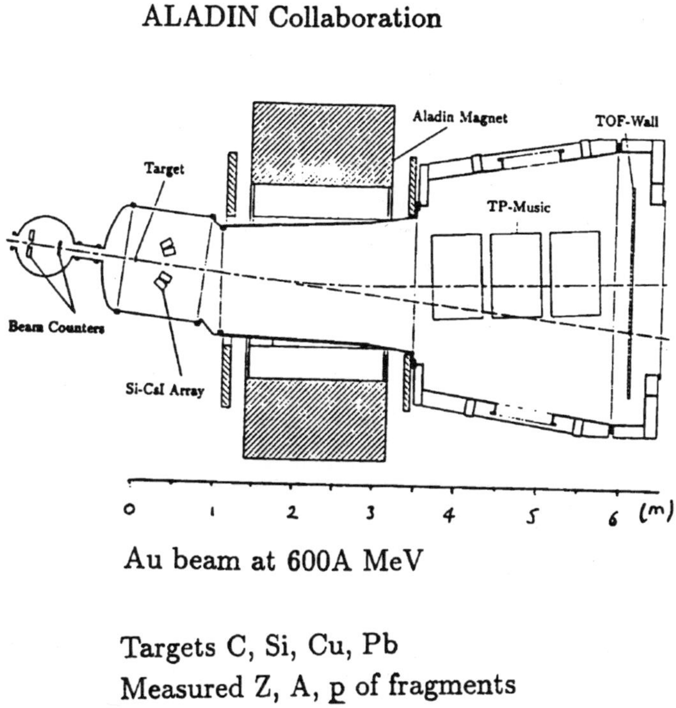

Fig. 24. ALADIN-spectrometer to study exclusively multifragmentation of excited projectile like objects by momentum analysis, Z-identification and time of flight analysis on all charged intermediate mass fragments emitted in forward direction.

Fig. 24 shows schematically the set up used to study exclusively intermediate mass fragment formation from highly excited projectile like fragments produced by bombardment of a series of targets such as C, Si, Cu and Pb with 600 MeV/u Au beams. Use is made of the kinematic focussing of the projectile like intermediate mass fragments in the beam direction. Essentially all fragments are momentum analyzed using the large acceptance spectrometer ALADIN. In combination with a time of flight measurement A of the fragments may be determined. Z-identification on was done with a MUSIC-chamber, which was operated in a time projection mode for track reconstruction.

Fig. 25 shows the main results of the first measurements of the average value of multiplicities of intermediate mass fragments with $Z > 2$ as function of a scaling parameter Z_{bound}, which is the sum of the charges of all projectile like fragments with $Z > 2$. Note that the data points for all targets measured lay on an universal curve with increasing multiplicity up to a maximum of $Z \sim 38$ to decrease again towards $Z_{\text{bound}} \sim 79$. The explanation of this observation, which was termed "rise and fall" of multifragmentation is very simple. Z_{bound} seems to be a measure of the energy per nucleon deposited into the projectile like nucleus.

If Z_{bound} is small, the large amount of energy deposited is used to fragment the nucleus into nucleons. At small deposited energy Z_{bound} is large. The scale for the en-

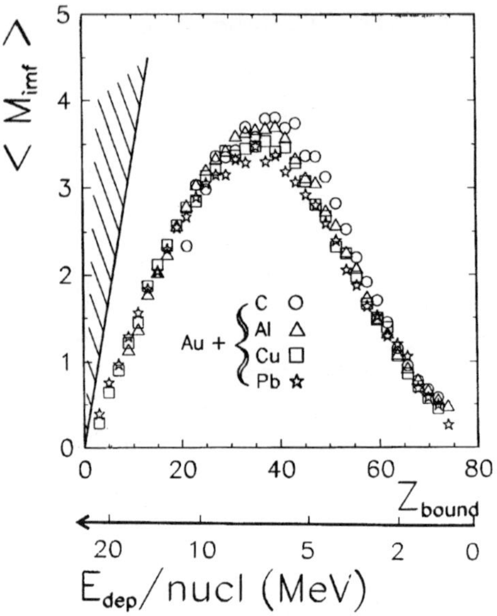

Fig. 25. Rise and fall of multifragmentation. Multiplicity of intermediate mass fragments (IMF) as function of Z_{bound} · $Z_{\text{bound}} = \sigma Z_i$ (Z_i>2)· Z_{bound} can be related to the deposited energy per nucleon in frame of a model. The scale shown on the figure is gained with the BUU-model.

ergy deposition originates from a model calculation. It suggests that multifragmentation reaches a maximum value for an energy deposited of about 8 MeV/u and drops again at higher energies. It would be of course very interesting to measure the deposited energy directly. Attempts are under way to measure the neutron multiplicity associated with the scaling parameter Z_{bound}. A fluctuation analysis performed with the data indeed suggests that a sort of phase transition has been observed. Nuclei heated to 8 MeV/u do not boil, but seem to fragment into intermediate size pieces.

5.3 Radial flow or explosion of compressed nuclear matter?

One of the main motivation to study central heavy ion collisions is the possible production of a compressed and heated piece of nuclear matter combined with the possibility to study the equation of state (EOS) of nuclear matter at elevated densities. As any observable in a collision integrates over all collision phases (interpenetration, high density and expansion) one can investigate strictly only the dynamical behavior of nuclei at higher energy densities. In order to receive an informative picture of the collision one should measure the momentum flow of all particles produced in a central collision. This is a very difficult task. A fair approximation to it seems to be the measurement of the momentum flow of all identified charged particles. For this purpose a 4π-detector for charged particles was constructed at GSI. A schematic drawing of this detector is shown in Fig. 26. It consists of a central detector with a large superconducting solenoid, tracking chambers, a time of flight scintillator barrel and Čerenkov threshold detectors, a forward detector

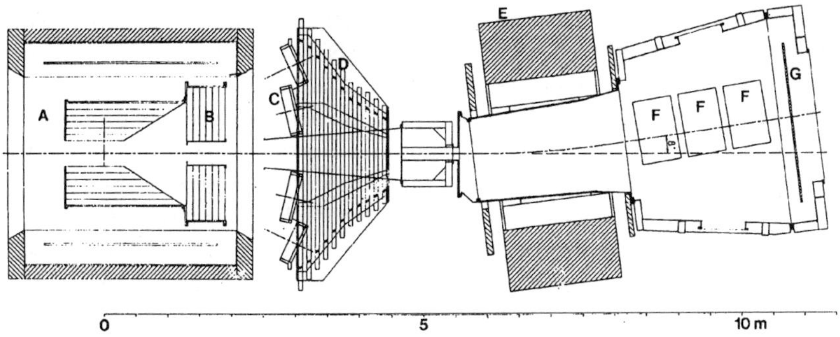

Fig. 26. Schematic layout of the 4π detector at GSI. It consists of a central drift chamber (A) and a helitron drift chamber (B) housed in a large superconducting coil, a forward wall with ionization chambers (C) and plastic scintillators (D) and a forward spectrometer (E) with position sensitive multiple sampling ionization chambers (F) and a time of flight wall (G).

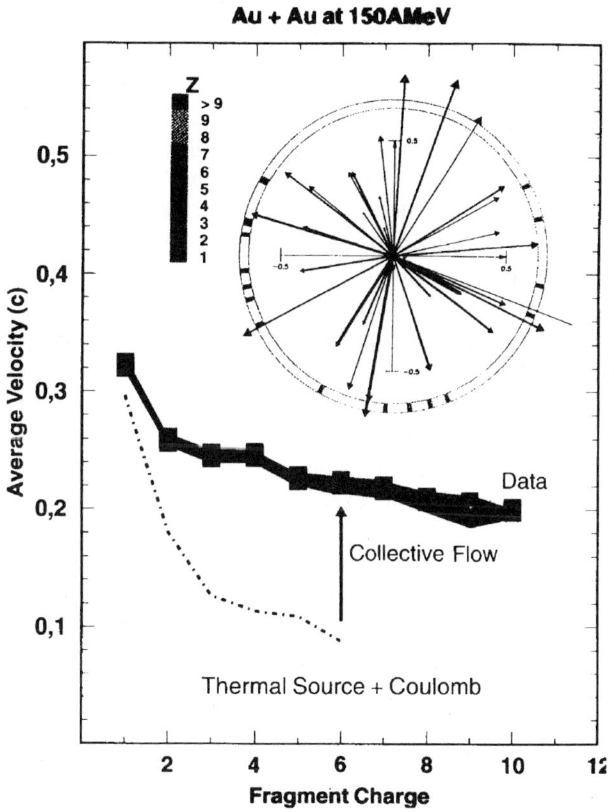

Fig. 27. Collective flow of fragments from a 150 MeV/u central Au-Au collision. The average velocity of the fragments is plotted versus the fragment charge. The insert shows the velocity vectors of various particles perpendicular to the beam direction.

with tracking chambers, ionization chambers for energy loss measurements and a time of flight wall consisting of an octahedral arrangement of scintillator bars. The whole detector system was recently taken into operation. Using the combination of momentum, time of flight and energy loss measurements all charged particles including pions and hopefully also kaons can be identified and their momentum flow may be determined.

In Fig. 27 we show one result of a first investigation using a part of the forward detector. Central Au-Au-collisions at 150 MeV/u energy were selected by a special trigger requiring uniform energy flow a all charged particles. The insert of Fig. 27 shows the perpendicular velocity vectors of particles with various charges marked by different colours. The main figure shows the average velocities of the fragments and their fluctuations as function of the average charge up to 10 units. Note, that there exists a strong radial velocity flow nearly independent on the fragment mass. More than 50% of the kinetic energy of the system is converted into radial kinetic energy. In central collisions nuclei explode into fragments of various sizes having all approximately similar velocities. This indicates the formation of a compressed system, which expands and breaks up in pieces with similar radial velocities.

5.4 Formation and study of properties of resonance matter.

Central nucleus-nucleus collisions at still higher energies (1-2 GeV/u) are expected to produce hot and dense hadronic matter with densities $\rho \approx$ 2-3 ρ_0 and $T \approx$ 60-90 MeV. Hadronic matter means that an appreciable fraction of the nucleons will be excited to their lowest resonances (Δ and N*). We like to call this new type of matter "resonance matter". The motivation of the experiments to be described are the study of the creation of resonance matter by investigations of subthreshold or high p_t-particle production. In medium modifications of properties of hadrons in dense matter, especially of their effective masses would very interesting to look for.

The measurement of excitation functions in subthreshold particle production offers itself as a good method to study resonance matter formation. In thermalized resonance matter, energy is stored which may be used to create high mass or high p_t-particles. Its production cross section will be enhanced and density increases will enhance the yields with high power of the density. The excitations functions which are not sensitive to the absorption of the created particles are expected to show flatter slopes due to an effective lowering of the thresholds.

Fig. 28 shows an analysis of π° p_t-spectra, measured with a two arm photon spectrometer (TAPS) at GSI for 1 GeV/u and 1.5 GeV/u Ar + Ca collisions, taking into account contributions from the Δ (1232), the N (1440) and N (1535) resonances. In order to reproduce the transverse momenta spectra excitation probabilities for the Δ-resonance of 7.6 % and 15.2 % and of the N (1440) resonance of 1.4 % and 4.4 % at 1.0 and 1.5 GeV/u respectively have to be taken into account. In order to fit the low p_t part of the spectra a shift of the Δ-resonance by -40 MeV to lower masses has to be included in the analysis. The average $\langle p_t \rangle$ values deduced from the π° spectra seem to reach a plateau at bombarding energies higher than 1 GeV/u.

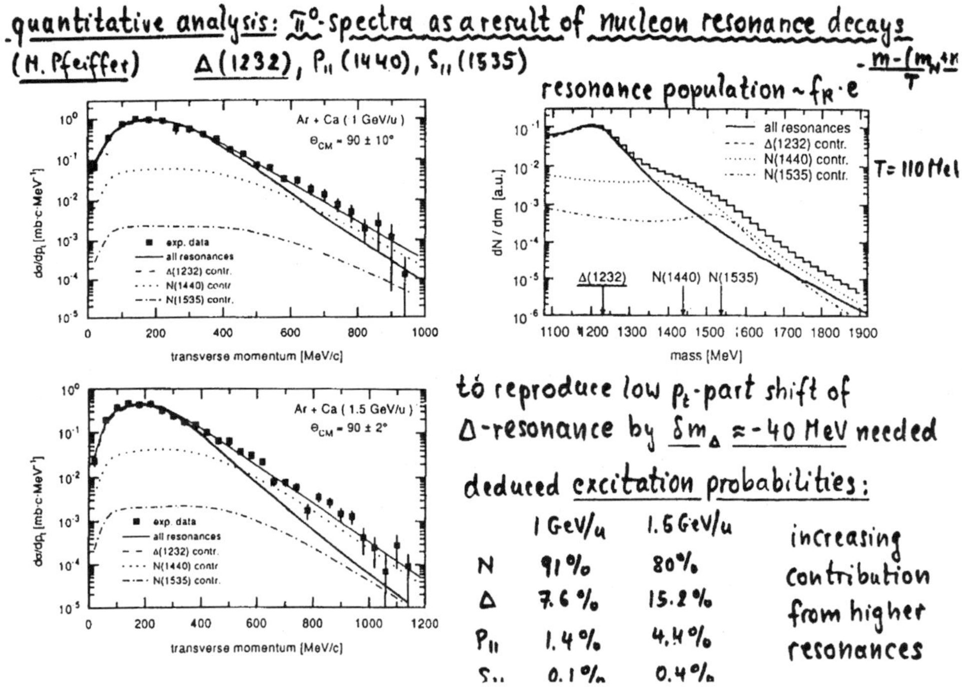

Fig. 28. Quantitative analysis of $\pi^o p_t$-spectra from Ar+Ca collisions at 1 GeV/u and 1.5 GeV/u bombarding energies. The table lists the deduced excitation probabilities of nucleon and Δ-resonances.

A very attractive subthreshold particle production reaction is the antiproton creation at energies around 2 GeV/u. This energy is far below the N-N-threshold, so that Fermi motion has a negligible contribution to the production mechanism and is already high enough for production of resonances and compression effects. The disadvantage of p̄-production studies is the importance of strong absorption and very small cross sections.

Fig. 29 shows the experimental set up used at GSI for antiproton production studies using ^{20}Ne and ^{58}Ni beams in the energy range between 2.0 and 1.6 GeV/u. It consists of the FRS operated in a dispersion free mode which allowed an increase of the solid angle to 3 msr and the momentum acceptance to $\Delta p/p$ = 6 %. The p̄ were identified by repeated momentum and time of flight measurements. The necessary high π- and K-suppression was reached by adding two type of threshold Cherenkov detectors (lucite and aerogel) at midplane and at the final focus of the spectrometer. Fig. 30 shows time of flight identification spectra with the aerogel detector giving vetos (a) and with both the aerogel-as well as the lucite-detector producing veto signals. Background free antiproton identification is indicated in spectrum b.

Fig. 31 shows excitation functions for p̄ production for various collision systems as function of the available energy in the cm system. Large production cross sections are indicated with slope parameters between 40 and 60 MeV/u indicating a unexpectedly

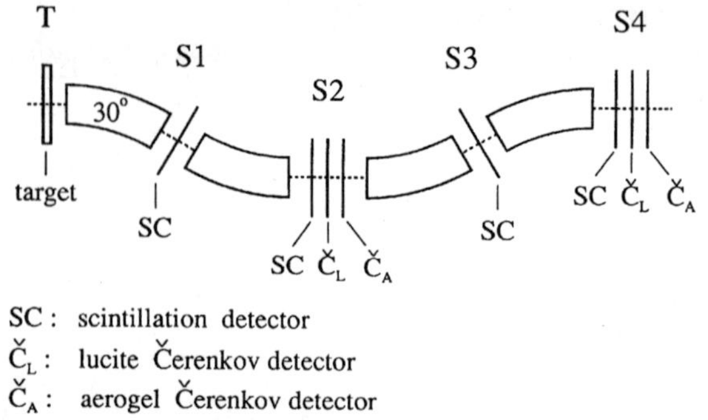

Fig. 29. Schematic lay out of the spectrometer to study forward production of antiprotons, K^- and π^- in heavy ion collisions. The length of the spectrometer is 72 m.

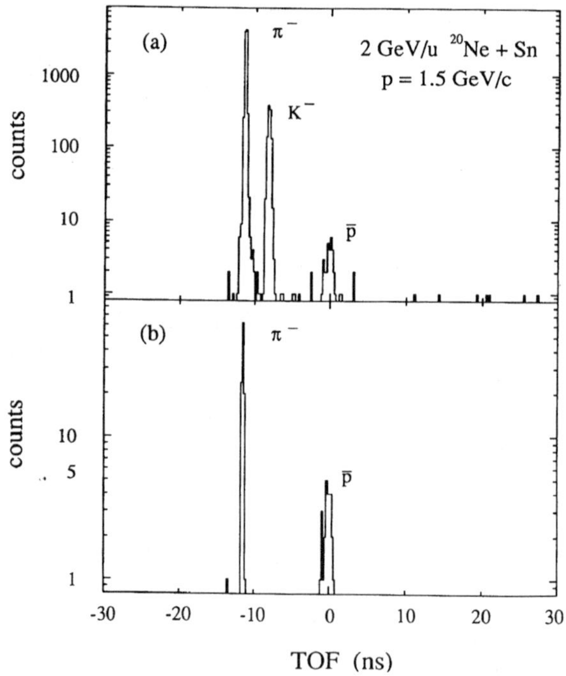

Fig. 30. Time of flight spectrum of negatively charged particles from the reaction of 2 GeV/u ^{20}Ne ions with Sn for the segment S_2 -S_3 - at p = 1.5 GeV/c. In spectrum (a) only events with no signal in the Cherenkov counter at S_2 were registered. In spectrum (b) in addition the veto signal from the lucite Cherenkov detector was required. The corresponding total number of pions detected in the focal plane S_3 was 1.3×10^8.

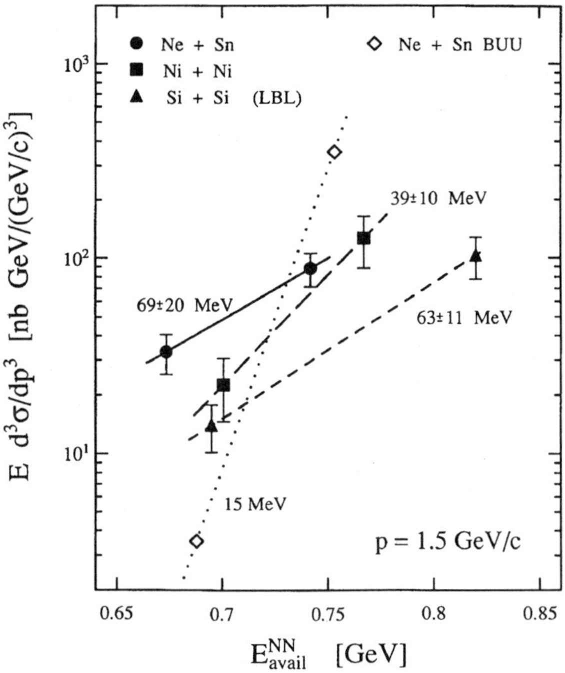

Fig. 31. Excitation functions for subthreshold $\bar{p}$ production with respect to the available energy in the N-N-system for several collision systems and p = 1.5 GeV/c. The dotted curve comes from a BUU-calculation. Its slope is too steep to explain the data with slope parameters between 39 and 69 MeV.

flat rise of the excitation function in this low bombarding energy regime. In fact the measured slope parameters are 4 times smaller than those predicted by BUU-calculation which take already account of Δ-Δ collision contributions. So an additional mechanism to raise the cross sections at lower energies seems to be effective. It is proposed that a decrease of the effective mass of the antiproton by 150 MeV in the dense hadronic matter could explain the flat rise of the excitation function and the magnitude of the observed cross section (Fig. 31). It is the first time that properties of antiprotons inside at dense medium have been studied.

In order to examine in more detail in medium effects on the masses of vector mesons, especially the ρ-mesons, due to partial restoration of chiral symmetry in dense matter, the construction of a High Acceptance Di Electron Spectrometer HADES is proposed at GSI.

Fig. 32 shows a schematic drawing of the proposed spectrometer with a toroidal field produced by 6 superconducting coils. It is able to analyse (e^+e^-)-pairs emitted into the forward hemisphere (Θ = 20°-80°). A RHICH-detector, with a large mirror, which focuses the Čerenkov light cones as rings on an annular position sensitive detector, surrounds the target. The RHICH is supposed to provide a pion suppression of 10^{-4} to 10^{-5}. Mini drift chambers are places as tracking detectors in front and behind

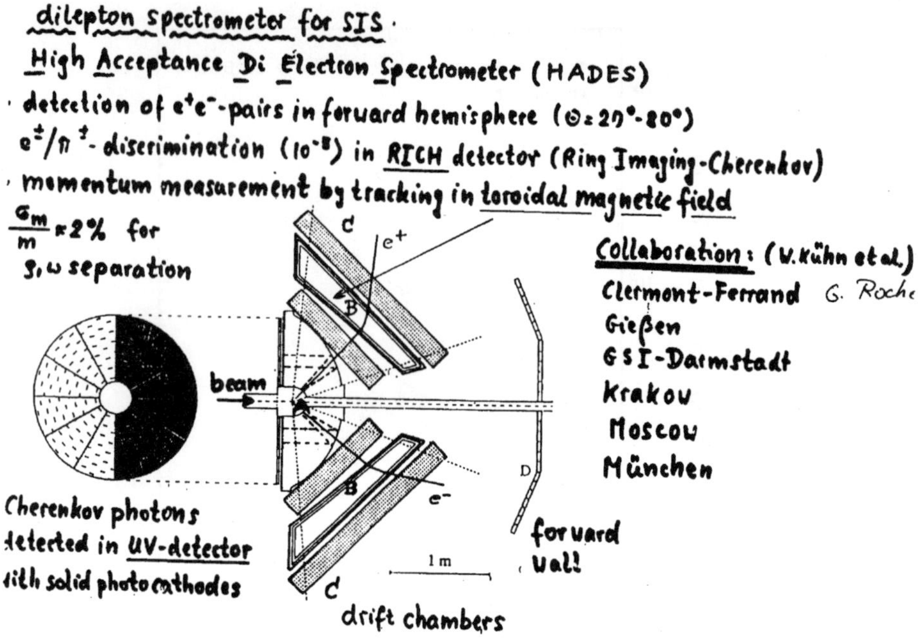

Fig. 32. Shows a schematic drawing of the high acceptance di electron spectrometer HADES to detect e^+e^- pairs in the forward hemisphere of 2 GeV/u heavy ion collisions. A is a RICH-detector for e/π discrimination, B are superconducting coils for the toroidal field, C are tracking detectors and D are particle hodoscopes and shower detectors for a first level trigger.

the magnetic coils, which give a momentum kick in transverse direction to the particles. We expect to achieve a mass resolution of better than 2 % for the reconstruction and the separation of the ρ- and ω-mesons. For trigger purpose a particle hodoscope combined with an electromagnetic shower detector will be placed behind the outer tracking chambers.

Fig. 33 shows a BUU-prediction for the invariant dilepton production cross section normalized with the projectile and target area as function of the invariant mass for ^{197}Au-^{197}Au collisions at 1 GeV/u bombarding energy. Various processes contribute to the total cross-section as indicated such as, p-n Bremsstrahlung, η-Dalitzpairs Δ-decays, dileptons from π^+-π^--annihilation and finally ρ and ω-decays. The φ contribution was estimated by assuming an density dependent ρ form factor as indicated in the upper part of Fig. 33.

It would be also very interesting to study the ρ-ω interference, due to its sensitivity on even small in medium mass shifts.

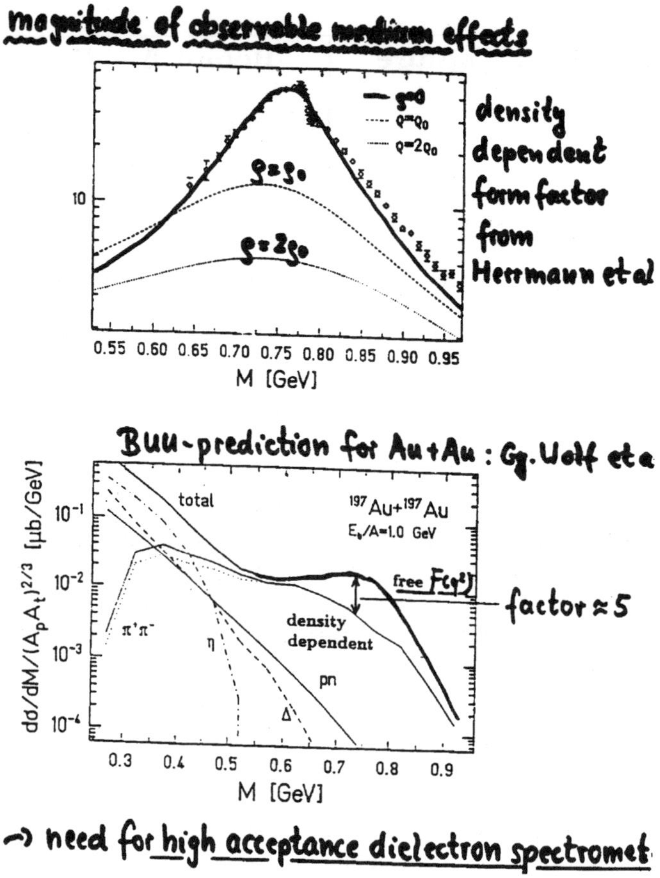

Fig. 33. Upper part: Density dependence of the ρ-meson form factors as predicted by Herrmann et al. Lower part: BUU-prediction of the various contributions to the e^+e^- cross section as function of the invariant mass. Note that the ρ-mass region is very sensitive to density effects.

6 Boundstate β-decay at the Interface between Nuclear and Atomic Physics

Boundstate β-decay was first discussed by Daudet, Jean and Lecoin in 1947. While bound β-decay, in which the decay electron gets captured in an atomic state and the antineutrino is emitted mono energetically, is of minor importance for neutral atoms it may become the dominant or even the only decay channel for highly ionized atoms, thus being an important process for element synthesis in stellar plasmas. With the availability highly ionized and stored heavy ion beams it can be studied for the first time in a laboratory. As indicated in Fig. 34, $^{163}_{66}$Dy is stable as a neutral atom, but as a fully stripped $^{163}_{66}\mathrm{Dy}^{66+}$ ion it can decay into $^{163}_{67}\mathrm{Ho}^{66+}$ by bound β-decay into the K and L shells of the daughter atom. For the experimental observation of this decay mode up to 10^8 $^{163}\mathrm{Dy}^{66+}$ ions at 290 MeV/u were accumulated, stored, and cooled in the ESR. Dur-

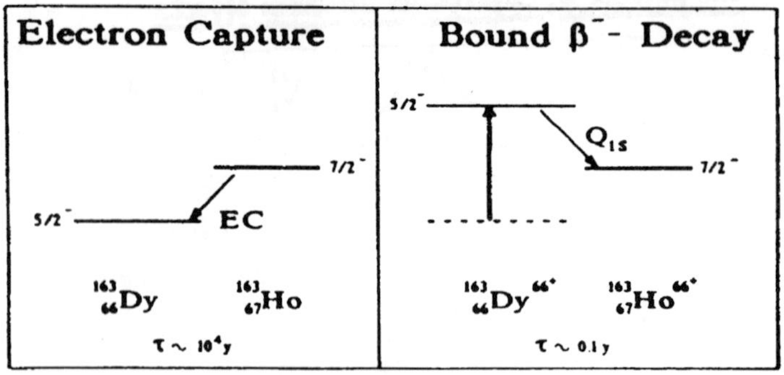

Fig. 34. Level schemes of the neutral pair ^{163}Dy-^{163}Ho and the ionized one including their decay modes.

ing the storage the bound β-decay daughters, stable hydrogen-like $^{163}\mathrm{Ho}^{66+}$ ions were continuously created. But due to the nearly identical charge to mass ratio of $^{163}\mathrm{Dy}^{66+}$ and $^{163}\mathrm{Ho}^{66+}$, a direct identification during the storage time was not possible. The experimental procedure to measure the $^{163}\mathrm{Ho}^{66+}$, is illustrated in Fig. 35. First, $^{163}\mathrm{Dy}^{66+}$ ions were accumulated in the ring for a typical time of 30 min. (Fig. 35 "accumulation"). Then an internal Ar gas jet, which vertically crossed the beam, was turned on for about 500 s. By this most of the $^{163}\mathrm{Ho}^{66+}$ daughters produced during accumulation were removed from the closed orbit via electron capture or electron loss ("purging"). After the gas jet was turned off, the ions-the primary $^{163}_{66}\mathrm{Dy}^{66+}$ as well as the decay products $^{163}\mathrm{Ho}^{66+}$ were stored and cooled for a variable time t_s("storage") ranging from 10 to 85 min. The electron cooler was in operation during all stages of the experiment.

The detection and identification of these $^{163}\mathrm{Ho}^{66+}$ daughters were based on the fact that only they could be further ionized, thus reducing their magnetic rigidity. Therefore, in the last step ("detection") the gas jet was turned on again in order to strip off the K electron in $^{163}_{67}\mathrm{Ho}^{66+}$ and to detect the bare $^{163}_{67}\mathrm{Ho}^{67+}$ with a position-sensitive particle counter placed in the deflection path behind the gas jet target.

From rate equations which describe the ratio $N_{\mathrm{Ho}}(t_s)/N_{\mathrm{Dy}}(t_s)$ as a function of the storage time t_s, the bound β-decay constant in the rest frame $\lambda_{\beta b}$, can be derived:

$$\frac{N^*_{\mathrm{Ho}}(t_s)}{N_{\mathrm{Dy}}(t_s)} \approx \frac{\lambda_{\beta b}}{\gamma} t_s \tag{12}$$

Using the above equation one gets from a linear fit to the experimental data, displayed in Fig. 36, a bound β-decay constant in the ion rest frame of $\lambda_{\beta b}$ = (1.72~0.17) $\times\ 10^{-7}\ \mathrm{s}^{-1}$ or a half life of $T^{\beta b}_{1/2}$ = 47~5 d.

A further, independent confirmation of our results came from the frequency analysis of the Schottky noise spectra which was performed for the stored ions during and after their interaction with the gas jet (the particle detectors were removed in that case). In addition to the intense primary beam, a significant signal from stored $^{163}\mathrm{Ho}^{67+}$ ions has been observed after storage and turning on the gas jet as shown in Fig. 37. This

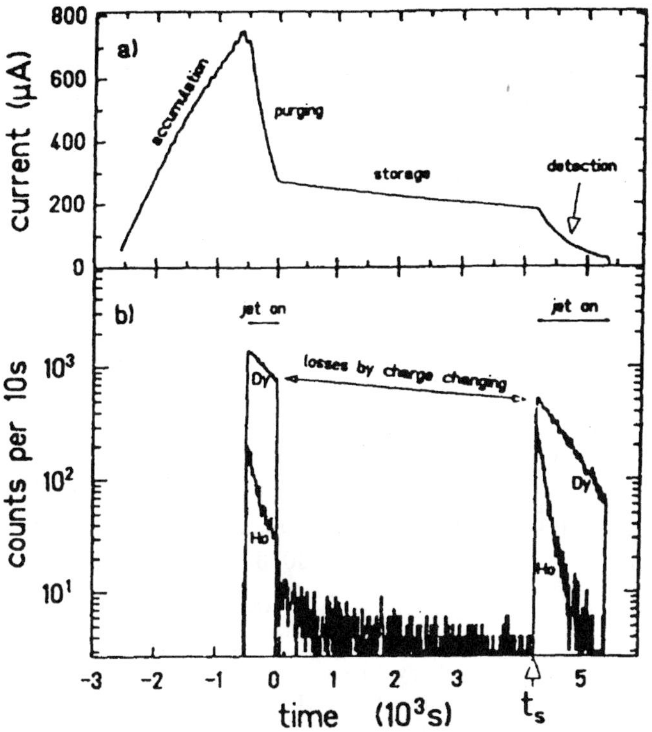

Fig. 35. (a) Current of stored bare $^{163}_{66}Dy^{66+}$ ions in the storage ring ESR during the various stages of the bound β-decay experiment. (b) Rates of Dy^{65+} ions (labeled "Dy", scaled down by 2^8) and of Ho^{67+} ions ("Ho") detected in the outer and inner position-sensitive detectors, respectively, which are moved into the ring chamber after the accumulation stage. Note the strong increase of the Ho^{67+} rate by more than an order of magnitude at t_s with respect to $t = 0$.

signal increases in proportion to t_s in contrast to the much smaller signals from nuclear reaction products being not proportional to t_s. At this point it is interesting to note that all secondary beams which have a slightly different $A/q-$ratio as the primary Dy^{66+} beam show a very narrow width of the Schottky noise frequency spectra. Because the ESR is even in the straight sections slightly dispersive, there is no spatial overlap of the secondary beams with the primary one. Electron cooling seems to be very effective for these beams with high Z and low particle numbers. So it looks like as if all particles in a certain beam have the same circulation frequency. There is some discussion whether this observation is caused by a disk like ordering of the particles in the secondary beams.

As a further interesting aspect, the result for $\lambda_{\beta b}$ along with the known value $\lambda_{\mathrm{EC}}^{\mathrm{M1}}$, $\lambda_{\mathrm{EC}}^{\mathrm{M2}}$ for electron capture from the M1 and M2 shell, respectively, allow to set bounds for both the Q^{EC} value of neutral ^{163}Ho and for the electron neutrino mass m. The ratio

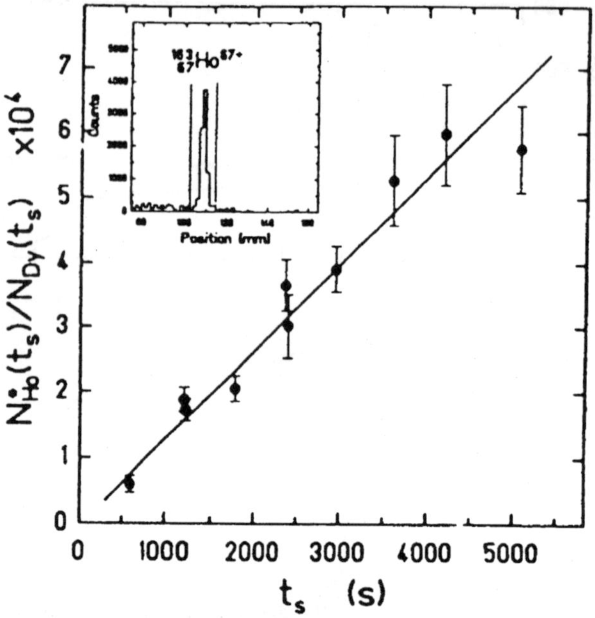

Fig. 36. Ratio of the number of β_b decay daughter atoms $^{163}_{67}Ho^{66+}$ $[N_{Ho}(t_s)]$ to primary $^{163}_{66}Dy^{66+}$ ions $[N_{Dy}(t_s)]$ as a function of the storage time t_s. From a linear fit to the data the decay constant $\lambda_{\beta b}$ can be derived. The inset shows the peak of Ho^{67+} ions recorded by the inner particle detector.

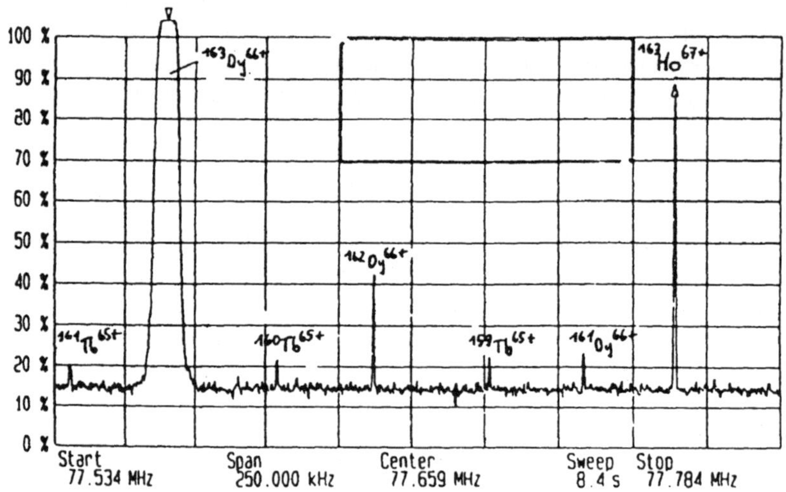

Fig. 37. Schottky noise frequency spectrum of the primary $^{163}Dy^{66+}$ beam and various secondary products circulating in the ESR after switching on the gas target. The signal for the bound β decay product increases in proportion to the storage time t_s.

of $\lambda_{\beta b}^{\mathrm{K}}/\lambda_{\mathrm{EC}}^{\mathrm{M1,2}}$ is given by

$$\frac{\lambda_{\beta b}^{\mathrm{K}}}{\lambda_{\mathrm{EC}}^{\mathrm{M1,2}}} = \frac{1}{2} \cdot \frac{2I_{\mathrm{Ho}}+1}{2I_{\mathrm{Dy}}+1} \cdot \frac{q_\nu W_\nu^{\mathrm{K}}}{q_\nu W_\nu^{\mathrm{M1,2}}} \cdot \frac{|\psi_{1s}^{66+}(R)|_{\mathrm{Ho}}^2}{|\psi_{\mathrm{M1,2}}^{0}(R)|_{\mathrm{Dy}}^2} \cdot \frac{B_{1s}^{66+}(\mathrm{Ho})}{B_{\mathrm{M1,2}}^{0}(\mathrm{Dy})} \tag{13}$$

where q_ν, W_ν are the momenta and total energies of the monochromatic neutrino and antineutrino, respectively, $|\psi_x^q(R)|^2$ are the electron densities at the nuclear surface for the corresponding daughter atom, B_x^q are the exchange and overlap correction factors, $I_{\mathrm{Ho}} = 7/2$, $I_{\mathrm{Dy}} = 5/2$ are the nuclear spins, and the factor 1/2 accounts for the (different) number of electrons involved in β_b-decay and in EC, respectively. Since bound β-decay and EC are determined by the same matrix elements for weak interaction, the latter quantity does not enter the above ratio. Inserting the measured and/or calculated values for the various quantities and using the approximations $W_\nu^{\mathrm{K}} \sim Q_{\beta,b}^{\mathrm{K}}$, $W_\nu^{\mathrm{M1,2}} \approx Q^{\mathrm{M1,2}} = Q_{\mathrm{EC}} - |B_{\mathrm{Dy}}^{\mathrm{M1,2}}|$, and $q_{-\nu} = \sqrt{W_{\bar\nu}^2 - m_{\bar\nu}^2} \approx W_{\bar\nu}$ (neglecting the antineutrino mass in the nominator), one gets the following constraints for the neutrino mass and the Q_{EC} value for neutral ^{163}Ho:

$$m_\nu < 410\ \mathrm{eV}; \quad 2616\ \mathrm{eV} < Q_{\mathrm{EC}} < 2694\ \mathrm{eV}. \tag{14}$$

With this experiment the bound β-decay mode could be observed for the first time. After the successful run with ^{163}Dy we will now start a second experiment to investigate the bound β-decay of $^{205}\mathrm{Tl}^{81+}$ to the first excited state of $^{205}\mathrm{Pb}^{81+}$ with a predicted half life of 120 d. As a further candidate the astrophysically interesting process of $^{187}\mathrm{Re}^{75+}$ decaying to the first excited state of ^{187}Os should be addressed. Since the predicted half life for $^{187}\mathrm{Re}^{75+}$ is 14 y, this experiment will be much more difficult.

Acknowledgements

I like to thank the organizers of this Asian-Spring-School especially my friend T. Yamazaki for their invitation. It was a very interesting experience for me to learn that the East Asian scientific cultures are so different. The atmosphere of the school was determined by the unsurmountable Japanese hospitality. It was hard work for me because I just changed from GSI Darmstadt to TU Munich, but I enjoyed it very much. The manuscript of my lecture became also a bit of a burden, but with the help of Petra Zeitlhofer, I also managed to write up things what I was telling in a relaxed atmosphere.

Additional reading material:

Electron cooling:

H. Poth, Physics Report **196**, 136 (1990)

B. Franzke Nucl. Instr. Meth. **B24**/25, 18 (1987)

M. Steck et al Contr. to 3rd European Conference on Particle Accelerators (EPAC) Berlin 1992

Synthesis of heavy elements:

G. Münzenberg Rep. Prog. Phys. **51**, 57 (1988)

Proton radioactivity:
S. Hofman GSI-93-04 preprint (to be published Handbook of Nuclear Decay modes CRC Press, Boca Raton, Florida)

General:
Directions of Heavy Ion Physics P. Kienle Springer Proceedings in Physics Volume **57** 1991

Lecture given at the JSPS-INS International Spring School, March 1993 at Shimoda

Non-Perturbative Aspects of Hadron Structure Examined Through Deep-Inelastic Scattering

A.W. Thomas and W. Melnitchouk ⋆

Department of Physics and Mathematical Physics, The University of Adelaide
Adelaide S.A. 5005 AUSTRALIA

1 Introduction

Understanding the structure of the nucleon is clearly one of the most challenging problems in modern physics. Apart from its intrinsic interest, there is by now a broad appreciation of the role its internal structure plays in determining the properties of finite nuclei. While lattice devotees continue to work on the brute force solution of QCD, the range and sophistication of phenomenological models of hadron structure continues to grow. On the other hand, the tests usually applied to those models tend to be rather indirect. It is really only in the last ten years or so that deep-inelastic scattering data, which is our *only* direct view of the quarks inside hadrons, has been taken seriously as a source of useful information [1, 2].

In these lectures we shall outline some of the progress that has been made in using deep-inelastic scattering (DIS) to refine our knowledge of hadron structure. Of course, in order to be reasonably self-contained we need to devote the first few sections to the kinematics of DIS, the parton model and the standard machinery required to treat DIS within QCD – namely the operator product expansion and the renormalisation group. In sect.4 we turn to recent work which successfully connects the MIT bag model with the measured parton distribution functions. We shall see that a number of features of those distributions can be understood in terms of familiar physics required in almost all quark models. We also examine a recent development which treats the spinor structure of the composite nucleon seriously.

There has recently been quite a bit of interest in the role of the meson cloud of the nucleon in DIS. This is explained in detail in sect.5. Our discussion includes the reasons for preferring the infinite momentum frame for the calculation as well as a detailed comparison with recent data. In sect.6 we turn to some unexpected surprises which DIS experiments have revealed – including the apparent violation of the Gottfried sum rule, the proton "spin crisis", and the nuclear European Muon Collaboration (EMC) effect. All of these phenomena are rich in information about the non-perturbative structure of the nucleon, free or bound. Finally in sect.7 we make some concluding remarks.

2 Basics of Deep-Inelastic Scattering

The elementary ideas of DIS can be found in many text books and reviews [3–10]. In this section our aim is simply to define our notation and collect together the essential formulae.

⋆ Address from 1 September 1993: Institut für Theoretische Physik, Universität Regensburg, D-93040 Regensburg, GERMANY

Consider the inclusive scattering of a high energy lepton (initial energy E, final energy E', and scattering angle θ) from a hadronic target (mass M and initial four-momentum p) as illustrated in Fig.1. The spacelike four-momentum transferred to the target is denoted q. For an unpolarised target the laboratory differential cross-section for electromagnetic scattering is calculated by contacting the lepton tensor $L_{\mu\nu}$ with the hadronic tensor $W^{\mu\nu}$:

$$W^{\mu\nu} = \tilde{g}^{\mu\nu} W_1 + \frac{\tilde{p}^\mu \tilde{p}^\nu}{M^2} W_2. \tag{1}$$

Here

$$\tilde{g}^{\mu\nu} = -g^{\mu\nu} + \frac{q^\mu q^\nu}{q^2} \tag{2}$$

and

$$\tilde{p}^\mu = p^\mu - \frac{p \cdot q}{q^2} q^\mu, \tag{3}$$

and the two arbitrary functions W_1 and W_2 contain all the information we can learn about the target from such experiments. After contracting the tensors and integrating over phase space we find:

$$\frac{d^2\sigma}{dE'd\Omega} = \frac{4\alpha^2 (E')^2}{q^4} \left[\cos^2 \frac{\theta}{2} \frac{F_2}{\nu} + 2 \sin^2 \frac{\theta}{2} \frac{F_1}{M} \right]. \tag{4}$$

All of the information concerning the structure of the target is now contained in the structure functions F_1 (= MW_1) and F_2 (= νW_2, with $\nu = E - E'$ the photon energy in the laboratory frame) which can depend on at most two variables. It is most usual to choose those to be the Lorentz invariant quantities $Q^2(= -q^2 > 0)$ and Bjorken x $(= -q^2/2p \cdot q = Q^2/2M\nu)$.

For $\nu(\bar{\nu})$ scattering from an unpolarised target we find a third structure function, F_3, associated with parity violation:

$$\frac{d^2\sigma^{\nu(\bar{\nu})}}{dE'd\Omega} = \frac{G_F^2 (E')^2}{2\pi^2} \left(\frac{M_W^2}{M_W^2 + Q^2} \right)^2 \left[\cos^2 \frac{\theta}{2} \frac{F_2^{\nu(\bar{\nu})}}{\nu} + 2 \sin^2 \frac{\theta}{2} \frac{F_1^{\nu(\bar{\nu})}}{M} \pm \frac{E + E'}{M\nu} \sin^2 \frac{\theta}{2} F_3^{\nu(\bar{\nu})} \right] \tag{5}$$

Finally, for scattering of a polarised electron (or muon) from a polarised, spin-1/2 target there are (at least in principle) two more structure functions (g_1 and g_2) which can be measured. Denoting beam and target helicity with arrows top and bottom respectively we find:

$$\frac{d^2\sigma}{dE'd\Omega} \left(\overset{\rightarrow}{\rightarrow} - \overset{\rightarrow}{\leftarrow} \right) = \frac{4\alpha^2 E'}{M\nu Q^2 E} \left[(E + E' \cos\theta) g_1 - 2Mx g_2 \right]. \tag{6}$$

(We note in passing some recent theoretical arguments [11] which imply a relationship between g_1 and g_2, but we shall not discuss these further here.) In the deep-inelastic regime Q^2 and ν are both very large ($Q^2 > 2\,\mathrm{GeV}^2$, $\nu > 1$ GeV) but $x \in (0, 1)$. Clearly the second term on the r.h.s. in equ.(6) will be negligible if g_1 and g_2 are of the same order. For this reason only g_1 has been measured so far. To determine g_2 one would need to work with a longitudinally polarised beam and a transversely polarised target. However, even then the cross-section is of order $(1/Q)$ times that given in equ.(6):

$$\frac{d^2\sigma}{dE'd\Omega}\left(\vec{\uparrow} - \vec{\downarrow}\right) = -\frac{4\alpha^2(E')^2}{M\nu Q^2 E}\left[g_1 + \frac{2Eg_2}{E+E'}\right]\sin\theta. \tag{7}$$

In the late 1960s tremendous excitement was generated by the discovery at SLAC that the structure functions were almost independent of Q^2 over a very wide range. That is, they were functions of the single variable — Bjorken x. It is very easy to see that this is what one would expect if the nucleon contained a collection of elementary constituents (initially called partons by Feynman but later identified with quarks) with low mass, which do not interact strongly during the DIS collision.

For simplicity it is usual to consider this problem in a so-called infinite momentum frame – e.g. one where the nucleon has momentum $P \gg M$ in the z-direction so that its 4-momentum is $p = (P; 0, 0, P)$. Suppose a constituent with 4-momentum $yp = (yP; 0, 0, yP)$, where y is the "momentum fraction" of the nucleon carried by the constituent, absorbs the photon. Its final invariant mass squared will be:

$$(yp+q)^2 = y^2p^2 - Q^2 + 2y\, p\cdot q. \tag{8}$$

But $p^2 \ll Q^2$ and $p\cdot q$, and the invariant mass squared of the parton must be small (≈ 0) by assumption. Then we find $y = Q^2/2p\cdot q$ which was called Bjorken x above. Thus we see that under the assumptions of the parton model, only a parton with fraction x of the momentum of the nucleon can absorb the exchanged photon (or W-boson). In the case that the impulse approximation is valid, DIS structure functions then measure the number density of partons in the nucleon with momentum fraction x.

It is usual to define distributions $q^{\uparrow\downarrow}(x)$ which give the number density of quarks in the target with helicity parallel or anti-parallel to that of the target. For example, $u(x)x dx$ gives the fraction of momentum of u quarks *in the proton* with momentum between xP and $(x + dx)P$ in the infinite momentum frame (and with either helicity). By charge symmetry u also gives the distribution of d quarks in the neutron.

The structure functions mentioned earlier are directly related to these distribution functions. For an electromagnetic probe one finds:

$$F_1(x) = \frac{1}{4}\sum_q e_q^2(q^\uparrow(x) + q^\downarrow(x)) \tag{9}$$

$$F_2(x) = 2xF_1(x) \tag{10}$$

$$g_1(x) = \frac{1}{2}\sum_q e_q^2(q^\uparrow(x) - q^\downarrow(x)) \tag{11}$$

with e_q the charge in units of e of quark flavor q. Equation (10) is the Callan-Gross relation and relies on the partons having spin 1/2 and no transverse momentum (in the infinite momentum frame). In general we have

$$F_2 = 2xF_1\frac{1+R}{1+2Mx/\nu} \tag{12}$$

where R is the ratio of cross-sections for absorbing a longitudinal to that for a transverse photon. Experimentally R is small [12] ($\lesssim 0.1$) for all x, for $Q^2 \gtrsim 5$ GeV^2.

For neutrino scattering from an isoscalar target one finds

$$F_2^{(\bar{\nu})} = (u + \bar{u} + d + \bar{d} + s + \bar{s}), \tag{13}$$

which measures the total quark content of the proton. Even more important, by combining ν and $\bar{\nu}$ data one can measure the combination

$$F_3^{(\nu,\bar{\nu})} = (u - \bar{u} + d - \bar{d}), \tag{14}$$

which isolates the excess of quarks over antiquarks – i.e. the valence quark distribution of the nucleon. Clearly we would expect the sum rule (due to Gross and Llewellyn Smith)

$$\int_0^1 dx\; F_3^{(\nu,\bar{\nu})}(x) = 3 \tag{15}$$

to be obeyed. It will also be useful to define the n'th moment of a structure function like xF_3, F_2 or xF_1 as, e.g.

$$M_{3n} = \int_0^1 dx\; x^{n-2}\, [xF_3(x)]. \tag{16}$$

Initially the major experimental activity in this field was at SLAC, but for the past 10-15 years, the emphasis has shifted to the muon and neutrino beams of CERN and Fermilab. For a thorough summary of the experiments at these laboratories we refer to the recent review by Morfin [13]. While most of data has been been accumulated for Q^2 between 5 and 20 GeV^2 and x between 0.1 and 0.65, Q^2 as high as 200 GeV^2 and x as low as 10^{-4} (at HERA) have now been obtained. (The latter is particularly relevant for certain sum rules as we shall see.) Figure 2 illustrates the results typically obtained — "this experiment" is the CDHS neutrino experiment [14]. We note that the antiquarks, which form half of the sea of virtual pairs in the nucleon, are concentrated at low x ($x \lesssim 0.3$). The valence quarks (c.f. equ.(14)) dominate the large x region. There is impressive agreement between the weak and electromagnetic experiments once one allows for the appropriate charges. Actually the situation is a little worse than Fig.1 might suggest. Because of systematic errors, different muon data sets on the same target may differ by as much as 10–20%. These differences are typically within the quoted systematic errors. Again such differences can be important whenever an absolute measurement is required — e.g. in the spin sum rule.

It is clear from the analysis of the experimental data that even in the Bjorken region the structure functions have a weak Q^2-dependence, and therefore so do the distribution functions which we write as $q(x, Q^2)$. If one sticks to any one data set in order to (partially) avoid systematic errors, this variation of the structure functions (scaling violation) is essentially logarithmic. In order to understand it one must go beyond the naive parton model to QCD.

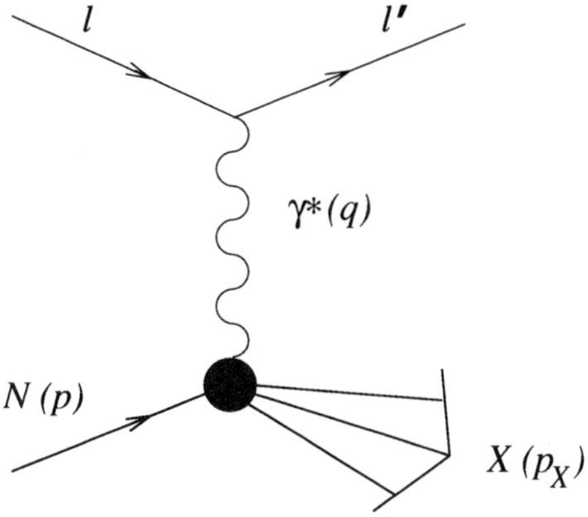

Fig. 1. The deep-inelastic cross-section involves a sum over all (unobserved) final states, X.

3 Perturbative QCD

In developing a theoretical description of the deep-inelastic process shown in Fig.1 it is usual to assume that the wavefunction of the target has no high-momentum components (i.e. $\mathbf{p}_T^2 \ll Q^2$). Thus any Q^2-dependence can only come from the lepton-quark scattering process. Scaling results if the quark is treated as point-like and the trivial Q^2-dependence of the Mott cross-section is factored out. On the other hand, in an interacting field theory, the lepton-quark scattering amplitude will involve radiative corrections, some of which add coherently (e.g. wave function and vertex renormalisation) while others are incoherent (e.g. bremmstrahlung). It is well known that such radiative processes lead to corrections which vary logarithmically with the appropriate cut-off scale — in this case Q^2.

For explicit calculations of these radiative corrections, we refer to some excellent texts [3-10]. It is particularly important from the point of view of application to other systems, that one can develop a very physical interpretation of this Q^2 variation. This is perhaps best expressed through Close's "onion skin" picture, whereby every time we increase Q^2, we increase the resolution at which we observe the structure of the target — hence revealing more and more of its previously virtual quarks and gluons. The mathematical description of the variation of the structure function of Q^2 is given by the Altarelli-Parisi equations. If one sticks to any one of the data sets mentioned above (in order to avoid systematic errors), the Q^2 variation of the structure functions is well described by these equations. However there are discrepancies between data sets and difficulties have also been encountered when trying to make a consistent fit to the EMC data on different nuclear targets.

The most rigorous approach to the calculation of structure functions, and the description of their Q^2 variation, comes through the operator product expansion and the renormalisation group. As these are also discussed in many texts, we highlight only those features needed for our main consideration, namely the prediction of structure functions from particular quark models.

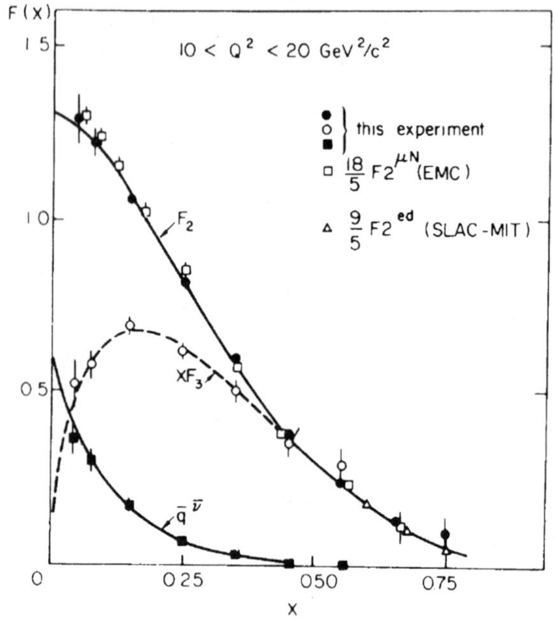

Fig. 2. Summary of structure function data taken with neutrinos – from the CDHS group [14] – in comparison with data from electromagnetic probes.

3.1 The Operator Product Expansion

As electromagnetic DIS involves a total cross-section for lepton scattering with a single photon exchanged, the structure functions are proportional (through the optical theorem) to the imaginary part of the forward Compton amplitude for a photon of 4-momentum q. The Compton amplitude is written:

$$T_{\mu\nu} = i \int d^4z\, e^{iq\cdot z}\, \langle N|T(j_\mu(z)j_\nu(0))\,|N\rangle, \tag{17}$$

where j_μ is the electromagnetic current operator. The essential idea of the operator product expansion (OPE) is that one can expand the time-ordered product of the currents in what is essentially a generalisation of the familiar Taylor series. That is, one writes it as an infinite series, whose terms each involve a (possibly singular) function of z^2 times a local operator, in general involving $\bar{\psi}(0)$ and derivatives of $\psi(0)$, contracted with products of z^λ . It is crucial that this is an expansion of the operator which is therefore target independent.

Dropping Lorentz indices, for purposes of illustration, one has schematically:

$$T(j_\mu(z)j_\nu(0)) = \sum_{i,n} \widetilde{C}_{i,n}(z^2)\, z_{\mu_1} \cdots z_{\mu_n}\, O_{i,n}^{\mu_1\cdots\mu_n} \tag{18}$$

where for example one might have an operator $O_{i,n}^{\mu_1\cdots\mu_n}$ of the form:

$$\bar{\psi}(0)\gamma^{\mu_1} D^{\mu_2} \cdots D^{\mu_n} \psi(0) \tag{19}$$

$$\bar{\psi}(0)\gamma^{\mu_1}D^{\mu_2}\psi(0)\bar{\psi}(0)\gamma^{\mu_3}\cdots D^{\mu_n}\psi(0) \tag{20}$$

and so on. Now ignoring renormalisation ψ has dimensions of $[\text{energy}]^{3/2}$, which is usually called simply dimension-3/2. Therefore the l.h s. of equ.(18) has dimension 6. Suppose the operator $O_{i,n}$ has dimension N_i, then $\widetilde{C}_{i,n}$ must have dimension $6+n-N_i$. After Fourier transformation to give the Compton amplitude (equ.(17)) it will therefore behave like Q^{6+n-N_i}. Since $Q^2 \to \infty$ in the Bjorken limit, the dominant operators will be those with the largest exponent, or the smallest value of $(N_i - n)$ — which is usually called "twist". One can easily check that the operator in equ.(19) is twist-2, while that in equ.(20) is twist-4. With a little thought one can see that twist-2 is the best that can be done, and therefore *DIS in the Bjorken limit will be entirely determined by twist-2 operators*.

The operator given in equ.(19) is called a singlet, twist-2 operator because it involves a trace over flavors. The only other twist-2, singlet operator involves the gluons:

$$g_{\alpha\beta}F^{\mu_1\alpha}D^{\mu_2}\cdots D^{\mu_{n-1}}F^{\mu_n\beta}. \tag{21}$$

One can also write down a set of non-singlet operators

$$\bar{\psi}(0)\gamma^{\mu_1}D^{\mu_2}\cdots D^{\mu_n}\lambda_i\psi(0), \tag{22}$$

where λ_i are the flavor-SU(3) matrices. For simplicity we shall concentrate on the non-singlet operators from now on.

In general the matrix elements of these operators must be of the form

$$\langle N|O_i^{\mu_1\cdots\mu_n}|N\rangle = p^{\mu_1}\cdots p^{\mu_n}\,\langle N|O_{i,n}|N\rangle. \tag{23}$$

Returning to equ.(17) we realise that the Fourier transform of z_μ is essentially q_μ/Q^2, which contracted with p^μ gives $1/x$. Thus we find that in the large-Q^2 limit, schematically (i.e. corresponding to equ.(18))

$$T \sim \sum_n C_n(Q^2)\,x^{-n}\,\langle N|O_n|N\rangle, \tag{24}$$

and we have incorporated an appropriate number of factors of Q^{-2} into the (n'th derivative of the) Fourier transform of $\widetilde{C}_n$ to give C_n. The latter is easily seen to be dimensionless.

Finally, for the full electromagnetic case one finds

$$\begin{aligned}
&i\int d^4z\, e^{iq\cdot z}\, T(j_\mu(z)j_\nu(0)) \\
&= \sum_{i,n}\Big[(g_{\mu\mu_1}g_{\nu\mu_2}Q^2 + g_{\mu_1\mu}q_\nu q_{\mu_2} + g_{\nu\mu_2}q_\mu q_{\mu_1} - g_{\mu\nu}q_{\mu_1}q_{\mu_2})C_{i,n}(Q^2) \\
&+ \left(g_{\mu\nu} - \frac{q_\mu q_\nu}{q^2}\right) q_{\mu_1}q_{\mu_2}C^L_{i,n}(Q^2) - i\epsilon_{\mu\nu\alpha\beta}g^\alpha_{\mu_1}q^\beta q_{\mu_2}C^3_{i,n}(Q^2)\Big] \\
&\times q_{\mu_3}\cdots q_{\mu_n}\left(\frac{2}{Q^2}\right)^n O_i^{\mu_1\cdots\mu_n}.
\end{aligned} \tag{25}$$

Here we have kept only the twist-2 operators, $C_{i,n}$ and $C^L_{i,n}$ correspond to the parity conserving transverse and longitudinal contributions respectively while $C^3_{i,n}$ is the parity violating term leading to F_3 in $\nu(\bar{\nu})$ scattering — c.f. equ.(5).

3.2 The Renormalisation Group

The arguments just presented must be modified in a field theory like QCD. The matrix element of the currents on the l.h.s. of equ.(25) must be renormalised, as must the matrix elements of the operators in the OPE on the r.h.s. This procedure introduces a new mass scale (or re-normalisation scale), μ^2, upon which no physical results can depend. It is very important that the μ^2 dependence of the coefficient functions $C_{i,n}(Q^2, \mu^2, g^2(\mu^2))$ be chosen such that equ.(25) is true after renormalisation. From the practical point of view it is crucial that the OPE is an operator relationship which holds independent of the target. One can therefore calculate the μ^2 dependence for a simple target, such as a free quark. As this too is found in many texts [3-8] we just review it briefly.

Assuming that the operator O_n is multiplicatively renormalised by $Z^{-1}_{O_n}$, we define

$$\beta(g) = \frac{\partial g(\mu^2)}{\partial \ln \mu^2} \tag{26}$$

and

$$\gamma_{O_n} = \frac{\partial \ln Z_{O_n}}{\partial \ln \mu^2}. \tag{27}$$

(Note that we have dropped the label i for convenience.) We shall show below that the product $C_n(Q^2, \mu^2, g^2)\langle N|O_n|N\rangle$ is measurable and therefore cannot depend on μ^2. (It is proportional to the n'th moment of the DIS structure function defined in equ.(16)). For the present we write

$$\mu \frac{d}{d\mu} C_n \langle N|O_n|N\rangle = 0, \tag{28}$$

and hence (in an appropriate gauge)

$$\left(\frac{\partial}{\partial \ln \mu^2} + \frac{\partial g}{\partial \ln \mu^2} \frac{\partial}{\partial g} - \gamma_{O_n} \right) C_n(Q^2, \mu^2, g^2) = 0. \tag{29}$$

However we saw above that for twist-2 C_n is dimensionless and therefore can only depend on Q^2 as Q^2/μ^2. Therefore we can replace $\partial/\partial \ln \mu^2$ by $-\partial/\partial \ln Q^2$ and equ.(29) becomes

$$\left(\frac{\partial}{\partial \ln Q^2} - \beta(g(Q^2)) \frac{\partial}{\partial g(Q^2)} + \gamma_{O_n} \right) C_n = 0. \tag{30}$$

The first two terms are easily identified as $-\beta(g(Q^2))d/dg(Q^2)$ so that equ.(30) implies

$$\frac{d \ln C_n}{dg(Q^2)} = \frac{-\gamma_{O_n}}{\beta}. \tag{31}$$

Finally we obtain:

$$C_n(Q^2, \mu^2, g^2(\mu^2)) = C_n(Q^2, Q^2, g^2(Q^2)) \exp\left(-\int_{g(\mu^2)}^{g(Q^2)} \frac{\gamma}{\beta} dg'\right). \tag{32}$$

In practice one has a series expansion for $\gamma(g')$ and $\beta(g')$ to only a few terms. In the so-called leading order we have

$$\beta(g) = -\frac{\beta_0 g^3}{16\pi^2}, \qquad \gamma(g) = \gamma_0^n \frac{g^2}{16\pi^2}, \tag{33}$$

and

$$C_n(Q^2, Q^2, g^2(Q^2)) = 1. \tag{34}$$

Then the integral in equ.(32) is easily performed and we find

$$C_n(Q^2, \mu^2, g^2) \overset{L.O.}{=} \left(\frac{\alpha(Q^2)}{\alpha(\mu^2)}\right)^{d_0^n}, \tag{35}$$

and using the calculated value of β_0 (for N_f quark flavors) and γ_0^n, the anomalous dimension d_0^n is

$$d_0^n = \frac{\gamma_0^n}{2\beta_0} = \frac{4}{33 - 2N_f}\left(1 - \frac{2}{n(n+1)} + 4\sum_{m=2}^{n} m^{-1}\right). \tag{36}$$

3.3 The Moments of the Structure Functions

As we hinted above, there is a direct connection between the moments of the structure functions and the Compton amplitude which we have calculated so far. In fact as a function of ν for fixed Q^2, $T_{\mu\nu}$ has two cuts, (ν_{th}, ∞) corresponding to the physical region and (∞, ν_{th}) corresponding to crossed processes. In terms of x, again for fixed Q^2, the corresponding cuts run from (0,1) and (–1,0). Thus the dispersion relation for $T_{\mu\nu}$ at fixed Q^2 has the form

$$T_{\mu\nu}(x, Q^2) = \frac{1}{2\pi i}\left[\int_{-\infty}^{-\nu_{th}} + \int_{\nu_{th}}^{\infty}\right] \frac{d\nu'(disc)T_{\mu\nu}(x', Q^2)}{\nu' - \nu}. \tag{37}$$

Replacing ν' by x' and using the optical theorem to replace $\mathrm{Im}T_{\mu\nu}$ by the total cross-section, which is by definition the structure function, we find

$$T_{\mu\nu}(x, Q^2) = \sum_n x^{-n} \int_0^1 dx'(x')^{n-1} W_{\mu\nu}(x', Q^2). \tag{38}$$

Note that for the various terms in $T_{\mu\nu}$ (see equ.(25)) the sum over n is restricted to even or odd values depending on the crossing properties of the corresponding piece of $W_{\mu\nu}$.

Comparing equ.(38) with equ.(24) we see that the product of the coefficient function C_n and the operator matrix element is measurable. Indeed it is equal to the moment of the appropriate structure function as defined in equ.(16). Using the result of the above analysis based on the OPE and the renormalisation group, we see that the Q^2 variation of the moments of the structure functions is given by perturbative QCD. To leading order one finds:

$$M_n(Q^2) = M_n(Q_0^2)\left(\frac{\alpha(Q^2)}{\alpha(Q_0^2)}\right)^{d_0^n}. \tag{39}$$

For fixed Q_0^2 it is then easily shown that

$$\ln M_n(Q^2) = \frac{d_0^n}{d_0^m}\ln M_m(Q^2) + \text{constant}, \tag{40}$$

and therefore a log-log plot of any two moments should be a straight line whose slope is predicted by QCD.

All of the above discussion of Q^2 evolution involves non-singlet operators. The Q^2 evolution of the operators given in equs.(19) and (21) is more complicated because they mix under renormalisation. While the corresponding analysis is not much more difficult (it involves a 2×2 matrix), it would divert us too much to explain it here. Instead we refer to the appropriate texts [3-8] - — for example there is a concise summary in Table 2 of the review by Altarelli [9].

3.4 The Inverse Mellin Transform

Given an analytic continuation of a set of moments, $M_n(Q^2)$, there is a standard method for reconstructing the corresponding function — this is the Inverse Mellin Transform (IMT):

$$xF_3(x,Q^2) = \frac{1}{2\pi i}\int_{C-i\infty}^{C+i\infty} dn\, x^{1-n} M_{3n}(Q^2). \tag{41}$$

(Here C is chosen so that the integral exists.) If the moments can be written as a product, as in equ.(24) or (25):

$$M_{2n}(Q^2) = C_n(Q^2,\mu^2)\,\langle N|O_n(\mu^2)|N\rangle \tag{42}$$

then the IMT xF_3 is just a convolution of the IMT of C_n (denoted by $\mathcal{C}_3$) and $\langle N|O_n|N\rangle$ (denoted $\mathcal{F}_3$), viz:

$$xF_3(x,Q^2) = \int_x^1 \frac{dy}{y}\,\mathcal{C}_3(x/y,Q^2,\mu^2)\,(y\mathcal{F}_3(y,\mu^2)). \tag{43}$$

This is an extremely important result. In particular, $\mathcal{C}_3$ is totally independent of the structure of the target — a property known as factorisation. Clearly if we can evaluate the structure function of the target at any renormalisation scale μ^2, equ.(43) allows us to calculate it at all higher values of Q^2. Higher order QCD corrections do not alter this result in principle, they just make $\mathcal{C}_3$ harder to compute. For this reason μ^2 cannot be too low.

4 Relation to Simple Quark Models

At last we have collected all the results necessary to understand how to relate quark models to QCD. (Of course, if one could use non-perturbative QCD (e.g. on the lattice) to calculate $\langle N|O_n(\mu^2)|N\rangle$ this would be unnecessary. However this is not feasible for more than a few moments at the present time — see for example Lissia et al. [15] and Sachrajda [16].) Since the models are only "QCD motivated", the connection cannot be rigorous. On the other hand we know of no sensible alternative to what is proposed here.

Apart from lattice technology, the only known technique for solving a bound state problem in field theory is to make quantum corrections about a classical solution. That is, one calculates radiative corrections at some renormalisation scale using perturbation theory and then solves (non-perturbative) classical equations of motion. One can then systematically add quantum corrections to the classical solutions. We assume that whatever quark model we are using represents just such a solution, renormalised at some scale μ^2. Although physically measurable quantities cannot depend on μ^2, the classical approximation may be better for some value. We treat the value of μ^2 appropriate to a given model, which we shall call Q_0^2, as a free parameter. If one can evaluate the twist-2 target matrix elements within the model, then through equ.(25) and (32) (or equivalently (42) and (43)), one can calculate the twist-2 structure function at all Q^2. Even though the twist-2 contribution may not be dominant at Q_0^2, the general considerations presented earlier (see the discussion of higher twist below equ.(20)) ensure that at high enough Q^2 it will eventually dominate.

The most convenient practical method for evaluating the twist-2 moments of the structure function of some target follows from important work by Jaffe [17]. However, his article which was entitled "parton distribution functions" involved no discussion of renormalisation group corrections, and the calculations in it were only made in the Bjorken limit. It would therefore not be surprising if the student were confused as to the connection between his parton distributions (e.g. calculated for some model) and experimental data. We shall make that connection quite clear.

Following ref.[17] we define a function $H(\alpha)$ (in the $A^+ = 0$ gauge):

$$H(\alpha) = \frac{M}{2\pi}\int_{-\infty}^{\infty} dz e^{-iM\alpha z}\langle N|T(\psi_+^\dagger(\xi^-)\psi_+(0))|N\rangle_c. \tag{44}$$

Here we understand a sum over the spins of the target ($|N\rangle$, mass M), c denotes a connected matrix element and $\psi_+(\xi^-)$ is an abbreviated notation for $\frac{1+\alpha^3}{2}\psi(z;0,0,-z)$.

To understand why the second field operator is evaluated at a point on the light-cone with respect to the first we recall equ.(17). In the target rest frame we can choose the photon four-momentum q to be $(\nu;0,0,-\nu-Mx)$ with $\nu\to\infty$. (Clearly Q^2 is $2M\nu x$.) The argument of the exponential in equ.(17) is $iq\cdot z$ which becomes $i[\nu(z^0+z^3)/2 + Mxz^3]$. The rapid oscillations as $\nu\to\infty$ drive (z^0+z^3) to zero in the Bjorken limit and hence the process is light-cone dominated. (Causality implies that z_T^2 must be zero if we are to obtain a non-zero contribution to the connected matrix element.)

As a further matter of some practical importance, Jaffe has argued that the time-ordering in equ.(44) can be dropped. In particular, it was shown in ref.[17] that for a connected matrix element involving field operators separated on the light-cone one can write either

$$\langle N|T(\psi_+^\dagger(\xi^-)\psi_+(0))|N\rangle_c = \langle N|\psi_+^\dagger(\xi^-)\psi_+(0)|N\rangle_c \tag{45}$$

or equivalently

$$\langle N|T(\psi_+^\dagger(\xi^-)\psi_+(0))|N\rangle_c = -\langle N|\psi_+(0)\psi_+^\dagger(\xi^-)|N\rangle_c. \tag{46}$$

The only reason for preferring one form to the other is calculational simplicity, because equ.(45) has no semi-disconnected contributions for $\alpha > 0$ while the second form has none for $\alpha < 0$. In general one can show that $H(\alpha) = 0$ for $|\alpha| > 1$, or alternatively that $H(\alpha)$ has support on (–1,+1).

The next step in establishing the significance of $H(\alpha)$ is to show that its n'th moment, A_n:

$$A_n = \int_{-1}^{+1} d\alpha\, \alpha^{n-1}\, H(\alpha) \tag{47}$$

is just $\langle N|O_{2,n}|N\rangle$ (see sect. 3.2 of ref.[17]) where $O_{2,n}$ is the local twist-2 operator of order n associated with the structure function $F_2^{(\nu)}$. Finally one has

$$F_2^{(\nu)}(x) = x\ (H(x) - H(-x))\,, \tag{48}$$

while

$$F_3^{(\nu,\bar{\nu})}(x) = (H(x) + H(-x))\,. \tag{49}$$

Comparing with the parton model formulae in equs.(13) and (14) we identify $q(x) = H(x)$ and $\bar{q}(x) = -H(-x)$ for $x > 0$.

Since, as explained earlier, we view the quark model which we use to evaluate $H(\alpha)$ as an approximate solution of the QCD equations at a renormalisation scale Q_0^2, we add the Q_0^2 label to q and $\bar{q}$. Using the relations (45) and (46) to simplify the calculation, we find therefore:

$$q(x, Q_0^2) = \frac{M}{2\pi} \int_{-\infty}^{\infty} dz e^{-iMxz} \langle N|\psi_+^\dagger(\xi^-)\psi_+(0)|N\rangle_c \tag{50}$$

and

$$\bar{q}(x, Q_0^2) = \frac{M}{2\pi} \int_{-\infty}^{\infty} dz e^{-iMxz} \langle N|\psi_+(\xi^-)\psi_+^\dagger(0)|N\rangle_c. \tag{51}$$

(In the last equation we used the translational invariance of the field operators to shift the argument ξ^- in $\psi_+^\dagger$ to $-\xi^-$ in ψ_+. We then changed the integration variable from z to $-z$.)

Before describing some numerical results obtained from equs.(50) and (51), some remarks must be made. Whereas we followed ref.[17] in taking the Bjorken limit ($Q^2 \to \infty$) in order to obtain these parton distributions, the operator matrix elements needed in equ.(17) in order to reproduce data at some finite Q^2 should be evaluated at that Q^2. Although there is no rigorous proof yet, we believe that the difference between the exact results and the equations we use should be of order $1/Q^2$. (This is the case for the simple models which we have considered.) Effectively it amounts to a smearing of $\langle N|\psi_+^\dagger(z^0;0,0,z^3)\,\psi_+(0)|N\rangle$ about the point $z^0 = -z^3$ by an amount of order $1/\nu$ (and therefore of $O(1/Q^2)$).

Since we only intend to compare our twist-2 predictions with data at high Q^2, it is consistent to ignore this correction. Indeed, from this point of view, equs.(50) and (51) are the twist-2 quark/parton distributions whose evolution is governed by perturbative QCD. In order to establish this result it is crucial that the renormalisation scale is not a momentum cut-off (c.f. ref.[18]). It is of course also crucial that the OPE has allowed us to factorise the target dependence from the Q^2 dependence of the quark-probe interaction.

4.1 The MIT bag

Starting from the usual expression for the twist-two quark distribution in equ.(50), one inserts a complete set of energy and momentum eigenstates between the field operators. For the nucleon itself and for the intermediate states we use translationally invariant Peierls-Yoccoz states. These will be two-quark (with mass in the region of $3M/4$) and three-quark + one-antiquark (with mass of order $5M/4$) states.

In the calculation of the antiquark distribution $\bar{q}(x, Q_0^2)$ (for which ψ and $\psi^\dagger$ are interchanged in equ.(50)) the dominant contribution is from a four-quark intermediate state (again with mass of order $5M/4$). One novel feature of this calculation is that it is quite clear that the nucleon has an intrinsic sea [19] — even in a model with just valence quarks, like the three-quark bag. Furthermore, as a result of the Pauli exclusion principle, this intrinsic sea will not be flavor symmetric [19]. Indeed we will find more $d\bar{d}$ pairs in the sea. (This is because, with two spins and three colors one can insert d quarks into five different $1s$-states in a proton bag whereas there are only four states available for u quarks.) Clearly an asymmetry such as this will have important consequences for the Gottfried sum rule as discussed in sect.6.2.

The dominant piece of the valence quark distribution calculated from equ.(50) is that involving a two-quark intermediate state. This term is controlled by two parameters, the bag radius and the mass. For the latter it is important to take into account [20] the effect of gluon exchange which raises the mass of a pair of quarks with spin 1 and lowers that of a spin 0 pair so that the resultant splitting is 200 MeV.

Rather than using the model for the contribution to the valence distribution from $3q + \bar{q}$ intermediate states, we simply use a phenomenological term of similar shape (say $(1-x)^7$) with a normalisation chosen to ensure that we have three valence quarks. Under QCD evolution, this phenomenological term moves to small x so that there is no significant uncertainty for $Q^2 \gtrsim 5\,\text{GeV}^2$ and $x \gtrsim 0.1$ [21, 22]. It is also worth noting that at small x we are sensitive to long-distance physics (the important values of z in equ.(50) are roughly up to order $(Mx)^{-1}$) which is difficult to handle in any phenomenological quark model, so it will be difficult to do better in the near future.

In Fig.3 we show a comparison between the valence quark distribution of the proton calculated for a bag radius of 0.8 fm and various phenomenological fits which will be loosely referred to as data. A priori we have no way to specify the bag scale μ. Instead it is determined by seeing how far one must evolve until the agreement with data at 10 GeV^2 is optimal. Clearly the overall description of the data is rather good. Only at very

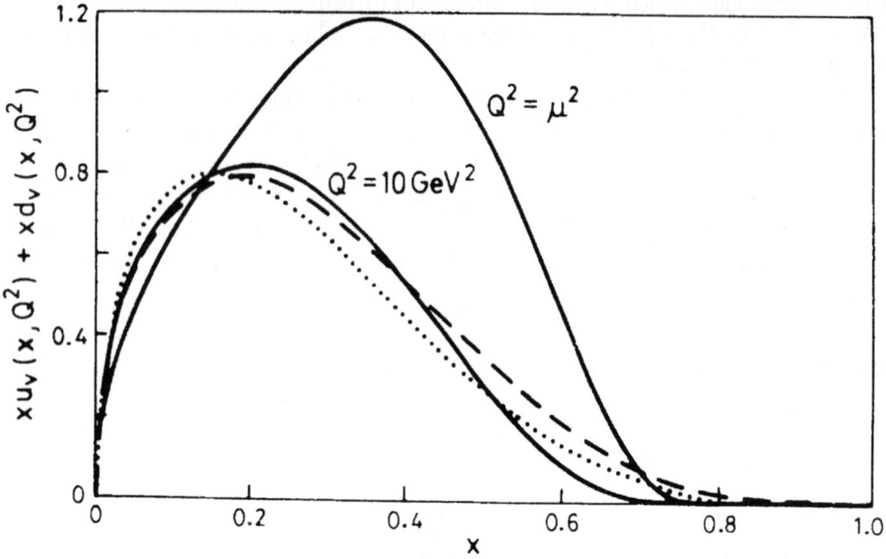

Fig. 3. The valence quark distributions for the proton in the bag model (R = 0.8 fm) at the bag scale μ^2 (0.25 GeV^2) and at 10 GeV^2 (solid lines) [21, 22]. The dashed and dotted lines are the Duke-Owens [23] and MRS parametrisations [24] at 10 GeV^2.

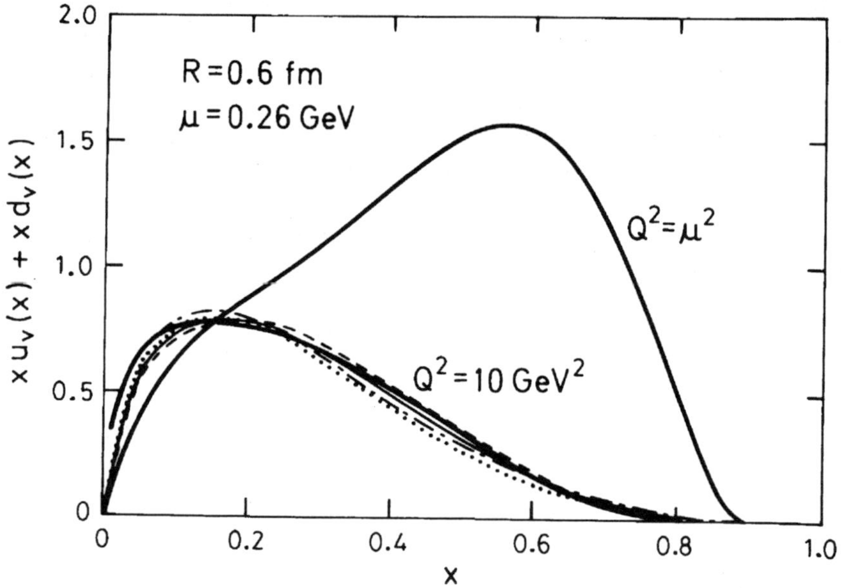

Fig. 4. The valence quark distribution of the proton as in Fig.3 but for R = 0.6 fm [21, 22]. We have also included the fits of EHLQ [25] and DFLM [26].

large values of x ($x \gtrsim 0.7$) is there a significant difference. At such values the struck quark will have a momentum greater than 1 GeV/c and one would expect to have to include the effect of correlations. There is an additional uncertainty associated with the use of leading order QCD, which may be less reliable for higher moments and hence large x.

On the other hand, the agreement with data for calculations with a bag radius of 0.6 fm are essentially perfect (see Fig.4). The improvement at large x is a consequence of the higher average momentum is the smaller cavity. Certainly it would be tempting to conclude that 0.6 fm is preferred. We choose not to draw that conclusion at this stage in view of the problems just cited. Instead we are content to observe that a bag radius in the range 0.6 to 0.8 fm gives a very good representation of the data. Particularly for the calculations at 0.6 fm the bag scale is rather low (e.g. 0.26 GeV in Fig.4). For Λ_{QCD} = 0.2 GeV, as used here, this gives a rather large value of $\alpha_s(\mu^2)$. Other phenomenological studies have used similar values in perturbative calculations of QCD evolution [27], but we would be more comfortable with μ closer to 0.7 or 0.8 GeV. This does seem to be a likely, desirable consequence of including the pionic corrections needed to preserve chiral symmetry [28, 29]. While we shall not pursue this discussion now, the effect of these pionic corrections on the Gottfried sum rule will be mentioned later.

Because the quark distributions measured in deep-inelastic scattering involve light-cone correlation functions, the energy of the struck quark is as important as its three-momentum. This is why, even for SU(6) wavefunctions for which the u and d quarks of all spin orientations have the same distribution of three-momentum, one finds important differences in $u^{\uparrow\downarrow}$ and $d^{\uparrow\downarrow}$ by including the first-order one-gluon-exchange corrections to the *energies* of the intermediate (diquark) states inserted in equ.(50) [20-22,30]. (The lowest mass diquark will give the hardest quark distribution.) In particular, in this unsophisticated model one can readily see that the ratio $d(x)/u(x)$ tends to zero for x going to one. As a consequence $F_{2n}/F_{2p} \rightarrow 1/4$ as $x \rightarrow 1$ [3]. Figure 5 shows the general agreement between the data for the d/u ratio and our calculations. Only at very large x is there any serious discrepancy and this may also be related to the absence of short-range correlations in the bag [31].

In Fig.6 we see that this same, simple physics, familiar from low energy spectroscopy, also leads to a quantitative understanding of the proton spin structure function, $g_{1p}(x)$. The crucial feature is that only a u quark with its spin parallel to that of the proton is accompanied by a low-mass spin-singlet pair of quarks (in an SU(6) proton). As a consequence $u^{\uparrow}$ is the dominant parton distribution at large x.

Of course, because we are using an SU(6) spin-flavor wavefunction, the integral of our g_{1p} agrees with the Ellis-Jaffe sum rule [33] — unlike the data [32]. A clear indication of this problem is the quantitative disagreement at intermediate x in Fig.6. One knows that a more sophisticated treatment of the proton wavefunction including gluonic and pionic corrections would improve the situation a little [34-36]. However, it is also known that the anomaly plays a critical role in the flavor singlet distribution [37-39] and this is still rather controversial [40]. We defer further discussion of this issue to sect.6.3.

Independent of the question of the Ellis-Jaffe sum rule, it would be extremely interesting to obtain data on the neutron spin-dependent structure function $g_{1n}(x)$. It is an exciting prediction of the bag model that g_{1n} should become positive at large x.

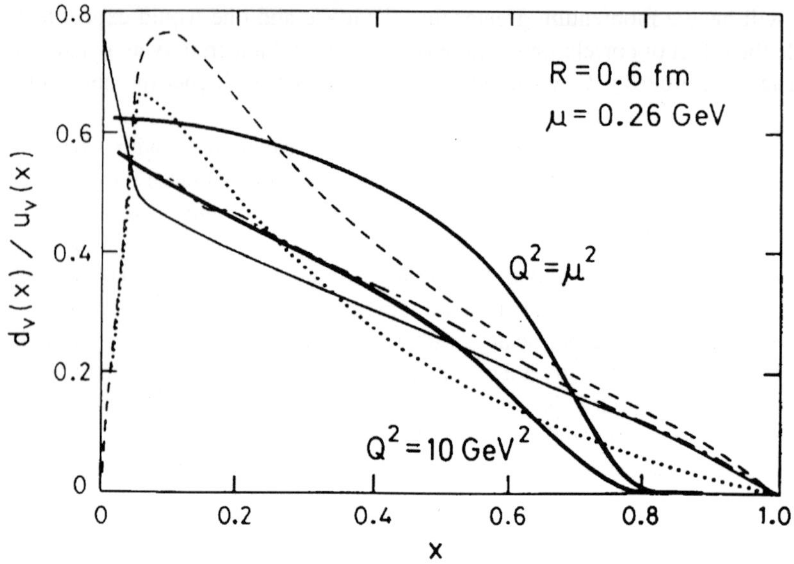

Fig. 5. The bag model prediction for the d/u ratio calculated including the energy shift arising from one-gluon-exchange (heavy solid curves). For comparison we also show four fits to world data. The parameters are as for Fig.4.

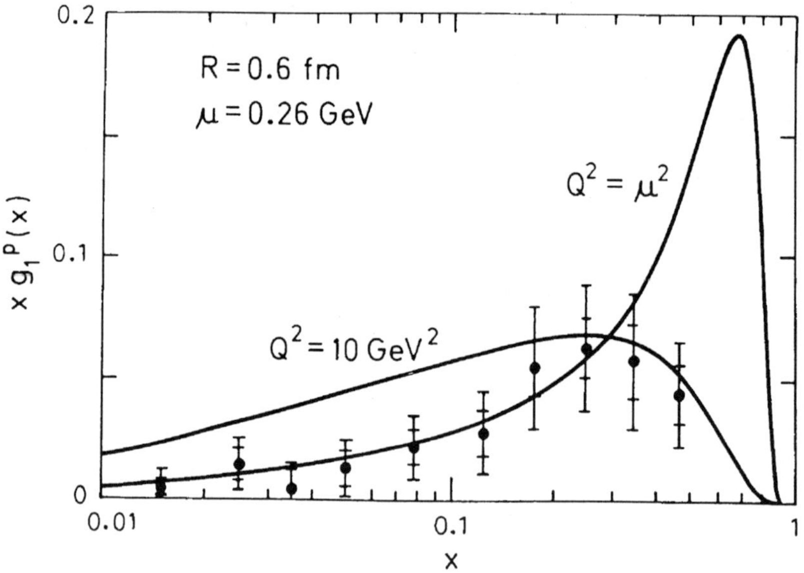

Fig. 6. The spin-dependent structure function of the proton, g_{1p}, calculated in the bag model, in comparison with the data of EMC [32].

However, a recent extension of that work to include pion corrections [29] has led us to question the sensitivity of this prediction to small changes in the model. Morever, if the discrepancy between quark model calculations (e.g. Fig.6 above) and the data arises because of the axial anomaly, there could even be a change in the sign of g_{1n} at large x [41, 42].

4.2 Relativistic Models

The advantages of a model such as the one just discussed is its clear connection to hadron spectroscopy and low energy properties. One drawback, however, is the fact that the nucleon wavefunction used in equ.(50) is essentially non-relativistic – i.e. we need to use a technique like Peierls-Yoccoz to construct a state of total momentum zero. Here we outline some recent work in calculating nucleon quark distributions within a relativistic, covariant approach. Further details can be found in ref.[43] (see also ref.[44]).

To see how one can formulate the problem in a relativistic manner, we need to go back to the hadronic tensor $W_{\mu\nu}$, and in particular delve deeper into the Lorentz and Dirac structure of the truncated nucleon tensor, that is, one which has its fermion legs amputated. Because of the additional spinor degree of freedom the structure of the truncated tensor will necessarily be more complicated than that of the full nucleon tensor, $W_{\mu\nu}$. However, once we identify the relevant structures that contribute to the physical tensor in the Bjorken limit, we will be able to use these in a fully relativistic, covariant calculation of the nucleon structure function. The formalism developed here can also be extended to the case of off-mass-shell nucleons, since the truncated tensor will generally depend upon p^2 as well as q^2 and $p \cdot q$.

We can firstly observe that the nucleon tensor can be written [45]:

$$M\ W^{\mu\nu}(p,q) = \frac{1}{2}\mathrm{Tr}\left[(\not{p} + M)\ \widehat{W}^{\mu\nu}(p,q)\right] \tag{52}$$

where we have explicitly separated the nucleon spinors from the remaining interaction. In general, $\widehat{W}^{\mu\nu}$ must be constructed from the Lorentz tensors (Dirac scalars) $g^{\mu\nu}, p^{\mu}, q^{\mu}$ and Dirac matrices $I, \gamma^{\mu}, \sigma^{\mu\nu}, \gamma^{\mu}\gamma_5$ and γ_5. By parity considerations terms involving $\gamma^{\mu}\gamma_5$ or γ_5 will not contribute to the spin-averaged tensor. Furthermore, terms with $\sigma^{\mu\nu}$ will not contribute to $W^{\mu\nu}$. Therefore we can write [43]

$$\widehat{W}^{\mu\nu}(p,q) = \tilde{g}^{\mu\nu}\left(I\widehat{W}_T^0 + \not{p}\widehat{W}_T^1 + \not{q}\widehat{W}_T^2\right) + \cdots \tag{53}$$

where each of the functions on the right hand side is a scalar function of $q^2, p \cdot q$ and p^2: $\widehat{W}_T^i = \widehat{W}_T^i(p,q) = \widehat{W}_T^i(p^2, p \cdot q, q^2)$. Substituting equ.(53) into (52) and comparing with equ.(1), we see that the transverse structure function W_T ($W_T = W_1 = F_1/M$) can be written as a linear combination of three independent terms,

$$M\ W_T(p,q) = 2\ M\ \widehat{W}_T^0 + 2\ M^2\ \widehat{W}_T^1 + 2\,p \cdot q\ \widehat{W}_T^2. \tag{54}$$

We can calculate the functions $\widehat{W}_T^i$ explicitly by considering the "handbag" diagram, which represents the impulse approximation for quarks. As in the operator product expansion, this amounts to separating out the q^2-dependent part of $\widehat{W}^{\mu\nu}$ from the

q^2-independent, non-perturbative part, which must be described by some model. To calculate the latter we need to consider quark–nucleon vertices that transform as scalars or vectors under Lorentz transformations (since the intermediate spectator diquark state will have either spin 0 or 1, and we need to make an overall Lorentz scalar). It is straightforward to identify the form of the vertices that are allowed by Lorentz, parity and time-reversal invariance, however the specific momentum dependence has to be determined within a model. In general there will be many independent scalar and vector vertex functions. For simplicity we choose a specific form for the scalar (say $I\Phi^S$) and vector ($\gamma_\alpha \Phi^V$) vertices. Calculating the functions $\Phi^{S,V}$ from first principles amounts to solving the relativistic, many-body, bound-state problem, which present day technology does not yet allow, so in practice we use phenomenological input to constrain their functional form. Denoting the quark four-momentum and mass by k and m_q, respectively, we can take the vertex functions as

$$\begin{aligned}\Phi^S(k^2) &= N_S \frac{(k^2 - m_q^2)}{(k^2 - \Lambda_S^2)^2} \\ \Phi^V(k^2) &= N_V \frac{(k^2 - m_q^2)}{(k^2 - \Lambda_V^2)^{7/2}},\end{aligned} \tag{55}$$

chosen to reproduce the correct large-x behaviour of the u and d quark distributions. The constants N_S and N_V are determined by the normalisation condition,

$$\int_0^1 dx\; q_0(x) \;=\; \int_0^1 dx\; q_1(x) \;=\; 1, \tag{56}$$

where $q_{0,1}$ are the distributions for a spin 0, 1 diquark state, respectively. Note that the above vertex functions incorporate the confinement mechanism, by removing the pole in the quark propagator which would have occurred at $k^2 = m_q^2$.

The result of this calculation is that each of the three functions $\widehat{W}_T^{0-2}$ contribute to W_T in the Bjorken limit [43]. This is extremely important if one wishes to discuss scattering from off-mass-shell nucleons. In particular, as shown in ref.[43], it is very unlikely that in scattering from a nuclear target one would obtain the same linear combination that occurs in the free case. As a consequence the usual approximation whereby the nuclear structure function is written as a convolution of a nucleon momentum distribution with the structure function of a free nucleon breaks down. For further details, including a fit to the free structure functions and application to the deuteron and nuclear matter, we refer to ref.[43].

5 Role of the Meson Cloud

Simply on the basis of the Heisenberg uncertainty principle we know that the long range structure of the nucleon must involve a pion cloud. For example, the non-zero value for the neutron charge radius can be easily understood in terms of the emission from a neutron of a light, negatively charged virtual pion, $n \to p + \pi^-$. Furthermore, from PCAC, and from the tremendous successes of chiral quark models [46–49] we expect that the nucleon should have a pion cloud. In addition, because there is no scale at which chiral symmetry can be ignored, the nucleon properties will have pionic corrections at all Q^2.

The possible relevance of the extended pionic structure of the nucleon in high energy processes, such as deep-inelastic scattering, was first suggested by Sullivan in the early 1970s [50]. It was shown that the contribution to the inclusive $\gamma^* N$ cross-section from pion exchange between the virtual photon and the nucleon scales in the Bjorken limit. The reason for this is that, in contrast to processes such as exclusive pion-production which are suppressed by $O(1/Q^2)$ form factors, here it is the inelastic structure function of the pion itself that is probed.

Using this picture of the physical nucleon, it was later noticed [51] that the pion cloud could be responsible for generating an asymmetry between the $\bar{u}$ and $\bar{d}$ quark content of the proton, through the preferred proton dissociation into a neutron and π^+. Furthermore, DIS data on the momentum fractions carried by antiquarks were used to obtain an upper limit on this non-perturbative pionic component. More recently it has been hypothesised [52-56] that this asymmetry could account for some of the apparent discrepancy between the naive parton model prediction for the Gottfried sum rule [57] and its recently determined experimental value [58].

Since it has by far the smallest mass, the pion was the first meson whose contributions to the nucleon structure function were investigated [59, 60]. However, just as other mesons can be included to give corrections to low energy nucleon properties, such as the electromagnetic nucleon form factors or magnetic moments [61], an extended mesonic structure of the nucleon may also be relevant in DIS.

In this section we shall give a detailed account of the calculation of the virtual meson and baryon contributions to the nucleon structure functions. Furthermore, we will use recent DIS data to examine the extent to which such a picture may be relevant in high energy reactions.

The basic hypothesis of this model, in which the nucleon has internal meson and baryon degrees of freedom, is that the physical nucleon state (in an infinite momentum frame) can be expanded (in the one-meson approximation) in a series involving bare nucleon and two-particle meson—baryon states

$$\begin{aligned} |N\rangle_{\text{phys}} = & \sqrt{Z}\,\{|N\rangle_{\text{bare}} \\ & + \sum_{MB} \int dy\, d^2\mathbf{k}_T\, g_{0_{MBN}}\, \phi_{MB}(y,\mathbf{k}_T)\, |M(y,\mathbf{k}_T); B(1-y,-\mathbf{k}_T)\rangle\, \} \end{aligned} \tag{57}$$

Here, $\phi_{MB}(y, \mathbf{k}_T)$ is the probability amplitude for the physical nucleon to be in a state consisting of a meson M and baryon B, having transverse momenta $\mathbf{k}_T$ and $-\mathbf{k}_T$, and carrying longitudinal momentum fractions y and $1 - y$, respectively. Z is the bare nucleon probability. Although we shall work in the one-meson approximation, we shall include higher order vertex corrections to the bare coupling constants $g_{0_{MBN}}$. Illustrated in Fig.7 is the DIS of the virtual photon from the two-particle state $|M; B\rangle$. In Fig.7(a) the photon interacts with a quark or antiquark inside the exchanged meson, while in Fig.7(b) the scattering is from a quark in the baryon component of the physical nucleon.

According to equ.(57), the probability to find a meson inside a nucleon with mo-

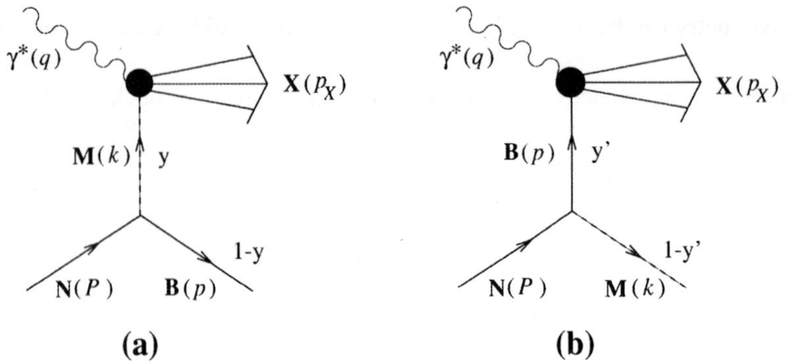

Fig. 7. DIS from the virtual (a) meson and (b) baryon components of a physical nucleon.

mentum fraction y (= $k \cdot q/p \cdot q = k_+/p_+$) is (to leading order in the coupling constant)

$$f_{MB}(y) \equiv Z\ g^2_{0MBN} \int d^2\mathbf{k}_T\ |\phi_{MB}(y, \mathbf{k}_T)|^2 . \tag{58}$$

This must also be the probability to find a baryon inside a nucleon with momentum fraction $1 - y$. The baryon distribution function, $f_{BM}(y')$, where $y' = p' \cdot q/p \cdot q$, is probed directly through the process in Fig.7(b), and should be related to the meson distribution function by

$$f_{MB}(y) = f_{BM}(1 - y) \tag{59}$$

for all y, if the above interpretation is valid. We also demand equal numbers of mesons emitted by the nucleon, $\langle n \rangle_{MB} = \int_0^1 dy\ f_{MB}(y)$, and virtual baryons accompanying them, $\langle n \rangle_{BM} = \int_0^1 dy'\ f_{BM}(y')$:

$$\langle n \rangle_{MB} = \langle n \rangle_{BM} . \tag{60}$$

This is just a statement of charge conservation. Momentum conservation imposes the further requirement that

$$\langle y \rangle_{MB} + \langle y \rangle_{BM} = \langle n \rangle_{MB} \tag{61}$$

where $\langle y \rangle_{MB} = \int_0^1 dy\ y\ f_{MB}(y)$ and $\langle y \rangle_{BM} = \int_0^1 dy'\ y'\ f_{BM}(y')$ are the average momentum fractions carried by meson M and the virtual baryon B, respectively. Equations (60) and (61), and in fact similar relations for all higher moments of $f(y)$, follow automatically from equ.(59).

In what follows we shall explicitly evaluate the functions f_{MB} and f_{BM}, and examine the conditions under which equ.(59) is satisfied. The results will be used to calculate the contributions to the nucleon structure function from the extended mesonic structure of the nucleon, which are expressed as convolutions of the functions $f(y)$ with the structure functions of the struck meson or baryon:

$$\delta^{(MB)}F_{2N}(x) = \int_x^1 dy\; f_{MB}(y)\; F_{2M}(x_M) \tag{62}$$

$$\delta^{(BM)}F_{2N}(x) = \int_x^1 dy'\; f_{BM}(y')\; F_{2B}(x_B) \tag{63}$$

with $x_M = x/y$ and $x_B = x/y'$. Note that equs.(62) and (63) are correct when physical (renormalised) meson—baryon coupling constants are used in the functions f_{MB} and f_{BM} (see sect.5.5 for a discussion on this point). By comparing against the experimental structure functions, we will ultimately test the reliability of the expansion in equ.(57), and also the relative importance of the states involving heavier mesons compared with the pion states.

5.1 Pions – Covariant Formulation

Let us firstly review the previous calculations of the contribution to F_{2N} from the pion cloud. Following the original method of Sullivan, the approach has been to simply treat the diagram in Fig.7(a) as a Feynman diagram. With a pseudoscalar πN coupling, $g_{\pi NN}\bar{u}(p)\; i\gamma_5 u(P)$, the contribution from this diagram to the hadronic tensor of the physical nucleon can be written

$$\delta^{(\pi N)}W^{\mu\nu}(P,q) =$$
$$\int \frac{d^3\mathbf{p}}{(2\pi)^3(2p_0)} \frac{g^2_{\pi NN}(k^2)}{(k^2-m_\pi^2)^2} \frac{1}{2}\mathrm{Tr}\left[(\not{P}+M)\, i\gamma_5\, W_\pi^{\mu\nu}(k,q)\, (\not{p}+M)\, i\gamma_5\right] \tag{64}$$

where the hadronic tensor for the virtual pion is expressed as

$$W_\pi^{\mu\nu} = \tilde{g}^{\mu\nu}W_{1\pi} + \frac{\tilde{k}^\mu\tilde{k}^\nu}{m_\pi^2}W_{2\pi} \tag{65}$$

and where $g^2_{\pi NN}(k^2)$ is the interaction strength. [2] It is customary to isolate the k^2 dependence of $g^2_{\pi NN}(k^2)$ into the πNN form factor: i.e. $g^2_{\pi NN}(k^2) = g^2_{\pi NN}\; \mathcal{F}_{\pi N}(k^2)$, where $g_{\pi NN}$ is now the coupling constant at the pion pole ($\mathcal{F}_{\pi N}(-m_\pi^2) = 1$). We can obtain the contribution to the nucleon structure function W_1 (or W_2, by the Callan-Gross relation) by collecting $\tilde{g}_{\mu\nu}$ (or simply $g^{\mu\nu}$) terms on both sides of equ.(64) [62] to obtain an expression like that in equ.(62).

Performing the elementary trace gives a factor $2P\cdot p - 2M^2 = -k^2$, so that the distribution function of a virtual pion accompanied by a recoiling nucleon is [51, 60]

$$f_{\pi N}(y) = \frac{3g^2_{\pi NN}}{16\pi^2}\; y \int_{-\infty}^{k^2_{max}} dk^2\; \frac{\mathcal{F}^2_{\pi N}(k^2)(-k^2)}{(k^2-m_\pi^2)^2}. \tag{66}$$

[2] Since only tree diagrams are ever considered, the pseudoscalar interaction is equivalent to that with a pseudovector coupling $(f_{\pi NN}/m_\pi)\; \bar{u}(p)\; i\gamma_\alpha\gamma_5\; u(P)\; k^\alpha$, providing the coupling constants are related by $f_{\pi NN} = g_{\pi NN}\; (m_\pi/2M)$.

Here, $k^2 = k^2_{max} - k^2_T/(1-y)$ is the 4-momentum squared of the virtual pion, with a kinematic maximum given by $k^2_{max} = -m^2y^2/(1-y)$, and k^2_T is the pion transverse momentum squared. We have also included a factor 3 by taking account of the different charge states of the nucleon (namely 2 for the dissociation process $p \to n\pi^+$ and 1 for $p \to p\pi^0$). In a covariant formulation the form factor, $\mathcal{F}_{\pi N}$, parameterising the πNN vertex, at which only the pion is off-mass-shell, can only depend on k^2. In the literature this is most often parameterised by a simple monopole or dipole function,

$$\mathcal{F}_{\pi N}(k^2) = \left(\frac{\Lambda^2_{\pi N} - m^2_\pi}{\Lambda^2_{\pi N} - k^2}\right)^n \tag{67}$$

for $n = 1$ and 2, respectively.

Because we integrate over the recoiling particle's momentum, in principle we could also have contributions from processes where a baryon other than a nucleon (e.g. a Δ isobar) is left in the final state in Fig.7(a). It is expected that contributions from the higher mass baryons will be suppressed relative to the nucleon, since the maximum value of k^2 for which energy and momentum can be conserved when a higher mass baryon is produced decreases rapidly as the mass of the baryon increases. Nevertheless, the importance of the Δ-resonance is well known in pion physics. In any quark model the coupling to the N and Δ would enter on the same footing.

The process where the nucleon emits a pion and leaves behind an on-shell Δ was previously calculated in refs.[56, 63, 64], using the effective interaction $(f_{\pi N\Delta}/m_\pi)\, \bar{u}_\alpha(p)\, k^\alpha\, u(P)$, where $u_\alpha(p)$ is the spin-3/2 Rarita-Schwinger spinor-vector [65], which can be formed by combining the spin-1/2 Dirac spinor $u(p,s)$ with vectors $\epsilon_\alpha(\lambda)$:

$$u_\alpha(p,S) = \sum_m \left\langle \frac{3}{2}S \,\middle|\, 1m, \frac{1}{2}s \right\rangle \epsilon_\alpha(m)\, u(p,s). \tag{68}$$

The vectors ϵ_α can be parameterised, in a frame where $p = (p_0;\ \cos\varphi \sin\theta\, |\mathbf{p}|,\ \sin\varphi \sin\theta\, |\mathbf{p}|$ $\cos\theta\, |\mathbf{p}|)$, by

$$\begin{aligned} \epsilon_\alpha(0) &= \frac{1}{M_\Delta}(|\mathbf{p}|;\ \cos\varphi \sin\theta\, p_0,\ \sin\varphi \sin\theta\, p_0,\ \cos\theta\, p_0) \\ \epsilon_\alpha(\pm 1) &= \frac{1}{\sqrt{2}}(0;\ \mp\cos\varphi\cos\theta + i\sin\varphi,\ \mp\sin\varphi\cos\theta - i\cos\varphi,\ \pm\sin\theta). \end{aligned} \tag{69}$$

The energy projection operator for the Rarita-Schwinger spinor-vector is [66]

$$\sum_S u_\alpha(p,S)\bar{u}_\beta(p,S) = \Lambda_{\alpha\beta}(p) \tag{70}$$

where

$$\Lambda_{\alpha\beta}(p) = (\not{p} + M_\Delta)\left(-g_{\alpha\beta} + \frac{\gamma_\alpha\gamma_\beta}{3} + \frac{\gamma_\alpha p_\beta - \gamma_\beta p_\alpha}{3\, M_\Delta} + \frac{2\, p_\alpha p_\beta}{3\, M^2_\Delta}\right). \tag{71}$$

Equation (71) can be verified by using the explicit parameterisation in equ.(69). Using this projection operator, we can therefore proceed to evaluate the $\pi\Delta$ trace factor, which in this case is

$$\frac{1}{2}\mathrm{Tr}\left[\Lambda_{\alpha\beta}(p)\, k^\alpha k^\beta\, W^{\mu\nu}_\pi(k,q)\, (\not{p} + M)\right] \tag{72}$$

and arrive at the distribution function for a pion with a Δ recoil:

$$f_{\pi\Delta}(y) = \frac{4}{3}\frac{f^2_{\pi N\Delta}}{16\pi^2\, m^2_\pi}\, y \int_{-\infty}^{k^2_{max}} dk^2\, \frac{\mathcal{F}^2_{\pi\Delta}(k^2)}{(k^2 - m^2_\pi)^2}\, \frac{\left[(M+M_\Delta)^2 - k^2\right]^2 \left[(M-M_\Delta)^2 - k^2\right]}{6\, M^2_\Delta} \quad (73)$$

where now the kinematic upper limit on k^2 is $k^2_{max} = -(M^2_\Delta - (1-y)\, M^2)\, y/(1-y)$. Note that a dipole function for the $\pi N \Delta$ form factor is necessary to suppress contributions from large $|k^2|$.

Contributions from higher-mass baryon resonances can all be computed from the formulae given above because the lower lying states all have spin 1/2 or 3/2. For the (spin 1/2) Roper resonance, which with a mass M_R = 1440 MeV is the next heaviest state after the Δ, the trace factor is

$$\frac{1}{2}\mathrm{Tr}\left[(\not{P} + M)\, i\gamma_5\, (\not{p} + M_R)\, i\gamma_5\right] = -k^2 + (M_R - M)^2. \quad (74)$$

With a $\pi N R$ coupling constant of $g^2_{\pi NR}/4\pi \approx 5.4$ the integral over y of the pion distribution function for a recoil Roper resonance comes to about 10% of that with a nucleon recoil for the same cut-off parameter. Furthermore, the pion distribution function with a Roper recoil appears at somewhat smaller y than $f_{\pi N}$ or $f_{\pi\Delta}$, which means that the convolution in equ.(62) with the Roper distribution function will only be potentially relevant at very small x. Furthermore, because the Roper has the same quantum numbers as the nucleon, its inclusion as an incoherent contribution is somewhat less justified. In what follows we shall therefore restrict ourselves to the nucleon and Δ baryons only.

In order to conserve momentum and charge, we must also allow for the incident photon to scatter from the recoiling N or Δ after a pion has been emitted, Fig.7(b). Previous attempts at calculating the contributions from these processes within a covariant framework were made by several authors, including Hwang et al. [55], Mulders et al. [44], and Dmitrašinović et al. [67], although all obtained different results. Partly because there is less phenomenological experience with so-called sideways form-factors (where the nucleon, rather than the pion, is off-mass-shell) some early work [52, 56, 68] simply defined $f_{N\pi}(y')$ through equ.(59). However, this is unsatisfactory from a theoretical point of view, and ideally we should be able to verify explicitly that within our model the functions $f_{\pi N}$ and $f_{N\pi}$ satisfy equ.(59). In historical terms, it was the careful examination of this process that opened up a whole Pandora's box of problems, and led to the realisation of the terminal shortcomings of the covariant convolution model. This issue was dealt with more deeply in ref.[43], but let us briefly summarise the origin of the problem.

Clearly the treatment of DIS from an interacting nucleon is considerably more involved than that from a free nucleon. As we saw in the previous section, the truncated nucleon tensor $\widehat{W}^{\mu\nu}_N$. which enters this calculation can be written as a linear combination of three independent terms. Initial calculations [44] assumed that only the term involving the operator $\not{q}$ was relevant. For pointlike nucleons this operator would indeed be the only one present, just as it is for a point-like quark inside a nucleon [69]. Treating the diagram in Fig.7(b) as a Feynman diagram, the contribution to the on-shell nucleon tensor from DIS off the virtual (structured) nucleon with a pion in the final state can be written

$$\delta^{(N\pi)}W^{\mu\nu}(P,q) = g_{\pi NN}^2 \int \frac{d^3\mathbf{k}}{(2\pi)^3(2k_0)} \frac{\mathcal{F}_{N\pi}^2(p^2)}{(p^2-M^2)^2} \times \frac{1}{2}\mathrm{Tr}\left[(\not{P}+M)\, i\gamma_5\, (\not{p}+M)\, \widehat{W}_N^{\mu\nu}(p,q)\, (\not{p}+M)\, i\gamma_5\right] \quad (75)$$

with the tensor $\widehat{W}_N^{\mu\nu}(p,q)$ as defined in equ.(53). Using only the operator $\not{q}$ leads to the virtual nucleon distribution function of ref.[44], namely

$$f_{N\pi}(y') = \frac{3g_{\pi NN}^2}{16\pi^2}\, y' \int_{-\infty}^{p_{max}^2} dp^2\, \frac{\mathcal{F}_{N\pi}^2(p^2)}{(p^2-M^2)^2} \left(-m_\pi^2 - \frac{1-y'}{y'}(p^2-M^2)\right) \quad (76)$$

where $p^2 = p_{max}^2 - p_T^2/(1-y')$ is the 4-momentum squared of the virtual nucleon, with the upper limit now given by $p_{max}^2 = M^2y' - m_\pi^2 y'/(1-y')$, and p_T^2 denotes the nucleon's transverse momentum squared. Apart from possible differences in the form factors, equs.(66) and (76) are clearly related by an interchange $y' \leftrightarrow 1-y$.

The large-$|p^2|$ suppression for the $N\pi N$ vertex is introduced by the form factor $\mathcal{F}_{N\pi}$, which is usually parameterised by a monopole or dipole function [44, 70, 71]

$$\mathcal{F}_{N\pi}(p^2) = \left(\frac{\Lambda_{N\pi}^2 - M^2}{\Lambda_{N\pi}^2 - p^2}\right)^n \quad (77)$$

for $n = 1$ and 2, respectively. However to satisfy equ.(60), the cut-off parameter $\Lambda_{N\pi}$ will in general have to be different from the cut-off $\Lambda_{\pi N}$ regulating the πNN vertex form factor in equ.(66), and a different $\Lambda_{N\pi}$ again to satisfy equ.(61). Furthermore, because the k_T^2 and p_T^2 dependence in the form factors in equs.(67) and (77) are clearly different, the calculated distribution functions $f_{\pi N}$ and $f_{N\pi}$ will in general not satisfy equ.(59). In Fig.8 we plot $f_{\pi N}(y)$ and $f_{N\pi}(1-y)$ for dipole form factors, and cut-offs $\Lambda_{\pi N} = 1$ GeV and $\Lambda_{N\pi} = 1475$ MeV, respectively (to give the same values for $\langle n\rangle_{\pi N}$ and $\langle n\rangle_{N\pi}$, namely ≈ 0.235. Clearly the shapes are quite different, the most obvious difference being that $f_{N\pi}(1-y)$ is finite at $y = 1$.

By using only one operator $\not{q}$ in equ.(75) we are of course assuming that the entire structure function of the virtual nucleon can be represented by the function $\widehat{W}_T^2$ in equ.(53). In the model calculation of the nucleon structure function in ref.[43] it was shown (using the simple quark–nucleon relativistic vertex functions described in sect.4.2), that generally one has non-zero scaling contributions from other functions as well. Furthermore, choosing a different operator form for $\widehat{W}_N^{\mu\nu}$ can also lead to unphysical results. For example, with an operator involving I rather than $\not{q}$ the trace factor in equ.(75) is proportional to $-m_\pi^2$ (i.e. negative) [67].

Problems also arise for the emission of scalar mesons, for which the trace factor in $f_{N\sigma}(y')$ for the structure $\not{q}$ is $4M^2 - m_\sigma^2 + (M^2-p^2)(1-y')/y'$, which is clearly related to the trace in $f_{\sigma N}(y)$ (namely k^2+4M^2) when written in terms of the transverse momentum squared. For an operator I, the trace factor in $f_{N\sigma}(y')$ is $2p^2+2M^2-m_\sigma^2$, which not only violates baryon number conservation but also leads to an unphysical (negative) cross-section. For the DIS from a virtual Δ component, these same difficulties

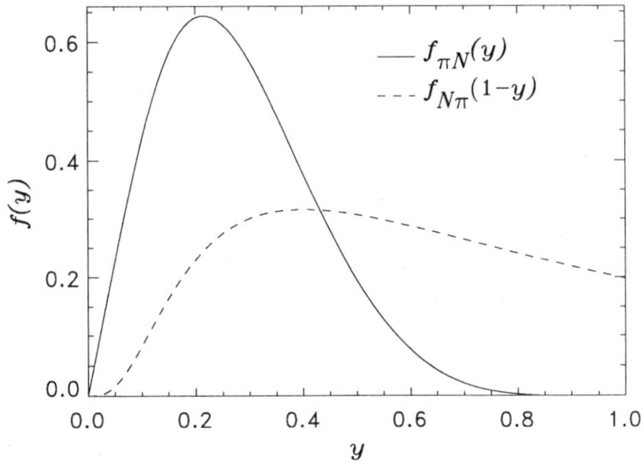

Fig. 8. Distribution functions $f_{\pi N}(y)$ and $f_{N\pi}(1-y)$ with dipole form factors, and cut-offs $\Lambda_{\pi N}$ = 1 GeV and $\Lambda_{N\pi}$ = 1475 MeV chosen to give $\langle n\rangle_{\pi N} = \langle n\rangle_{N\pi} = 0.235$.

will also be present, since the Δ hadronic tensor will have a non-trivial spinor structure, similar to that for the nucleon.

These are the first hints of problems with the covariant approach to calculating DIS processes involving virtual nucleons. Indeed, the convolution formula in equ.(63) appears to be a very special case that cannot be easily obtained from the above considerations. The prescription of ignoring some of the structures in $\widehat{W}_N^{\mu\nu}$ is clearly unsatisfactory, as in principle all should be used. Another important assumption in the covariant convolution model is that the dependence of the virtual meson and baryon structure functions in equs.(62) and (63) on the particles' invariant mass squared is negligible. The argument usually made is that the vertex form factor suppresses contributions from the far off-mass-shell configurations (i.e. for $|k^2| \gtrsim 10\ M^2$ [56]). However, strictly speaking, in this approach even the identification of the off-shell structure functions themselves is not very clear. Some suggestions about how to relate the off-shell functions to the on-shell ones were made [72, 73] in the context of DIS from nuclei, although these were more *ad hoc* prescriptions rather than theoretical derivations. More importantly, a covariant treatment of DIS from virtual nucleons essentially involves both nucleon and antinucleon degrees of freedom. In contrast to this, the Fock state expansion in equ.(57), and in particular the interpretation of $f(y)$ as meson and baryon probability functions, is only meaningful in the IMF. Thus, simply put, the difficulties encountered in trying to obtain sensible results from the covariant calculation of $f(y)$ result from an incompatibility of the covariant formalism with the initial hypothesis that the physical nucleon state can be expanded as in equ.(57). The challenge is therefore to formulate the problem self-consistently, using a single formalism. Since we would like to study the relevance of the virtual meson cloud of the nucleon, the most economical solution would be to keep the Fock state expansion in equ.(57), and reformulate the rest of the problem in time-ordered perturbation theory (TOPT), where equ.(57) is well defined. In fact, an early calculation of the function $f_{\pi N}(y)$ in TOPT was performed some time ago by Güttner et al. [74], in the context of pion electroproduction. More recently the merits of this approach were expounded by Zoller [75], who demonstrated that the distribution functions for the πN and $\pi\Delta$ states calculated in this fashion satisfied equ.(59).

5.2 Pions – TOPT in the IMF

An alternative to the use of covariant Feynman diagrams, in the form of "old-fashioned" time-ordered perturbation theory in the IMF, was proposed some time ago by Weinberg [76] for scalar particles. This was later extended by Drell, Levy and Yan [77] to the πN system in DIS. The main virtues of this approach are that off-mass-shell ambiguities in the structure functions of virtual particles can be avoided, and that the meson and baryon distribution functions can be shown to satisfy equ.(59) exactly. We firstly review the results for the pion cloud, and then compare these with the previous, covariant calculations.

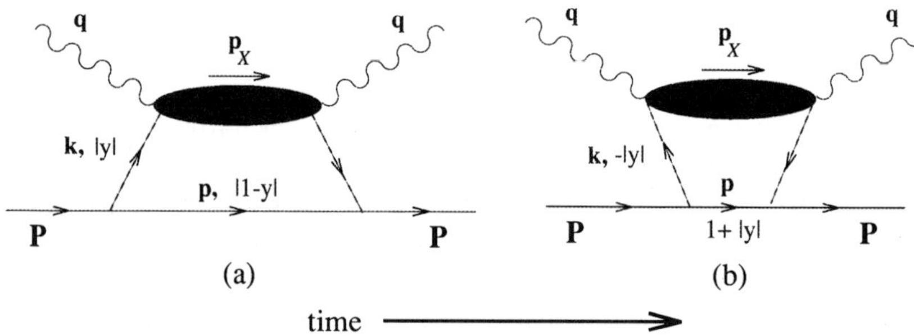

Fig. 9. Time-ordered diagrams for pions moving (a) forwards and (b) backwards in time. Time is increasing from left to right.

In the time-ordered theory the analogue of Fig.7(a) will now involve two diagrams in which the pion moves forwards and backwards in time, Fig.9. However, in a frame of reference where the target nucleon is moving fast in the z direction with longitudinal momentum $P_L(\to\infty)$, only that diagram involving a forward moving pion gives a non-zero contribution. In the IMF the target nucleon of momentum $\mathbf{P} = (0_T, P_L)$ has energy

$$P_0 = P_L + \frac{M^2}{2P_L} + O\left(\frac{1}{P_L^2}\right). \tag{78}$$

Following Weinberg [76] we write the pion 3-momentum as

$$\mathbf{k} = y\,\mathbf{P} + \mathbf{k}_T \tag{79}$$

where $\mathbf{k}_T \cdot \mathbf{P} = 0$, and conservation of momentum demands that the recoil nucleon momentum is

$$\mathbf{p} = (1-y)\,\mathbf{P} - \mathbf{k}_T. \tag{80}$$

Since all particles are on their mass shells the energies of the intermediate meson and baryon must be

$$k_0 = |y|\ P_L + \frac{k_T^2 + m_\pi^2}{2\ |y| P_L} + O\left(\frac{1}{P_L^2}\right) \tag{81}$$

$$p_0 = |1-y|\ P_L + \frac{k_T^2 + M^2}{2\ |1-y| P_L} + O\left(\frac{1}{P_L^2}\right). \tag{82}$$

For forward moving particles, Fig.9(a), y and $1-y$ are positive, and applying the rules of TOPT [76] the contribution to the hadronic tensor of the physical nucleon can be written

$$\delta^{(\pi N)} W^{\mu\nu}(P,q) = \int \frac{d^3\mathbf{k}}{(2\pi)^3 (2p_0)(2k_0)^2} \frac{g_{\pi NN}^2(k)}{(P_0 - p_0 - k_0)^2} \times \frac{1}{2} \mathrm{Tr}\left[(\not{P} + M)\, i\gamma_5\, W_\pi^{\mu\nu}(p,q)\, (\not{p} + M)\, i\gamma_5\right]. \tag{83}$$

The energy denominator in equ.(83) can be rewritten as $(P_0 - p_0 - k_0) = (M^2 - s_{\pi N})/2P_L$, where

$$s_{\pi N} = s_{\pi N}(k_T^2, y) = (p_0 + k_0)^2 - (\mathbf{p} + \mathbf{k})^2 = \frac{k_T^2 + m_\pi^2}{y} + \frac{k_T^2 + M^2}{1-y} \tag{84}$$

is the centre of mass energy squared of the intermediate πN state. Changing the variables of integration from $d^3\mathbf{k}$ to dy and dk_T^2, all powers of P_L are seen to cancel when combined with the vertex factors, $(2p_0)^{-1}$ and $(2k_0)^{-2}$, leading to a result that is P_L-independent. Equating coefficients of $g_{\mu\nu}$, we find that the π distribution function with an N recoil is

$$f_{\pi N}(y) = \frac{3 g_{\pi NN}^2}{16\pi^2} \int_0^\infty \frac{dk_T^2}{(1-y)\, y} \frac{\mathcal{F}_{\pi N}^2(s_{\pi N})}{(M^2 - s_{\pi N})^2} \left(\frac{k_T^2 + y^2 M^2}{1-y}\right) \tag{85}$$

which means that the result of equ.(66) is reproduced, form factor aside. Obviously because here all particles are on-mass-shell, we cannot use the same k^2-dependent form factor as in the covariant case. In the time-ordered calculation, it is quite natural to choose the form factor to be a function of the centre of mass energy squared of the πN system, $s_{\pi N}$, as was done by Zoller [75]. For the functional form of $\mathcal{F}_{\pi N}(s_{\pi N})$ we choose a dipole parameterisation,

$$\mathcal{F}_{\pi N}(s_{\pi N}) = \left(\frac{\Lambda^2 + M^2}{\Lambda^2 + s_{\pi N}}\right)^2 \tag{86}$$

normalised so that the coupling constant $g_{\pi NN}$ has its standard value at the pole ($\mathcal{F}(M^2) = 1$). Previously, in refs.[75, 78] an exponential function was used

$$\mathcal{F}_{\pi N}(s_{\pi N}) = \exp\left(\frac{M^2 - s_{\pi N}}{\Lambda_e^2}\right), \tag{87}$$

although ref.[75] in addition followed an unconventional normalisation.

For a backward moving meson, Fig.9(b), y is negative, and in this case the energy denominator becomes $(P_0 - p_0 - k_0) = 2yP_L + O(1/P_L)$. Therefore in the $P_L \to \infty$ limit this time-ordering is suppressed by a power of $1/P_L^4$, and so does not contribute.

For an interacting nucleon with a pion recoil, the contribution to the nucleon hadronic tensor is

$$\delta^{(N\pi)}W^{\mu\nu}(P,q) = \int \frac{d^3\mathbf{p}}{(2\pi)^3(2p_0)^2(2k_0)} \frac{g^2_{\pi NN}(p)}{(P_0 - p_0 - k_0)^2} \times \frac{1}{2}\mathrm{Tr}\left[(\not{P} + M)\, i\gamma_5\, (\not{p} + M)\, \widehat{W}_N^{\mu\nu}(p,q)\, (\not{p} + M)\, i\gamma_5\right]. \quad (88)$$

The kinematics here are similar to those described above, namely the nucleon and pion move with 3-momenta

$$\mathbf{p} = y'\,\mathbf{P} - \mathbf{k}_T \quad (89)$$

$$\mathbf{k} = (1 - y')\,\mathbf{P} + \mathbf{k}_T \quad (90)$$

and have energies

$$p_0 = |y'|\, P_L + \frac{k_T^2 + M^2}{2\,|y'|P_L} + O\left(\frac{1}{P_L^2}\right) \quad (91)$$

$$k_0 = |1 - y'|\, P_L + \frac{k_T^2 + m_\pi^2}{2\,|1 - y'|P_L} + O\left(\frac{1}{P_L^2}\right) \quad (92)$$

respectively. Then direct evaluation of the trace in equ.(88) gives

$$2\,(2P\cdot p \;-\; 2M^2)\left[\widetilde{g}^{\mu\nu}\;\left(2M\widehat{W}_T^0 \;+\; 2M^2\widehat{W}_T^1 \;+\; 2p\cdot q\,\widehat{W}_T^2\right) \;+\; ...\right]$$
$$= 2\,(2P\cdot p \;-\; 2M^2)\,\widetilde{g}^{\mu\nu}\; W_{1N}(p,q) \;+\; \cdots \quad (93)$$

where now the exact on-shell nucleon structure function appears, and automatically factorises.

For a backward moving nucleon, y' is negative, and $2P \cdot p - 2M^2 = -4y'P_L^2 + O(1/P_L)$, so that the numerator becomes large in the $P_L \to \infty$ limit. Technically this is due to the "badness" of the operator γ_5, which mixes upper and lower components of the nucleon spinors. The energy denominator here is $(P_0 - p_0 - k_0) = 2y'P_L + O(1/P_L)$, and when squared and combined with the $1/P_L^2$ from the integration and vertex factors, the contribution from this diagram vanishes when P_L is infinite. Therefore we need only evaluate the diagram with the forward moving nucleon, which gives the result

$$f_{N\pi}(y') = \frac{3g^2_{\pi NN}}{16\pi^2} \int_0^\infty \frac{dk_T^2}{(1 - y')\,y'\,(M^2 - s_{N\pi})^2} \frac{\mathcal{F}^2_{N\pi}(s_{N\pi})}{} \left(\frac{k_T^2 + (1 - y')^2 M^2}{y'}\right) \quad (94)$$

with

$$s_{N\pi}(k_T^2, y') \;=\; s_{\pi N}(k_T^2, 1 - y') \;=\; \frac{k_T^2 + M^2}{y'} + \frac{k_T^2 + m_\pi^2}{1 - y'}. \quad (95)$$

Notice that the integrand is identical to that in equ.(76), when p^2 there is written in terms of p_T^2 (or k_T^2), except perhaps for the form factor. It was shown in [75] that within this approach there is an explicit symmetry between the processes in which the intermediate pion and the intermediate nucleon are struck if the form factor in $f_{N\pi}$ is taken to be

$$\mathcal{F}_{N\pi}(s_{N\pi}) = \mathcal{F}_{\pi N}(s_{\pi N}). \qquad (96)$$

Then as long as the same cut-off mass parameter is used in both vertex functions, equ.(59) is automatically satisfied. [3] In Fig.10 we plot the function $f_{\pi N}(y)$ evaluated in the IMF, with both the y-dependent exponential, equ.(87), and dipole, equ.(86), form factors, and compare this with the function evaluated in the covariant approach, with the k^2-dependent dipole form factor in equ.(67). In order to make the comparison meaningful the cut-offs have been chosen to yield the same pion multiplicity $\langle n\rangle_{\pi N}$ ($\simeq 0.235$), for which the cut-offs are $\Lambda_{\pi N}$ = 1 GeV, Λ_e = 1380 MeV and Λ = 1425 MeV. With the y-dependent exponential form factor $f_{\pi N}(y)$ is a little broader and peaks at around $y = 0.3$, compared with $y \simeq 0.2$ for the covariant convolution model with a dipole form factor. The y-dependent dipole form factor yields a distribution which is a little broader still. The consequence of this will be that the convolution of $f_{\pi N}(y)$ with F_{2M} for the y-dependent form factors will have a slightly smaller peak and extend to marginally larger x (see sect.5.4).

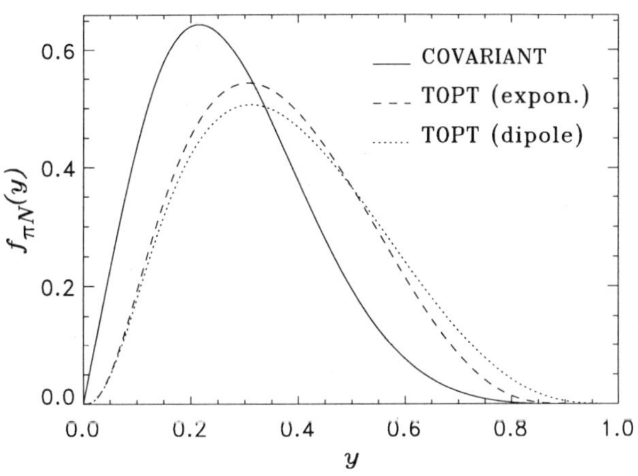

Fig. 10. Distribution functions $f_{\pi N}(y)$ evaluated using covariant and time-ordered perturbation theory. The covariant function is as in Fig.8. The TOPT functions are evaluated with exponential and dipole form factors, with cut-offs Λ_e = 1380 MeV and Λ = 1425 MeV, respectively, to give the same value for $\langle n\rangle_{\pi N} = 0.235$.

The processes involving DIS from $\pi\Delta$ states can also be calculated in the IMF, although some care must be taken when describing the $\pi N\Delta$ interaction vertex in TOPT.

[3] The form factor in equ.(86) may also reduce the number of free parameters in models of NN or $N\overline{N}$ scattering [79], where currently different form factors are necessary for the meson- and baryon-exchange diagrams.

Namely, in TOPT the relevant vertex is $\bar{u}_\alpha(p)\,(P-p)^\alpha\, u(P)$, rather than $\bar{u}_\alpha(p)\, k^\alpha\, u(P)$ as in the covariant theory, where of course the two are (trivially) identical. Using the same formalism as for calculating $f_{\pi N}$, and with the kinematics as given by equs.(78) to (82), but with $M \to M_\Delta$, we find that the pion distribution function with a Δ left in the final state is:

$$f_{\pi\Delta}(y) = \frac{4}{3}\frac{f_{\pi N\Delta}^2}{16m_\pi^2\pi^2}\int_0^\infty \frac{dk_T^2}{(1-y)\,y\,(s_{\pi\Delta}-M^2)^2}\frac{\mathcal{F}_{\pi\Delta}^2(s_{\pi\Delta})}{} \times\frac{\left[k_T^2+(M_\Delta-(1-y)M)^2\right]\left[k_T^2+(M_\Delta+(1-y)M)^2\right]^2}{6\,M_\Delta^2\,(1-y)^3} \qquad (97)$$

where $s_{\pi\Delta} = s_{\pi N}(M \to M_\Delta)$, and we take the same functional form for the $\pi N\Delta$ form factor as for the πNN form factor in equ.(86).

For an interacting Δ with a pion recoil we need additional information on the truncated Δ hadronic tensor, which in this case will involve additional Lorentz indices stemming from the fact that the Δ has spin 3/2. For an on-shell Δ the hadronic tensor can be represented as [80]:

$$W_\Delta^{\mu\nu}(p,q) = \frac{1}{4}\mathrm{Tr}\left[\Lambda_{\alpha\beta}(p)\,\widehat{W}_\Delta^{\mu\nu\alpha'\beta'}(p,q)\right] \qquad (98)$$

with $\Lambda^{\alpha\beta}(p)$ the Δ energy projector given in equ.(71). Assuming the simplest structure for the truncated Δ tensor, namely [75]

$$\widehat{W}_\Delta^{\mu\nu\alpha\beta}(p,q) = -g^{\alpha\beta}\,\widehat{W}_\Delta^{\mu\nu}(p,q) \qquad (99)$$

where $\widehat{W}_\Delta^{\mu\nu}$ has the same Dirac and Lorentz structure as the truncated nucleon tensor, gives the result

$$W_\Delta^{\mu\nu}(p,q) = 2\left(M_\Delta\widehat{W}_\Delta^0 + M_\Delta^2\widehat{W}_\Delta^1 + p\cdot q\widehat{W}_\Delta^2\right)\tilde{g}^{\mu\nu} + \cdots \qquad (100)$$

so that by comparing coefficients of $g^{\mu\nu}$ we can use the term in parentheses to define the on-shell Δ structure function. Utilising the form equ.(99) for the truncated Δ tensor, we can then evaluate the trace

$$\begin{aligned}&\frac{1}{2}\mathrm{Tr}\left[(\not{P}+M)\,(P-p)_\alpha\,\Lambda^{\alpha\alpha'}(p)\;\widehat{W}_{\mu\nu\alpha\beta}(p,q)\,\Lambda^{\beta'\beta}(p)\;(P-p)_\beta\right]\\ &= \frac{1}{6\,M_\Delta^2\,y'^3}\left(k_T^2+(M_\Delta-y'M)^2\right)\left(k_T^2+(M_\Delta+y'M)^2\right)^2\\ &\quad\cdot\left((2M_\Delta\widehat{W}_\Delta^0+M_\Delta^2\widehat{W}_\Delta^1+p\cdot q\widehat{W}_\Delta^2)\,\tilde{g}_{\mu\nu} + \cdots\right)\end{aligned} \qquad (101)$$

so that the on-shell Δ structure function factorises to give equ.(63), with the virtual Δ distribution function given by

$$f_{\Delta\pi}(y') = \frac{4}{3}\frac{f_{\pi N\Delta}^2}{m_\pi^2\,16\pi^2}\int_0^\infty \frac{dk_T^2}{(1-y')\,y'\,(s_{\Delta\pi}-M^2)^2}\,\mathcal{F}_{\Delta\pi}^2(s_{\Delta\pi}) \times\frac{\left[k_T^2+(M_\Delta-y'M)^2\right]\left[k_T^2+(M_\Delta+y'M)^2\right]^2}{6\,M_\Delta^2\,y'^3}. \qquad (102)$$

Clearly this is related to $f_{\pi\Delta}(y)$ by equ.(59) if $\mathcal{F}_{\pi\Delta}(s_{\pi\Delta}) = \mathcal{F}_{\Delta\pi}(s_{\Delta\pi})$, where $s_{\Delta\pi} = s_{\pi\Delta}(M \to M_\Delta)$.

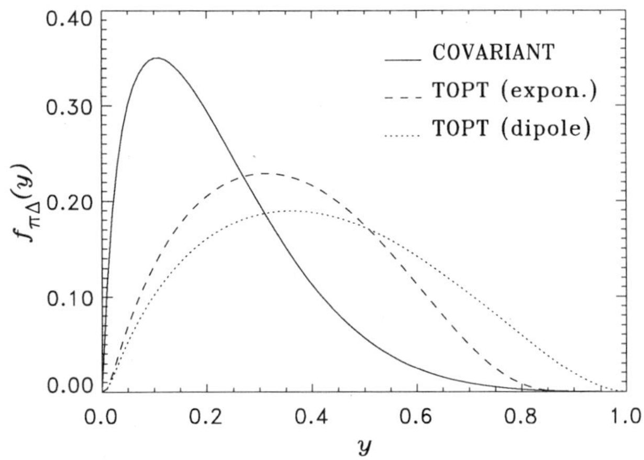

Fig. 11. Distribution functions $f_{\pi\Delta}(y)$ evaluated using covariant and time-ordered perturbation theory. The covariant function is evaluated with a dipole form factor with cut-off mass $\Lambda_{\pi\Delta}$ = 1 GeV. The TOPT functions have exponential and dipole form factors, with cut-offs Λ_e = 1565 MeV and Λ = 1512 MeV, respectively, to give the same value for $\langle n\rangle_{\pi N} = 0.114$.

In Fig.11 we compare the function $f_{\pi\Delta}(y)$, calculated in the IMF, with the function given by equ.(73). The k^2-dependent form factor in the covariant formulation is a dipole form ($\Lambda_{\pi\Delta}$ = 1 GeV), while the $s_{\pi\Delta}$-dependent form factors are dipole (Λ = 1512 MeV) and exponential (Λ_e = 1565 MeV), with all functions normalised to give the same $\langle n\rangle_{\pi\Delta}(\approx 0.114)$. Whereas for $f_{\pi N}(y)$ the y-dependent form factors produced a slight hardening of the distributions when compared with the covariant form factor, here we see a marked difference between the two calculations, in which the distributions calculated in the IMF are considerably broader and extend to larger y.

Having found a useful method for obtaining the pion distributions in a self-consistent manner, we next apply the TOPT/IMF formalism to non-pseudoscalar mesons. As we shall see the difficulties encountered in attempting to compute the contributions from vector mesons make the covariant approach to this problem very problematic, and from a technical point of view the vector mesons can only be handled adequately in the IMF. From a physical point of view, our aim will be to test the relevance or otherwise of the higher mass meson states in the physical nucleon. We focus primarily on the vector mesons, but also briefly re-examine the importance of kaons in the time-ordered formalism.

5.3 Heavier Mesons

The importance of vector mesons in nuclear physics is well known. In the context of meson exchange models of the NN force in nuclear physics, it has long been realised that vector mesons play a vital role [79,81-86]. For example, the isovector ρ meson provides some cancellation of the tensor force generated by π meson exchange. On the other hand, the isoscalar ω meson, through its large vector coupling, is responsible for the short range NN repulsive force, and also provides most of the spin-orbit interaction. Traditionally it has been necessary to use hard vector meson—nucleon form factors in order to fit the NN phase shifts [79]. However, alternative approaches have recently been developed in which the NN data can be fitted with quite soft form factors [86-88].

From another direction, the vector meson dominance model of the elastic electromagnetic nucleon form factors, in which an isovector photon couples to the nucleon via a ρ meson, provides a natural explanation of the dipole Q^2 behaviour of the γNN vertex function. Recent analyses [86] have shown that a ρNN vertex parameterised by a soft monopole form factor ($\Lambda_{\text{monopole}} \sim 800$ MeV) provides a good description of the Q^2 dependence of the Dirac and Pauli form factors. The effect of vector mesons upon nucleon electromagnetic form factors has also been explored [61, 89] in the cloudy bag model [47], and in various soliton models [90].

In previous calculations [55], the vector meson distributions were evaluated within a covariant framework, but with the assumption that the vector meson and nucleon intermediate states were on-mass-shell. In this section we extend the analysis of pions in sect.5.2 to the vector meson sector. Specifically, we shall demonstrate that the vector meson functions, calculated within the TOPT/IMF formalism, can be made to satisfy the relation equ.(59) exactly.

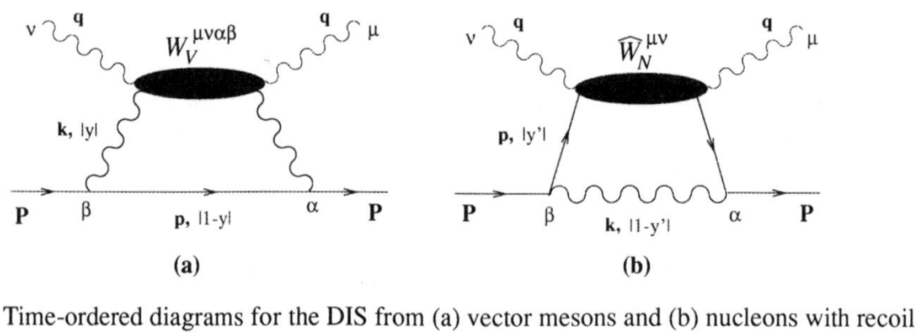

Fig. 12. Time-ordered diagrams for the DIS from (a) vector mesons and (b) nucleons with recoil vector mesons, that are non-zero in the IMF.

For the effective VNN interaction we include both a vector, $g_{VNN}\, \bar{u}(p)\gamma^\alpha \epsilon_\alpha u(P)$, and a tensor, $f_{VNN}/(4M)\, \bar{u}(p)\, i\sigma^{\alpha\beta}[(P_\alpha - p_\alpha)\epsilon_\beta - -(P_\beta - p_\beta)\epsilon_\alpha]u(P)$, coupling, where $V = \rho$ or ω, and $\epsilon_\alpha(\lambda)$ is the polarisation vector for a spin 1 meson with helicity λ. In the calculation of the vector meson distributions in ref.[78] the tensor coupling was taken to be $\sim \bar{u}(p)i\sigma^{\alpha\beta}k_\alpha\epsilon_\beta u(P)$ [79]. In our treatment of the $\pi\Delta$ states in the previous section, the derivative interaction was constructed from baryon momenta, $P_\alpha - p_\alpha$, rather than from the pion momentum k_α. For overall consistency in calculating contributions from all the meson–baryon states, we therefore use the above interaction for the tensor VNN vertex also [91]. In both cases, however, one can explicitly verify that the probability conservation condition, equ.(59), is satisfied [78].

The contribution from the diagram with backward moving vector mesons is suppressed in the IMF by the energy denominators, just as for pions. Therefore we only need to evaluate the diagram with forward moving vector mesons, which gives the following contribution to the nucleon hadronic tensor

$$
\begin{aligned}
\delta^{(VN)}W^{\mu\nu}(P,q) = \int \frac{d^3\mathbf{k}}{(2\pi)^3(2p_0)(2k_0)^2} \frac{1}{(P_0-p_0-k_0)^2} \\
\times \frac{1}{2}\sum_{\lambda,\lambda'} \mathrm{Tr}\left[(\not{P}+M)\left(g_{VNN}(k)\,\gamma^{\alpha'} + \frac{f_{VNN}(k)}{4M} 2i\sigma^{\alpha''\alpha'}\,(P_{\alpha''}-p_{\alpha''})\right)\right. \\
\cdot(\not{p}+M)\,\epsilon^*_{\alpha'}(\lambda)\epsilon_\alpha(\lambda)\;W_V^{\mu\nu\alpha\beta}(k,q)\;\epsilon^*_\beta(\lambda')\epsilon_{\beta'}(\lambda') \\
\left.\cdot\left(g_{VNN}(k)\,\gamma^{\beta'} + \frac{f_{VNN}(k)}{4M}2i\sigma^{\beta''\beta'}\left(P_{\beta''}-p_{\beta''}\right)\right)\right]
\end{aligned}
\tag{103}
$$

Evaluating the trace gives

$$
\left(g^2_{VNN}(k)\,A_{\alpha\beta} \;+\; f^2_{VNN}(k)\,B_{\alpha\beta} \;+\; g_{VNN}(k)\,f_{VNN}(k)\,C_{\alpha\beta}\right)
\tag{104}
$$

where

$$
\begin{aligned}
A_{\alpha\beta} &= (P-p)^2\,g_{\alpha\beta} \;+\; 2\,(P_\alpha p_\beta \;+\; p_\alpha P_\beta) \\
B_{\alpha\beta} &= (P-p)^2\,g_{\alpha\beta} \;-\; (P_\alpha P_\beta + p_\alpha p_\beta) \;-\; \frac{1}{2}(P_\alpha-p_\alpha)(P_\beta-p_\beta) \\
&\qquad + \frac{P\cdot p}{2M^2}(P_\alpha+p_\alpha)(P_\beta+p_\beta) \\
C_{\alpha\beta} &= 2\,(P-p)^2\,g_{\alpha\beta} \;--\; (P_\alpha-p_\beta)\,(P_\beta-p_\alpha)
\end{aligned}
\tag{105}
$$

are the VNN vertex trace factors for the vector, tensor and vector–tensor interference couplings, respectively. For an on-mass-shell vector meson, the spin 1 tensor $W_V^{\mu\nu\alpha\beta}$, symmetric under the interchange of $\mu \leftrightarrow \nu$ and $\alpha \leftrightarrow \beta$, is given by

$$
W_V^{\mu\nu\alpha\beta}(k,q) = \left(\tilde{g}^{\mu\nu}\;W_{1V}(k,q) \;+\; \frac{\tilde{k}^\mu\,\tilde{k}^\nu}{m_V^2}\;W_{2V}(k,q)\right)\;\tilde{g}^{\alpha\beta}.
\tag{106}
$$

This form guarantees that the vector current is conserved, $k_{\alpha(\beta)}W_V^{\mu\nu\alpha\beta} = 0 = q_{\mu(\nu)}W_V^{\mu\nu\alpha\beta}$. Furthermore, it reproduces the correct unpolarised on-shell spin 1 tensor when contracted with the meson polarisation vectors ($\epsilon_{\alpha(\beta)}$) and summed over the V helicity, λ [92]:

$$
\begin{aligned}
W_V^{\mu\nu}(k,q) &= \sum_\lambda \epsilon^*_\alpha(\lambda,k)\;\epsilon_\beta(\lambda,k)\;W_V^{\mu\nu\alpha\beta}(k,q) \\
&= \left(-g_{\alpha\beta} + \frac{k_\alpha k_\beta}{k^2}\right)\;W_V^{\mu\nu\alpha\beta}(k,q) \\
&\propto \tilde{g}^{\mu\nu}\;W_{1V}(k,q) \;+\; \frac{\tilde{k}^\mu\;\tilde{k}^\nu}{m_V^2}\;W_{2V}(k,q).
\end{aligned}
\tag{107}
$$

In the case of DIS from a vector particle emitted by a nucleon, we contract the spin 1 tensor $W_V^{\mu\nu\alpha\beta}$ with the VNN vertex trace factors in equ.(105). Equating coefficients of $g^{\mu\nu}$ in equ.(103), and using the same IMF kinematics as for the πN system, except with $m_\pi \to m_V$, together with the Callan-Gross relation for the nucleon and vector meson, enables the contribution to F_{2N} from vector mesons to be written as a convolution of the vector meson distribution function $f_{VN}(y)$ with the on-shell vector meson structure function $F_{2V}(x/y)$, as in equ.(62), where now

$$
\begin{aligned}
f_{VN}(y) = {} & \frac{c_V}{16\pi^2}\int_0^\infty \frac{dk_T^2}{(1-y)\,y\,(M^2-s_{VN})^2}\frac{\mathcal{F}_{VN}^2(s_{VN})}{} \\
\times & \left\{ g_{VNN}^2 \left(\frac{\left(k_T^2+y^2M^2+m_V^2\right)\left(k_T^2+y^2M^2+(1-y)^2m_V^2\right)}{y^2(1-y)m_V^2} + \frac{k_T^2+y^2M^2}{1-y} - 4M^2 \right) \right. \\
& + g_{VNN}\, f_{VNN} \left(- \frac{\left(k_T^2+y^2M^2-(1-y)m_V^2\right)^2}{2(1-y)^2)m_V^2} + \frac{4\left(k_T^2+y^2M^2\right)}{(1-y)} \right) \\
& + f_{VNN}^2 \left(\frac{(2-y)^2\left(k_T^2+y^2M^2+(1-y)m_V^2\right)^2\left(k_T^2+(4(1-y)+y^2)M^2\right)}{16y^2(1-y)^3M^2m_V^2} \right. \\
& \quad - - \frac{\left(k_T^2+y^2M^2\right)\left(k_T^2+(-4(1+y)+y^2)M^2\right)}{4(1-y)^2M^2} \\
& \quad \left.\left. - - \frac{\left(k_T^2+y^2M^2+(1-y)^2m_V^2\right)^2}{2y^2(1-y)^2m_V^2} - - \frac{\left(k_T^2+y^2M^2+m_V^2\right)^2}{2y^2m_V^2} \right) \right\}
\end{aligned}
\qquad (108)
$$

where $c_V = 1 + 2\delta_{V\rho}$ is the isospin factor (here δ is the Kronecker-δ symbol). The VN centre of mass energy squared is $s_{VN} = s_{\pi N}(m_\pi \to m_V)$, and we take the same form factor for both the vector and tensor couplings, $g_{VNN}(s_{VN}) = g_{VNN}\ \mathcal{F}_{VN}(s_{VN})$ and $f_{VNN}(s_{VN}) = f_{VNN}\ \mathcal{F}_{VN}(s_{VN})$, with $\mathcal{F}_{VN}(s_{VN})$ defined analogously to equ.(86).

For the vector meson recoil process, we evaluate the distribution function $f_{NV}(y')$ using the full spinor structure of $\widehat{W}_N^{\mu\nu}$ given in equ.(53):

$$
\begin{aligned}
\delta^{(NV)}W^{\mu\nu}(P,q) = {} & \int \frac{d^3\mathbf{p}}{(2\pi)^3(2p_0)^2(2k_0)}\frac{1}{(P_0-p_0-k_0)^2} \\
\times \frac{1}{2}\sum_\lambda \mathrm{Tr} & \left[(\not{P}+M)\left(g_{VNN}(p)\,\gamma^\alpha + \frac{f_{VNN}(p)}{4M}2i\sigma^{\alpha'\alpha}(P_{\alpha'}-p_{\alpha'}) \right)(\not{p}+M)\,\widehat{W}_N^{\mu\nu}(p,q)\,(\not{p}+M) \right. \\
& \left. \cdot \left(g_{VNN}(p)\,\gamma^\beta + \frac{f_{VNN}(p)}{4M}2i\sigma^{\beta'\beta}(P_{\beta'}-p_{\beta'}) \right)\epsilon_\alpha^*(\lambda)\epsilon_\beta(\lambda) \right] .
\end{aligned}
\qquad (109)
$$

Performing the contractions over the indices α, β leads to the convolution integral of equ.(63), with the nucleon distribution function $f_{NV}(y')$ with a vector meson recoil given by equ.(108) but with $y \to 1-y'$, with $s_{NV}(k_T^2, y') = s_{VN}(k_T^2, 1-y')$. Again, we have evaluated only the diagram with forward moving nucleons which is non-zero in the IMF. Thus we can prove that the probability distributions for the VN intermediate states are related by $f_{NV}(y') = f_{VN}(1-y')$.

One should observe that the trace factor inside the braces in $f_{VN}(y)$ is divergent in the limit $y \longrightarrow 0$. To illustrate one of the problems with the covariant approach to calculating f_{VN}, consider a form factor which behaves like $\exp\left[y\,(M^2 - s_{VN})\right]$. Since $k^2 = -(k_T^2 + M^2y^2)/(1-y)$, this form factor corresponds to a k^2-dependent covariant form factor proportional to $\exp\left[k^2 - m_V^2\right]$. With such a form factor, $\delta^{(VN)}F_{2N}(x)$ would approach a finite value as $x \longrightarrow 0$, much like a perturbative sea distribution. However, there are several problems with accepting such a result, the most obvious of which is that it would violate charge and momentum conservation very badly, since $f_{NV}(y') \longrightarrow 0$ for $y' \longrightarrow 1$ and $\longrightarrow$ constant as $y' \longrightarrow 0$ when the same form factor is used for the diagram where the nucleon is struck (i.e. a form factor which in the covariant formalism corresponds to $\exp\left[p^2 - M^2\right]$). Furthermore, it would lead to a gross violation of the Adler sum rule, which integrates the flavor combination $u - \bar{u} - d + \bar{d}$, and such a violation has not been observed in the range $1 < Q^2 < 40\,\mathrm{GeV}^2$ [93]. This gives further evidence for the preference of the IMF approach together with the y-dependent form factor in equ.(86) or (87).

To complete our discussion of vector mesons, we give the results for the functions describing the $V\Delta$ states. We saw in the previous section that the contributions from the $\pi\Delta$ states were certainly not negligible in comparison with the πN components of the physical nucleon. For the vector mesons, we would also like to examine whether the Δ isobar is of any importance. Since the ω meson is isoscalar, the only vector meson able to couple to a nucleon and Δ is the ρ, and for this we use a pseudovector coupling [79], $(f_{\rho N\Delta}/m_\rho)\,\bar{u}(P)\,i\gamma_5\gamma^\alpha u^\beta(p)\,((P-p)_\alpha\epsilon_\beta - (P-p)_\beta\epsilon_\alpha)$. Again, we drop those diagrams which give order $1/P_L^4$ contributions in IMF (i.e. for backward moving ρ mesons or Δ). The contribution from scattering from a ρ meson with a Δ recoil is obtained by evaluating the following trace:

$$
\begin{aligned}
&\frac{1}{2}\sum_{\lambda,\lambda'}\mathrm{Tr}\Big[(\not{P}+M)\,i\gamma_5\gamma^\alpha\,\Lambda^{\alpha'\beta'}(p)\,\big((P-p)_\alpha\epsilon^*_{\alpha'}(\lambda)\epsilon_{\alpha''}(\lambda) - (P-p)_{\alpha'}\epsilon^*_\alpha(\lambda)\epsilon_{\alpha''}(\lambda)\big)\\
&\quad\cdot W_\rho^{\mu\nu\alpha''\beta''}(k,q)\,i\gamma_5\gamma^\beta\,\big((P-p)_{\beta'}\epsilon^*_{\beta''}(\lambda')\epsilon_\beta(\lambda') - (P-p)_\beta\epsilon^*_{\beta''}(\lambda')\epsilon_{\beta'}(\lambda')\big)\Big]
\end{aligned}
\tag{110}
$$

where $W_\rho^{\mu\nu\alpha\beta}$ is the ρ meson structure tensor, as given in equ.(106). The resulting probability distribution to find a ρ in the physical nucleon with a Δ recoil is therefore

$$
\begin{aligned}
f_{\rho\Delta}(y) &= \frac{4}{3}\frac{f_{\rho N\Delta}^2}{m_\rho^2\,16\pi^2}\int_0^\infty \frac{dk_T^2}{(1-y)\,y}\frac{\mathcal{F}_{\rho\Delta}^2(s_{\rho\Delta})}{(M^2-s_{\rho\Delta})^2}\\
&\times\Big\{--\frac{4\,MM_\Delta}{3}\left(2M_\Delta^2+MM_\Delta+2M^2\right) -- \frac{4\,MM_\Delta}{3m_\rho^2}((P-p)\cdot k)^2\\
&\quad --\frac{4}{3m_\rho^2}\left(M_\Delta^2(P\cdot k)^2+M^2(p\cdot k)^2\right)+\frac{4\,P\cdot p}{3}\left(2M_\Delta^2+4MM_\Delta+M^2\right)\\
&\quad +\frac{4\,P\cdot p}{3m_\rho^2}(p\cdot k)^2\left(1-\frac{M^2}{M_\Delta^2}\right) -- 4\,(P\cdot p)^2\left(1-\frac{2P\cdot k\,p\cdot k}{3m_\rho^2M_\Delta^2}-\frac{P\cdot p}{3M_\Delta^2}\right)\Big\}
\end{aligned}
\tag{111}
$$

with the kinematics as given in equs.(78) through (82), except with $m_\pi \longrightarrow m_\rho$ and $M \longrightarrow M_\Delta$.

For the DIS off a virtual Δ with a ρ meson recoil, we need to evaluate the trace

$$\frac{1}{2}\sum_{\lambda} \mathrm{Tr}\left[(\not{P}+M)\, i\gamma_5\gamma_\alpha \left((P-p)^\alpha \epsilon^{\alpha'*}(\lambda) - (P-p)^{\alpha'}\epsilon^{\alpha*}(\lambda)\right)\right.$$
$$\cdot \Lambda_{\alpha'\alpha''}(p)\, \widehat{W}_{\Delta}^{\mu\nu\alpha''\beta''}(p,q)\, \Lambda_{\beta''\beta'}(p)$$
$$\left.\cdot\, i\gamma_5\gamma_\beta \left((P-p)^{\beta'}\epsilon^{\beta}(\lambda) - (P-p)^{\beta}\epsilon^{\beta'}(\lambda)\right)\right] \qquad (112)$$

It's then straightforward to show, using the kinematics of equs.(89) to (92) that $f_{\Delta\rho}(y') = f_{\rho\Delta}(1-y)$, when the form factors satisfy $\mathcal{F}_{\rho\Delta}(s_{\rho\Delta}) = \mathcal{F}_{\Delta\rho}(s_{\Delta\rho})$.

In Fig.13 we show the vector meson distribution functions $f_{\rho N}$, $f_{\omega N}$ and $f_{\rho\Delta}$ as a function of y, for the dipole form factor of the form in equ.(86), with $\Lambda = 700$ MeV in all cases. The dominant contributions come from the tensor (derivative) couplings, which is reflected in the larger ρN and $\rho\Delta$ distributions in comparison with the ωN. Also, the vector distribution functions tend to peak at slightly larger y values ($y \sim 0.5$) in comparison with the πN and $\pi\Delta$ functions.

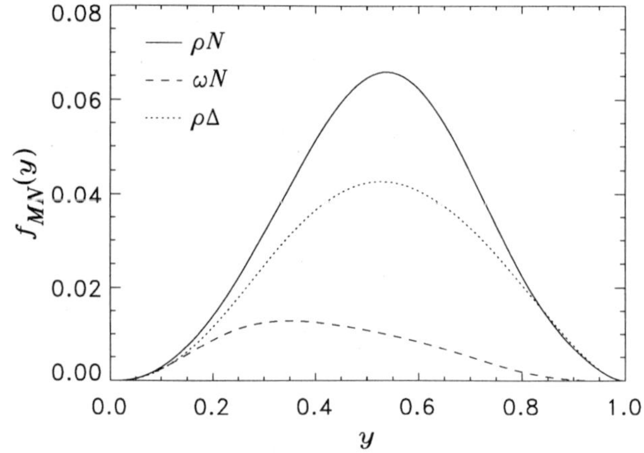

Fig. 13. Vector meson distribution functions in the nucleon. The (dipole) form factor cut-off is $\Lambda = 700$ MeV for all curves.

To conclude this discussion of heavy mesons we consider the DIS process involving the kaon cloud of the nucleon using the time-ordered formalism in the IMF (noting, however, that the role of kaons was first examined by Signal and Thomas [94] within a covariant approach). Through the proton dissociation processes $p \to K^0\Sigma^0(\Lambda^0)$ and $p \to K^+\Sigma^-$, the virtual photon will probe the quark structure of the virtual strange mesons and hyperons. Such a process will naturally generate a non-perturbative strange quark component of the nucleon, as well as a different antistrange sea, thereby breaking SU(3) flavor symmetry of the sea in the process.

Taking a pseudoscalar coupling for the KNH vertex, where the hyperon $H = \Sigma$ or Λ, the kaon distribution function is similar to the pion distribution function $f_{\pi N}(y)$, except the mass of the recoil state is now different,

$$f_{KH}(y) = c_H \frac{g_{KNH}^2}{16\pi^2} \int_0^\infty \frac{dk_T^2}{(1-y)\, y\, (M^2 - s_{KH})^2} \frac{\mathcal{F}_{KH}^2(s_{KH})}{} \times \left(\frac{k_T^2 + y^2 M^2 + (M_H - M)(M_H - (1-2y)M)}{1-y} \right) \quad (113)$$

where the isospin factor is $c_H = 1 + 2\delta_{H\Sigma}$. Similarly for the DIS from a strange baryon with a spectator K, repeating the calculation of sect.5.2, we find that the hyperon distribution function $f_{HK}(y')$ is given by the right hand side of equ.(113) with $y \to 1 - y'$, thereby satisfying equ.(59) exactly. Numerically, the kaon distributions are much smaller than the vector meson distributions, and can for practical purposes be dropped from the analysis.

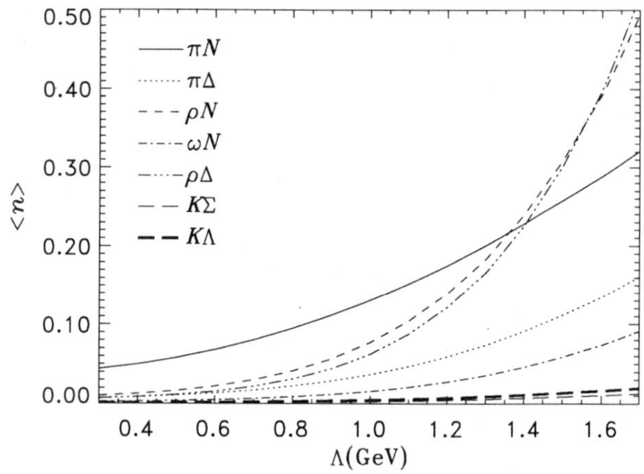

Fig. 14. Calculated average meson multiplicities in the nucleon, as a function of the (dipole) meson-nucleon form factor cut-off.

The relatively small size of the kaon contributions is also clear from Fig.14, where we compare the average number of all mesons considered, $\langle n \rangle_{MB}$, as a function of the dipole form factor cut-off Λ. For relatively small cut-off masses, $\Lambda \lesssim 0.7$ GeV, the dominant contribution is from the πN component. However, the rapid growth with Λ of the ρ meson multiplicities and momentum fractions means that for large Λ ($\gtrsim 1.2 - 1.3$ GeV) the vector mesons become as important numerically as pions. In fact, the strong k_T dependence in $f_{\rho N}(y)$ and $f_{\rho\Delta}(y)$ implies that for $\Lambda \gtrsim 1.4$ GeV $\langle n \rangle_{\rho N}$ actually exceed $\langle n \rangle_{\pi N}$.

Note that for the πN component, $\Lambda = (600, 1000, 1400)$ MeV corresponds to an exponential cut-off $\Lambda_e \simeq (580, 1130, 1360)$ MeV, and a covariant dipole form factor cut-off $\Lambda_{\pi N} \simeq (590, 760, 980)$ MeV for the same $\langle n \rangle_{\pi N}$. In many nuclear physics calculations quite hard form factors of the k^2-dependent type are often used, for example in NN potential models, where cut-offs of the order of 1.5—2 GeV are typical. Clearly such large cut-offs would imply an extremely large number of pions and an even larger number of vector mesons. Whether or not it is reasonable to accept such large heavy meson components in nucleon DIS is debatable, but obviously we would like some data to tell us whether this is so.

Up until now the cut-off Λ has been a free parameter. Indeed, because quantities such as the average number of mesons in the nucleon are not directly (or unambiguously) extracted from experiment, we cannot draw any conclusions about the size of Λ from the functions $f_{MB}(y)$ alone. However, we may be able to restrict the range of allowable values of Λ by comparing the calculated meson and baryon contributions with the experimental structure functions, or quark distributions. This is where we turn our attention next.

5.4 Nucleon Quark Distributions

With the functions f_{MB} and f_{BM} now calculated, we are able to compute the contributions to the quark and antiquark distributions of the proton from the DIS from its virtual meson and baryon components. The total contribution to a quark distribution in the proton from this process is

$$\delta q(x) = \sum_{M,B} \left(\delta^{(MB)} q(x) + \delta^{(BM)} q(x) \right) \tag{114}$$

and similarly for the antiquark distribution. Using the Clebsch-Gordan coefficients for the various charge states of the meson–baryon combinations we can easily obtain the individual flavor distributions. For example, for DIS from virtual π or ρ mesons we have contributions

$$\begin{aligned} \delta^{(\pi/\rho\, N)} u(x) &= \int \frac{dy}{y} f_{\pi/\rho\, N}(y) \left(\frac{1}{3} u^{\pi^0}(x_M) + \frac{2}{3} u^{\pi^+}(x_M) \right) = \int \frac{dy}{y} f_{\pi/\rho\, N}(y) \frac{5}{6} V^M(x/y) \\ \delta^{(\pi/\rho\, N)} d(x) &= \int \frac{dy}{y} f_{\pi/\rho\, N}(y) \left(\frac{1}{3} d^{\pi^0}(x_M) + \frac{2}{3} d^{\pi^+}(x_M) \right) = \int \frac{dy}{y} f_{\pi/\rho\, N}(y) \frac{1}{6} V^M(x_M) \\ &\text{etc.} \end{aligned} \tag{115}$$

Similar expressions can be deduced for other mesons based solely on SU(3) symmetry. For simplicity we have assumed here the same meson valence quark distribution $V^M(x_M)$ for all mesons (sea components of the meson distributions themselves are not included),

$$\begin{aligned} u^{\pi^+/\rho^+} = \bar{d}^{\pi^+/\rho^+} &= 2u^{\pi^0/\rho^0/\omega} = 2\bar{u}^{\pi^0/\rho^0/\omega} = 2d^{\pi^0/\rho^0/\omega} = 2\bar{d}^{\pi^0/\rho^0/\omega} \\ &= u^{K^+} = d^{K^0} = \bar{s}^{K^+} = \bar{s}^{K^0} = V^M \end{aligned} \tag{116}$$

and have used SU(3) flavor symmetry to obtain the others.

In practical applications, for V^M we can use the experimental pion valence distribution, which has been determined from Drell-Yan proton—pion scattering [95-97]. The pion valence quark distribution was found to be consistent with a behaviour $xV^M(x) \sim x^{0.44}(1-x)^{0.89}$ at $Q^2 \sim 4$ GeV^2 [95]. This is in fairly good agreement with the behaviour expected from large-x ($\sim (1-x)$) counting rules [98] and small-x ($\sim x^{1/2}$) Regge behaviour [99]. It was also found in [100] that the ratio of kaon to pion valence quark distributions was consistent with unity over most of the x range, although dropping slightly at large x, $V^K/V^\pi \sim (1-x)^{0.18\pm0.07}$. Unfortunately, the vector meson valence distribution has not yet been determined experimentally. As a first approximation it may seem reasonable to assume that its x-dependence resembles that of the π meson. Deviations from this may be expected on theoretical grounds, if one assumes that a spin flip for one of the quarks in a spin 1 meson induces an additional power of $(1-x)$ at large x [98].

For the diagrams with a meson spectator the contributions from DIS from the virtual baryon B can be obtained in a similar way. For example, the change in the u quark distribution of the proton is:

$$\begin{aligned}\delta^{(N\,\pi/\rho)}u(x) &= \int \frac{dy'}{y'} f_{N\,\pi/\rho}(y') \left(\frac{1}{3}u^p(x_B) + \frac{2}{3}u^n(x_B)\right) \\ &= \int \frac{dy'}{y'} f_{N\,\pi/\rho}(y') \left(\frac{1}{3}u(x_B) + \frac{2}{3}d(x_B)\right) \\ \text{etc.} & \end{aligned} \tag{117}$$

Again, for simplicity, we relate all of the baryon quark distributions to those of the proton. For the neutron this is trivial if one assumes charge symmetry. Since the Δ has spin and isospin 3/2, from the SU(6) quark model we expect that the valence spectator diquark will always have spin and isospin of 1. Using this fact we can relate the valence quark distributions in the Δ to the d quark distribution in the proton (since the spectator uu diquark in the proton has the same quantum numbers), $u^{\Delta^{++}} = \frac{3}{2}u^{\Delta^+} = 3d^{\Delta^+} = 3d$, with the distributions for the other charge states obtained from isospin symmetry. Similarly for the Σ and Λ hyperons, according to SU(3) flavor symmetry we would expect $s^{\Sigma^+} = d$ and $u^{\Sigma^+} = 2q^{\Sigma^0} = 2q^{\Lambda} = u$.

For our numerical results we use experimentally determined coupling constants, all of which are referred to the nucleon pole. For the πNN coupling we use the recently determined value $g^2_{\pi NN}/4\pi = 13.6$ [101], which is marginally smaller than the "traditional" value [102]. The vector meson—nucleon couplings are obtained from analyses of πN scattering data, $g^2_{\rho NN}/4\pi = 0.55$, $f_{\rho NN}/g_{\rho NN} = 6.1$ [103], and $g^2_{\omega NN}/4\pi = 8.1$, $f_{\omega NN}/g_{\omega NN} = 0$ [104]. For the kaon–hyperon–nucleon couplings we use $g^2_{KN\Lambda}/4\pi = 13.1$ and $g^2_{KN\Sigma}/4\pi = 3.7$, as in ref.[94] (although this $KN\Sigma$ coupling is somewhat larger than the ones determined from KN forward dispersion relations [105] or in some hyperon—nucleon potentials [106], however even so the strange contributions are still very small). Finally, we use the quark model to relate the $\pi N\Delta$ and $\rho N\Delta$ couplings to other experimentally measured ones [107], $f^2_{\pi N\Delta} = (72/25) f^2_{\pi NN}$, and $f^2_{\rho N\Delta} = (f^2_{\pi N\Delta}/f^2_{\pi NN})\, g^2_{\rho NN}\, (m_\rho/2M)^2\, (1 + f_{\rho NN}/g_{\rho NN})^2$.

Apart from the coupling constants, the only other parameters in the model are the meson–baryon form factor cut-offs, Λ. The initial idea about how one might use DIS data to constrain Λ was to compare $\langle y \rangle_{MB}$ with the measured momentum fractions carried by the antiquarks [51]. Even more stringent constraints can be achieved by demanding that the shape of the meson exchange contributions to $\bar{q}(x)$ (i.e. $\delta^{(MB)}\bar{q}(x)$) are consistent with the shape of the experimental antiquark distribution.

As mentioned in the previous section, the fact that the old $f_{\pi N}(y)$ calculated in a covariant framework peaked more sharply and at smaller y compared with the $f_{\pi N}$ calculated in the IMF means that the quark distributions in equs.(115), (117) will also peak at smaller x for the covariant k^2-dependent form factor. Because the TOPT/IMF formulation generally gives broader antiquark distributions, the limits on the cut-offs will be more severe than for the covariant case, since at intermediate x ($x \gtrsim 0.2$) the TOPT/IMF distributions are still large compared with the experimental data.

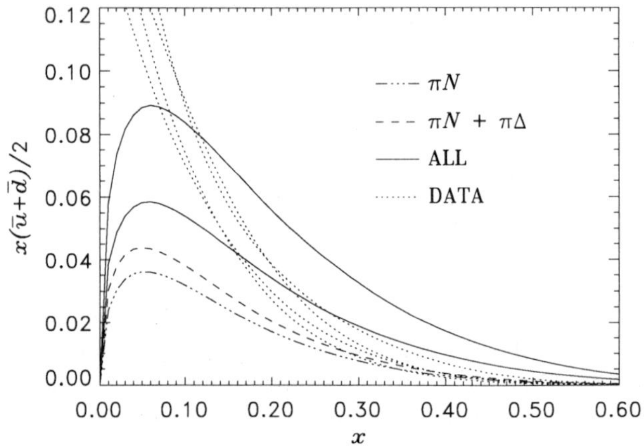

Fig. 15. Proton SU(2) flavor antiquark distributions for DIS on the various meson–baryon components of the nucleon. The dot-dashed and dashed curves represent the contributions from πN and $\pi N + \pi\Delta$ states, respectively, for Λ = 700 MeV. The solid curves are the total contributions from all meson–baryon states, for Λ = 700 MeV (lower curve) and 900 MeV (upper curve). The data, indicated by the dotted lines, are from refs.[108, 109].

Figure 15 shows the contributions to the SU(2) antiquark distribution $x(\bar{u}+\bar{d})/2$ from all of the meson–baryon components of the nucleon, for Λ = 700 and 900 MeV – in equ.(86). This is compared with recent empirical data (as parameterised by Morfin and Tung [109], Owens [108], Eichten et al. [25] and Diemoz et al. [26]) for $(\bar{u}+\bar{d})/2$ at Q^2 = 4 GeV^2. Also shown are the calculated results (for Λ = 700 MeV) for the πN and $\pi N + \pi\Delta$ states alone. Clearly the SU(2) $\bar{q}$ content of the nucleon is well saturated for $\Lambda \approx$ 700 MeV in the intermediate-x region when all meson–baryon components are included. The main contributions in this region come from the ρN and $\rho\Delta$ states, since the distribution functions $f_{\rho N(\Delta)}(y)$ generally extend to larger y compared with the pion distributions. As mentioned above, one uncertainty in the treatment of the vector meson contributions arises from the fact the structure function for a spin 1 meson may deviate at large x from the behaviour observed for the pion structure function. The effect on the ρN contribution to $x(\bar{u}+\bar{d})/2$ of including an extra power of $(1-x)$ in the ρ

meson structure function is a slightly softer distribution, so that this would allow for a marginally larger cut-off mass when comparing against the data in Fig.15. If only the πN states are included, slightly harder form factors could also be accommodated, with $\Lambda \approx 1$ GeV. In either case, for the πNN vertex this corresponds to a dipole form factor cut-off in the covariant formulation of $\Lambda_{\pi N} \sim 700 - 800$ MeV (to give the same value of $\langle n \rangle_{\pi N} \sim 0.10 - 0.15$), which is still considerably smaller than that used by many authors.

5.5 Renormalisation, Incoherence

A subtle, but nonetheless important, point that needs to be made concerns the renormalisation of the total quark distributions in the presence of mesons. The meson and baryon exchange diagrams in Fig.7 describe physical processes (inclusive baryon and meson leptoproduction) whose cross-sections involve physical (renormalised) coupling constants. When integrated over the recoil particles' momenta these yield the inclusive DIS cross-sections, which are proportional to the total quark (and antiquark) distributions

$$q(x) = Z\ q_{\text{bare}}(x) + \sum_{M,B} \left(\delta^{(MB)}q(x) + \delta^{(BM)}q(x)\right). \tag{118}$$

Therefore $\delta^{(MB)}q(x)$ and $\delta^{(BM)}q(x)$, and the convolution integrals in equs.(62) and (63), are expressed in terms of renormalised coupling constants contained in the functions $f_{MB}(y)$ and $f_{BM}(y')$. In equ.(118) we also identify the bare nucleon probability

$$Z = 1 - \sum_{M,B} \langle n \rangle_{MB} \tag{119}$$

chosen such that the baryon number and momentum sum rules are satisfied. We emphasise that all quantities in equs.(118) and (119) are evaluated using renormalised coupling constants.

We could, of course, choose to work at a given order in the bare coupling constant, and explicitly verify that the various sum rules are satisfied. For example, to lowest order (g_0^2) the total quark distributions would be [29]

$$q(x) = Z \left\{ q_{\text{bare}}(x) + \sum_{M,B} \left(\delta^{(MB)}q_{(0)}(x) + \delta^{(BM)}q_{(0)}(x)\right) \right\} \quad (\text{order } g_0^2) \tag{120}$$

with

$$Z = \left(1 + \sum_{M,B} \langle n_{(0)} \rangle_{MB}\right)^{-1} \quad (\text{order } g_0^2) \tag{121}$$

where the subscript (0) indicates that the functions $f(y)$ are evaluated using bare couplings. Equations (118) and (119) are easily recovered since the bare couplings, to this order, are defined by $g_0^2 = g_{\text{ren}}^2/Z$. It would, however, be inconsistent to use (120) and (121) with renormalised coupling constants, especially with large form factor cut-offs.

As long as the form factors are soft, the difference between the bare and renormalised couplings is not very large. However, with large cut-off masses the bare couplings would need to be substantially bigger than the physical ones. (In fact, the form factor cut-off dependence of the bare πN coupling constant in the cloudy bag model [48] showed some 40% difference for very hard form factors — or small bag radii, $\sim$ 0.6 fm.) In addition, with large values of Λ the higher order diagrams involving more than one meson in the intermediate state would become non-negligible, and the initial assumption that the series in equ.(57) can be truncated at the one-meson level would be seriously in doubt. Fortunately, we need not consider the multi–meson contributions, since Fig.15 clearly demonstrates the difficulty in reconciling the empirical data with quark distributions calculated with such large cut-offs.

Finally, we need to make some additional comments regarding the justification of our calculation in terms of an incoherent summation of cross-sections (rather than amplitudes) for the various meson exchange processes. A possible breakdown of incoherence may arise when there are different exchange processes leading to the same final state. For nucleon final states, because of the pseudoscalar nature of the πNN vertex, there will be no interference between π meson and vector meson exchange. Furthermore, no mixing will take place between the ω and ρ exchange configurations due to their different isospins. In fact, all of the meson exchange processes with a recoil N considered in this analysis can be added incoherently. For a Δ recoil, the only mesons coupling to N and Δ are the π and ρ, but since they have different G-parities, interference effects from these will again be excluded. However, the possibility exists in the pion exchange process that the decay products of a Δ recoil, namely π and N, may mix with the state containing an N recoil together with a π from the hadronic debris X of the shattered exchanged pion. Interference between the πN and $\pi\Delta$ states could therefore occur if the π from the debris had very low momentum, enabling the combined system to have an invariant mass squared $\sim M_\Delta^2$. However the vast majority of semi-inclusive meson events in lepton–nucleon DIS are those with high momentum mesons (slow hadrons are almost exclusively baryons), so that the probability of interference arising from such processes will not be large. A similar argument can be given for the potential interference from hyperon decay into $N\pi$.

For the baryon exchange processes, the requirement of the same recoil meson eliminates interference from most states, except from DIS off N and Δ with π (or ρ) in the final state, and from DIS off a Σ^0 and Λ^0 with a K^+ recoil. For the latter, the different isospin quantum numbers of the $\Lambda(I = 0)$ and $\Sigma(I = 1)$ rule out interference, just as for the ρN and ωN states. A similar argument can be made for excluding interference contributions from $N(I = 1/2)$ and $\Delta(I = 3/2)$ exchange. Once again, the fact that the decay products of the recoil mesons have low momentum, while the pions from the baryonic debris are fast, will again reduce the size of any interference effects.

Therefore we see that by considering only the lowest lying meson and baryon states (i.e. by excluding resonances having the same quantum numbers as the mesons and baryons considered here) we can avoid potential problems with interference, and certainly for the values of Λ allowed by the data, the only relevant states are those with the lowest masses.

6 Challenging Problems

It is a widely held belief that the best laid plans of mice and men often go astray! This is at least as true of DIS as of any other field of study. In this section we study several topical examples which have generated a great deal of excitement (and consternation!). The first concerns the nature of the sea of $q\bar{q}$ pairs in the nucleon. According to perturbative QCD these are generated from gluons which in turn carry no flavor information. As a consequence one expects to find equal numbers of $u\bar{u}$ and $d\bar{d}$ pairs. The Gottfried sum rule was built on this idea. Yet recent results from the New Muon Collaboration (NMC) at CERN have confirmed earlier hints that the sum rule is badly violated. Clearly this is more than a curiosity — it contains fundamental information about the non-perturbative structure of the nucleon. We explore some of the ideas proposed to explain the result in sects.6.1 and 6.2.

Spin-dependent measurements have often been the graveyard of otherwise successful theories. In DIS this would probably be a fair summary of the situation immediately after the discovery by the EMC of a violation of the Ellis-Jaffe sum rule. This result, which was initially christened the proton "spin crisis", has since led to important advances in our understanding of how the U(1) axial anomaly is realised. Although the dust has not yet settled, and a lot more experimental and theoretical work is needed, we outline the current situation in sect.6.3.

Our final outstanding problem, dealt with in sect.6.4, is the old nuclear EMC effect – the variation of nuclear structure functions. At various stages over the past decade it has seemed like the problem was understood. This is certainly not the case – even now. However, we have reached a stage of illumination where we can at least say that the physics required in a decent theoretical treatment is now known.

6.1 Flavor asymmetry

One of the more interesting observations that can be made within the meson model is that it predicts that contributions from DIS off virtual mesons and baryons to the u, d and s quark (and the corresponding antiquark) distributions in the proton (equs.(115), (117)) will be different. Because the contributions to s and $\bar{s}$ from DIS from kaons and hyperons are very much smaller than those from the non-strange mesons and baryons to the u and d distributions (mainly because $m_K \gg m_\pi$) we see that the meson model produces significant SU(3) flavor symmetry violation. Furthermore, it is apparent that the contributions to the u and d (and $\bar{u}$ and $\bar{d}$) quark distributions themselves are not the same, so that SU(2) flavor symmetry of the proton sea is also broken. In the case of the pion cloud, the simple origin of this is asymmetry is the predominance of the dissociation process $p \to n\pi^+$ over $p \to p\pi^0$. In the former, the π^+ valence quark content is $d\bar{u}$, while in the latter the ratio of $\bar{u}$ to $\bar{d}$ quarks is the same. This process certainly respects isospin symmetry, which simply says that the dissociation $p \to n\pi^+$ is as likely as $n \to p\pi^-$, or at the quark level, $u \to d(u\bar{d})$ is as likely in the proton as $d \to u(d\bar{u})$ in the neutron. But it clearly implies an excess of $\bar{d}$ quarks in the proton, and an equal excess of $\bar{u}$ quarks in the neutron.

If the masses of quarks were identical (i.e. SU(3) flavor symmetry limit) then the ratio of strange to non-strange antiquark distributions in the proton would be 1:2. From neutrino experiments (at $Q^2 \sim 4$ GeV^2) the measured ratio was found to be about 1:4 [14], which can be understood semi-quantitatively from the heavier mass of the strange quark. On the other hand, because charge symmetry is such a good symmetry in strong interaction physics, it was naively expected that SU(2) flavor symmetry of the sea would be an excellent approximation. Indeed, this expectation has been built into almost all of the analyses of the nucleon structure function data. The main reason for believing this has been the simple picture, motivated by perturbative QCD, in which the mechanism for producing antiquarks is gluon splitting into $q\bar{q}$ pairs. However, unless isospin symmetry is genuinely violated (by giving a non-zero mass difference between the u and d quarks), the perturbative process $g \to q\bar{q}$ should be SU(2) flavor symmetric, as the gluons of QCD are flavor–blind. Therefore a $\bar{d} - \bar{u}$ difference cannot be produced by perturbative QCD. Actually, this statement should be qualified by saying that at lowest order in α_S there is no asymmetry. A higher order perturbative QCD calculation of $\bar{d} - \bar{u}$ was performed some time ago by Ross and Sachrajda [110], who found a non-zero result for this difference, although the absolute value was very small. This means that the calculated $\bar{d} - \bar{u}$ difference will essentially be preserved in QCD evolution. But the fact that we get a non-zero $\bar{d} - \bar{u}$ difference in the meson–baryon model is not surprising, since this is a non-perturbative model, and its predictions are not in conflict with QCD, nor with isospin symmetry.

In Figs.16 and 17 we plot the $\bar{d} - \bar{u}$ difference, and the ratio $(\bar{d} - \bar{u})/(\bar{d} + \bar{u})$, respectively, calculated within the meson-exchange model of the nucleon. Fig.16 shows that the inclusion of $\pi\Delta$ states (with Λ = 700 MeV) eliminates some of the $\bar{d}$ excess, since here the dominant process is $p \to \Delta^{++}\pi^-$, which at the quark level, $d \to u(d\bar{u})$, is seen to produce a $\bar{u}$ excess. However, adding the vector meson components (lower solid curve) restores the original $\bar{d}$ excess at small x, and enhances the excess at larger x. At larger values of Λ (= 900 MeV) the $\bar{d} - \bar{u}$ difference is larger still (upper solid curve), although the ratio $(\bar{d} - \bar{u})/(\bar{d} + \bar{u})$ is smaller, Fig.17. On similar grounds the processes $p \to K^0\Sigma^+$ and $p \to K^+\Sigma^0(\Lambda)$ will introduce not only a different strange quark content of the proton than a non-strange, but also a different $\bar{s}$ distribution as well. However, because of the very small magnitude of the strange contributions it will be difficult for these effects to be observed in DIS experiments in the near future.

Of course the idea of an asymmetric proton sea is not a new one. The earliest, and perhaps most obvious, suggestion for why we should expect $\bar{u} \neq \bar{d}$ was made by Feynman and Field [111]. As mentioned in sect.4.1, because the proton has an unequal number of valence u and d quarks, by the Pauli exclusion principle we would therefore expect creation of additional $q\bar{q}$ pairs inside the proton to be sensitive to the number of quarks of each flavor already in the proton. Since there are 2 valence u quarks in the proton compared with only 1 valence d quark, we therefore expect a larger $\bar{d}$ sea since $u\bar{u}$ pair creation will be suppressed relative to $d\bar{d}$. In ref.[111] the $\bar{d}$ and $\bar{u}$ distributions were parameterised by $x\bar{d} = 0.17(1-x)^7$ and $x\bar{u} = 0.17(1-x)^{10}$, see Fig.18. With these, the integrated difference is $\int dx(\bar{d} - \bar{u}) = 0.057$.

An early calculation of the u and d sea quark probabilities in the proton, incorporating the effects due to the Pauli principle, was made by Donoghue and Golowich [112] using the MIT bag model. However, this involved calculating the one-gluon-exchange

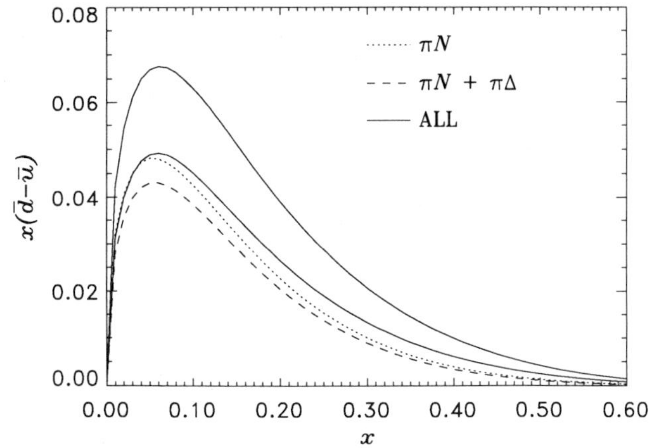

Fig. 16. Calculated $\bar{d}-\bar{u}$ difference from the various meson-baryon states. For the dotted, dashed and lower solid curves a dipole form factor cut-off of Λ = 700 MeV is used, while the upper solid curve is calculated with Λ = 900 MeV.

induced $q\bar{q}$ admixture in the proton wavefunction. This is quite a different effect from that discovered in the recent work of Signal and Thomas [19] when calculating the of quark distribution functions in the MIT bag (see sect.4.1). Their work suggested a quantitative method of calculating the intrinsic sea associated with the different vacuum structure of the bag from that in free space. They also showed that the Pauli exclusion principle implied a $\bar{d}-\bar{u}$ difference as well. In particular these intrinsic antiquark distributions arose only from 4 q intermediate bag states. Thus the $\bar{d}$ distribution required the intermediate state to consist of 2 u and 2 d quarks, while DIS from a $\bar{u}$ quark implied a $3u$ + $1d$ intermediate state, which, because of the Pauli principle, has a smaller probability. Furthermore, in ref.[19] the $\bar{d}$ excess associated with the Pauli effect was equal to the d excess and satisfied the condition

$$\int_0^1 dx\ \bar{d}_{\text{excess}}(x) = \int_0^1 dx\ (d_{\text{sea}}(x) - u_{\text{sea}}(x)) = \int_0^1 dx\ p_N(x) \equiv \mathcal{P}_N. \tag{122}$$

Here, $p_N(x)$ denotes the piece of the valence quark distribution associated with a four quark intermediate state (all in a $1s$ state), while $1 - \mathcal{P}_N$ is the integral over the distribution function associated with a two quark intermediate state. In ref.[19] the calculated distributions were found to peak at $x \sim 1 - M_{int}/M$, where M_{int} is the mass of the intermediate spectator state. Since the 4-quark intermediate states have mass greater than M, the antiquark distributions will peak at negative x. Therefore in the physical region ($x \geq 0$), $p_N(x)$ should resemble a typical sea quark distribution, namely be finite at x = 0 while dropping rapidly to zero by $x \sim 0.2 - 0.3$. For simplicity, we can parameterise the large-x behaviour by a $p_N(x) \propto (1-x)^7$ form [25, 56]. On theoretical grounds, we can also expect that due to the lack of Regge $f - A_2$ exchange degeneracy, at small x the $\bar{d}-\bar{u}$ difference should be proportional to $p_N(x) \sim x^{\alpha}$, where $\alpha(\simeq 0.5)$ is

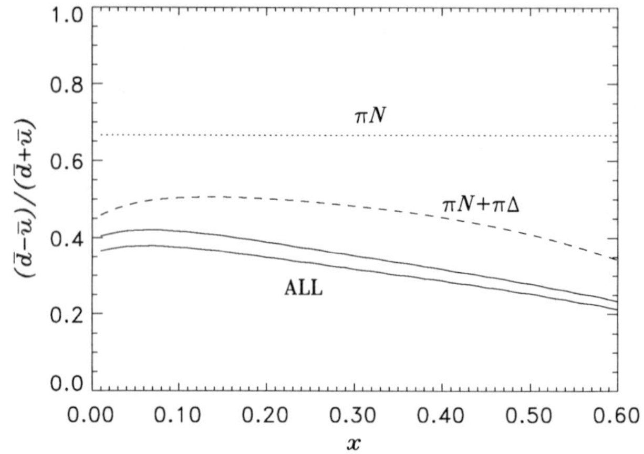

Fig. 17. Ratio of the difference and sum of the calculated SU(2) antiquark distributions in the proton. The cut-off for the dotted, dashed and upper solid curves is Λ = 700 MeV, while the lower solid curve is calculated with Λ = 900 MeV.

the Regge intercept [99, 113]. The overall normalisation $\mathcal{P}_N$ was calculated in ref.[52] to be less than about 0.25 for bag radii $R \lesssim 0.8$ fm.

In Fig.18 we compare the Feynman & Field parameterisation with the function $p_N(x)$, normalised so that $\mathcal{P}_N = 0.05, 0.15$ and 0.25. Also shown is the effect of using a slightly more singular small-x behaviour, $\alpha = 0.62$, as suggested by the NMC data on $F_{2p} - F_{2n}$ [58].

Unfortunately at the present time there are not sufficient data on $\bar{d}-\bar{u}$ to make definitive conclusions about these various mechanisms for SU(2) flavor symmetry breaking. However, there have been a number of interesting suggestions for experiments that could directly probe the light sea quark content of the proton, and we briefly review these now.

Recently it was suggested by Martin, Stirling and Roberts [114] that one could learn about the SU(2) sea by observing the W-boson asymmetries in $p\bar{p}$ collisions, $p\bar{p} \to W^{\pm}X$. The simple idea is that a $u(d)$ quark in a proton interacts with a $\bar{d}(\bar{u})$ antiquark in an antiproton to produce a $W^{+}(W^{-})$ boson. Because the $u(\bar{u})$ quark in the proton (antiproton) carries more momentum (lies at larger x) than the $d(\bar{d})$ quark, $W^{+}(W^{-})$ bosons will be predominantly produced in the proton (antiproton) direction. However, in addition there will be $W^{+}(W^{-})$-bosons produced by the annihilation of $\bar{d}(\bar{u})$ quarks in the proton with $d(u)$ quarks in the antiproton. Thus at large energies the asymmetry in the W-boson rapidity (y_W) distribution

$$A(y_W) = \frac{\sigma^+ - \sigma^-}{\sigma^+ + \sigma^-} = \frac{u(x_1)d(x_2) + \bar{d}(x_1)\bar{d}(x_2) - -d(x_1)u(x_2) - \bar{d}(x_2)\bar{d}(x_1)}{u(x_1)d(x_2) + d(x_1)u(x_2) + d(x_1)u(x_2) + \bar{d}(x_2)\bar{d}(x_1)}, \quad (123)$$

where $\sigma^{\pm} = d\sigma/dy_W(W^{\pm})$, would be sensitive to the antiquark distributions in the proton. Here, $x_{1,2} = M_W/\sqrt{s}\exp(\mp y)$, s is the centre of mass energy squared, and the W-boson rapidity is defined by $y_W = (1/2)\ \ln(q_-/q_+)$, with q the W-boson mo-

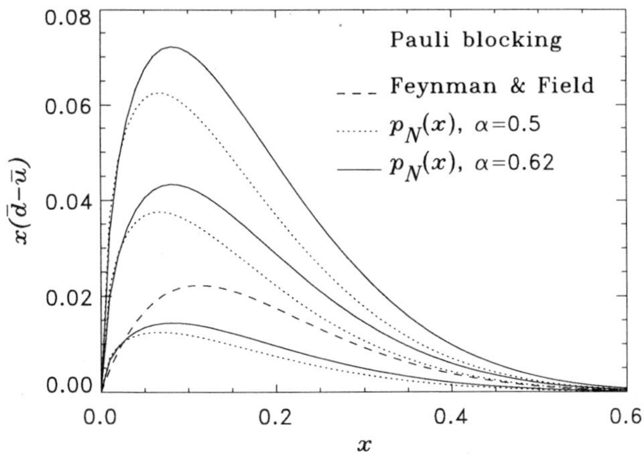

Fig. 18. Difference between $\bar{d}$ and $\bar{u}$ quark distributions in the proton, for the Feynman and Field parameterisation [111], and a parameterisation of the function $p_N(x) \sim x^{0.62}(1-x)^7$, with normalisation $\mathcal{P}_N = 0.05, 0.15$ and 0.25.

mentum. Furthermore, since only left-handed quarks (right-handed antiquarks) couple to W-bosons, in the resulting $W \to e\nu$ decay the electron (positron) distribution will generally follow the direction of the incoming proton (antiproton). It was suggested in [114] that the experimental $e^{\pm}$ asymmetry, $A(y_e) = (d\sigma/dy_{e^+}(y_{e^+}) - d\sigma/dy_{e^-}(y_{e^-}))/(d\sigma/dy_{e^+}(y_{e^+}) + d\sigma/dy_{e^-}(y_{e^-}))$, could then serve as an independent check on the $\bar{u}$ and $\bar{d}$ distributions in the proton. The claim in ref.[114] was that their existing parameterisations with no SU(2) flavor asymmetry are consistent with the data on $A_e(y_e)$ taken at the Collider Detector at Fermilab (CDF) [115]. However, the error bars in this experiment are quite large, and the data at present will have difficulty in discriminating between SU(2) flavor symmetric parameterisations, and those with a small $\bar{d}-\bar{u}$ difference, such as that suggested by the meson model in the previous section. On the other hand, a large $\bar{d}-\bar{u}$ difference, such as that arising from the meson model with large form factor cut-offs Λ, may well introduce a detectable difference.

In another experiment, performed some than 10 years ago by the E288 Collaboration at FNAL [116], the slope of the rapidity distribution for proton–nucleus Drell-Yan production was measured, and found to be sensitive to the $\bar{u}/\bar{d}$ ratio. In that experiment, the quantity

$$\frac{d}{dy} \ln \left(s \frac{d^2\sigma}{d\sqrt{\tau}dy} \right)_{y=0} \tag{124}$$

was measured as a function of $\sqrt{\tau}$, where $\tau = M^2_{l^+l^-}/s$. It was found that a parameterisation with $\bar{d} > \bar{u}$ improved the quality of the fit [117]. However, since the analysis of this experiment required the quark and antiquark distributions in the nucleus, any conclusions reached about the nucleon sea distributions were obviously dependent upon any nuclear assumptions made. In fact, it was later shown by Ericson and Thomas [118] that a similar improvement in the fit could be made by assuming a small difference between the $\bar{q}$ distributions in the nucleon and in a nucleus.

Proton–nucleus Drell-Yan production was also studied recently by the E772 Collaboration at Fermilab. It was found that by comparing the yield per nucleon in a proton collision with a neutron-rich target such as tungsten with that for an isoscalar nucleus, the resulting ratio would also be sensitive to the $\bar{d} - \bar{u}$ difference. However, it has since been argued by Eichten et al. [119] that this too may not be a sensitive enough experiment for a small non-zero $\bar{d} - \bar{u}$ difference to be discernable from no difference.

Perhaps the experiment that is most sensitive to the light sea quark distributions was that recently proposed by Ellis & Stirling [120], who suggested measuring the asymmetry between the pn and pp Drell-Yan production (i.e. $pN \to l^+l^-X$) cross-sections at zero rapidity, $A_{DY} = (\sigma^{pp} - \sigma^{pn})/(\sigma^{pp} + \sigma^{pn})$, where $\sigma^{pN} \propto d^2\sigma^{pN}/d\sqrt{\tau}dy|_{y=0}$. Neglecting terms involving annihilation of sea quarks in the (beam) proton and (target) nucleon, the cross-sections can be written

$$A_{DY} = \frac{(4\lambda_V - 1)(\lambda_S - 1) + (\lambda_V - 1)(4\lambda_S - 1)}{(4\lambda_V + 1)(\lambda_S + 1) + (\lambda_V + 1)(4\lambda_S + 1)} \tag{125}$$

where $\lambda_V = u_V/d_V$ and $\lambda_S = \bar{u}/\bar{d}$. The advantage of measuring this ratio is that it would be free from any nuclear dilution effects, and the complete asymmetry could be determined from ratios of valence and sea quark distributions alone. Since the d_V/u_V ratio is well determined, A_{DY} would then serve as an accurate indicator of λ_S. In Fig.19 we plot this Drell-Yan asymmetry as a function of $\sqrt{\tau} = \sqrt{Q^2/s}$ with $s \sim 1500$ GeV^2 (corresponding to a proton beam energy of about 800 GeV) calculated using the quark and antiquark distributions of the meson model (with $\Lambda = 700$ MeV). This is compared with the asymmetry arising from the parameterisation of Morfin & Tung [109] for the valence quarks (dotted curve), and from the d_V/u_V ratio fixed at $0.57(1 - x)$ [25] (dashed curve), with $\lambda_S = 1$ in both cases. It is clear that even small deviations of $\bar{u}/\bar{d}$ from unity will have a big impact upon A_{DY}.

An extension of this idea was discussed in ref.[121], where it was argued that one could directly measure the difference $\bar{d} - \bar{u}$ by going to large projectile momentum fractions x_1, but small target fractions x_2. In that case the term in A_{DY} involving the product of projectile sea and target valence distributions could be neglected and the asymmetry reduced to

$$A_{DY} \approx \frac{(4\lambda_V - 1)(\lambda_S - 1)}{(4\lambda_V + 1)(\lambda_S + 1)}. \tag{126}$$

Unfortunately, there are as yet no data on A_{DY}, although a proposal has been made [122] for an experiment to measure the Drell-Yan cross-sections for hydrogen and deuterium targets. Such data would be eagerly anticipated.

Finally, an interesting observation was made by Levelt, Mulders and Schreiber [123], who found that semi-inclusive charged-hadron production could be used to obtain information on the integrated $\bar{d} - \bar{u}$ difference. Following earlier work by Gronau et al. [124] and Field and Feynman [111] on the parton model for semi-inclusive DIS, Levelt et al. showed that the integrated difference should be proportional to the measured difference between the charged pion and kaon production rates from DIS on protons and neutrons. However, the available data from the EM Collaboration at CERN on semi-inclusive

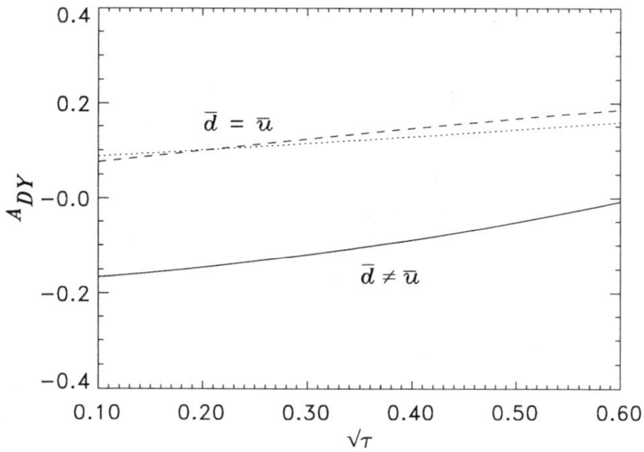

Fig. 19. Asymmetry for pp and pn Drell-Yan production. The $\bar{u} = \bar{d}$ predictions, with the valence quark parameterisation of Morfin and Tung [109] (dotted) and with a fixed valence ratio $d_V/u_V = 0.57(1-x)$ [25], are compared with the meson model calculation with $\bar{u} \neq \bar{d}$ (for Λ = 700 MeV).

charged-meson production [125] are not yet sufficiently accurate to discriminate between SU(2) flavor symmetry and asymmetry.

However, the most important impact on the question of SU(2) flavor symmetry in the proton sea, and certainly the stimulus for the close attention this question has received in recent times, has come from the measurement by the NMC of the difference between F_{2p} and F_{2n} [58], and the consequent determination of the Gottfried sum rule. We will now discuss the issues involved in this experiment more fully.

6.2 Gottfried Sum Rule

The Gottfried sum rule [57] is perhaps the most famous consequence of SU(2) flavor symmetry of the sea. Because this measures the x-integrated difference between the proton and neutron structure functions, it is sensitive only to the non-singlet SU(2) content of the nucleon. Let us firstly define the quantity

$$S_G(x,1) = \int_x^1 \frac{dx'}{x'} \left(F_{2p}(x') - F_{2n}(x')\right) . \tag{127}$$

Relating the proton and neutron structure functions to the quark distributions in the proton (i.e. using charge symmetry), we have

$$S_G(x,1) = \frac{1}{3}\int_x^1 dx' \left(u(x') + \bar{u}(x') - d(x') - \bar{d}(x')\right) \tag{128}$$

$$= \frac{1}{3}\int_x^1 dx' \left(u_V(x') - d_V(x')\right) - \frac{2}{3}\int_x^1 dx' \left(\bar{d}(x') - \bar{u}(x')\right) \tag{129}$$

where the valence quark distributions are defined by $q_V \equiv q - \bar{q}$. Since the number of valence quarks in a hadron does not change, we obtain the Gottfried sum rule

$$S_G \equiv S_G(0,1) = \frac{1}{3} \qquad \text{[QPM]} \qquad (130)$$

provided we make the additional assumption $\int_0^1 dx\, \bar{d} = \int_0^1 dx\, \bar{u}$, as would be expected in the simple quark-parton model (QPM).

The early experimental data for $S_G(x,1)$ did in fact suggest a value lower than 1/3, but with errors large enough to be consistent with it. However, armed with the theoretical expectation of SU(2) flavor symmetry, most authors believed that S_G would tend to 1/3 as the accuracy of the data improved. To the surprise of many, the recent, accurate determination of S_G by the NMC appears to support the idea that $\bar{u} \neq \bar{d}$ [58, 126]. Neglecting nuclear effects, the NMC found

$$S_G(x_{min}, 1) = 0.229 \pm 0.0157 \qquad (131)$$

where $x_{min} = 0.004$. Included in (131) is an extrapolation from $x = 0.8$ to 1, the contribution from which was estimated to be 0.002 ± 0.001 if a smooth extrapolation of F_{2p}/F_{2n} to 1/4 at $x = 1$ is assumed. From the unmeasured region ($x < 0.004$), using the extrapolation

$$F_{2p}(x) - F_{2n}(x) \longrightarrow \alpha\, x^{\beta} \qquad \text{as } x \to 0 \qquad (132)$$

with $\alpha = 0.21, \beta = 0.62$, the contribution was found to be $S_G(0, x_{min}) = (\alpha/\beta)\, x_{min}^{\beta} = 0.011$. With the conventional Regge theory assumption that $\beta = 0.5$, $S_G(0, x_{min})$ would be 0.014. The combined integral over the whole range of x was therefore

$$S_G = 0.240 \pm 0.016 \qquad \text{[NMC]} \qquad (133)$$

with errors added in quadrature. [4] Although not the only one, the most natural explanation for the smaller than expected value of S_G is that $\bar{d}(x) \neq \bar{u}(x)$ (see later for a discussion of other possibilities). Taken at face value, the NMC result would imply that

$$\int_0^1 dx\, (\bar{d}(x) - \bar{u}(x)) = 0.14 \pm 0.06. \qquad (134)$$

The various mechanisms discussed in sect.5.2 are then potential candidates for generating such a difference. Before turning to more exotic explanations, it seems more sensible that the simplest possibilities should be exhausted first. The most compelling, and most economical from the theoretical point of view, appear to be those based on the Pauli exclusion principle, and on the presence of a small pion (and perhaps other meson) cloud.

We therefore examine the consequences of both of these mechanisms on the $p - n$ structure function difference and the Gottfried sum rule. Firstly note that because

[4] From a very recent re-evaluation of the ratio F_{2n}/F_{2p} based on newly measured values of F_{2D} the NMC found a slightly larger value for S_G, namely $S_G = 0.258 \pm 0.017$ [127], although this is still consistent with the original value.

the non-strange baryon recoil contributions to the quark and antiquark distributions are related by

$$\delta^{(MB)}u(x) = \delta^{(MB)}\bar{d}(x), \qquad \delta^{(MB)}d(x) = \delta^{(MB)}\bar{u}(x) \qquad [M = \text{non} - \text{strange}] \tag{135}$$

the contributions to S_G from DIS from pions or vector mesons would cancel. Furthermore, since the strange content of the proton and neutron should be the same, contributions from kaons and hyperons to S_G should also vanish. However, the Pauli blocking effect should be present in DIS from virtual K^0, K^+ and Σ^+, since these contain unequal numbers of $u(\bar{u})$ and $d(\bar{d})$ quarks. It will not be present in DIS from Σ^0 or Λ^0. This raises the interesting possibility that we may pick up a non-zero strange quark contribution to S_G from the Pauli principle, if the Pauli effect in DIS from K with Σ recoil and in DIS from Σ with K recoil are different (and in principle they should be), which would spoil the cancellation of these components. Still, having seen that the role of strange mesons in the DIS process is negligible, we can be fairly confident that by dropping the strange contributions our results will not be significantly affected.

What may be more significant is the possibility that the shape of the Pauli $\bar{d} - \bar{u}$ contribution from DIS off a virtual Δ, with π or ρ recoil (labelled $p_\Delta(x)$) may differ from the shape of the Pauli difference from DIS off a nucleon with a π or ρ recoil, $p_N(x)$. In principle these should be different because the spins of the 4-quark intermediate states (which arise when the $\bar{u}$ or $\bar{d}$ quarks are probed) in the nucleon and Δ are different. This means that, for example, while a quark inserted into a (spin 1/2) proton can produce a state with spin 0 or 1, one inserted into a (spin 3/2) Δ^+ could produce either a spin 1 or spin 2 intermediate state. One way to make spin 0, 1 or 2 four-quark states is to construct them from spin 0 or spin 1 diquarks, and since a vector diquark is more massive than a scalar diquark (see ref.[20]), and therefore has a softer x-distribution, the result is that the Pauli blocking function $p_\Delta(x)$ should have a softer shape than $p_N(x)$. Furthermore, the integral over $p_\Delta(x)$ (denoted $\mathcal{P}_\Delta$) need not necessarily equal $\mathcal{P}_N$. Having said this, it is probably also true that the uncertainty introduced in taking these to be the same will be much smaller than the overall uncertainty in the absolute normalisation of $\bar{d} - \bar{u}$ due to Pauli blocking in the nucleon.

The final expression for $F_{2p} - F_{2n}$, including meson and Pauli effects, is

$$\begin{aligned} F_{2p}(x) - F_{2n}(x) = {} & \frac{Z}{3}\left(xu_V(x) - xd_V(x) \quad 2xp_N(x)\right) \\ & - \frac{1}{9}\int_x^1 dy' \; \left(f_{N\pi}(y') + f_{N\rho}(y') - 3f_{N\pi}(y')\right) \\ & \qquad \times \left(x_B u_V(x_B) - x_B d_V(x_B) - 2x_B p_N(x_B)\right) \\ & + \frac{5}{9}\int_x^1 dy' \; \left(f_{\Delta\pi}(y') + f_{\Delta\rho}(y')\right)\left(x_B d_V(x_B) - 2x_B p_\Delta(x_B)\right) \\ & + \frac{4}{9}\int_x^1 dy' f_{\Sigma K}(y')\left(\frac{1}{2}x_B u^{\Sigma^+}(x_B) - 2x_B p_\Sigma(x_B)\right) \\ & - \frac{1}{9}\int_x^1 dy \; \left(f_{K\Sigma}(y) - 3f_{K\Lambda}(y)\right)\left(x_M V^M(x_M) - 2x_M p_K(x_M)\right). \end{aligned} \tag{136}$$

Making the above approximations, we plot the resulting x distribution in Fig.20. Note that kaons and ω mesons contribute to the structure functions themselves, even though their contributions will cancel when the structure functions are integrated over x. (We include the K and ω contributions only for the sake of completeness, as dropping them altogether has numerically negligible consequences.)

The most noticeable consequence of the meson cloud is a decrease in the peak value of $F_{2p} - F_{2n}$ at $x \sim 0.3$. Since here the parameterisation clearly overestimates the NMC data, the effect of mesons is to move the curve in the right direction. At the same time, however, the structure function difference becomes larger for $x \lesssim 0.1$. Here, the action of the πN states is to decrease $F_{2p} - F_{2n}$ at small x, while adding the $\pi\Delta$ tends to do the opposite. However, it is only with the addition of the vector mesons that there is an increase over the parameterisation in this region. Because the parameterisation is already too large in this region compared with the NMC data, it's clear that mesons alone cannot improve the fit at small x.

At large x the meson-corrected curves consistently lie beneath the NMC data points. This is a consequence of the original parameterisation [109] underestimating the NMC $F_{2p} - F_{2n}$ results (in fact most other parameterisations [25, 108] also have this property). If we had a parameterisation which could better reproduce the large-x data, the quality of the fits for the corrected curves would naturally improve. We should add, however, that the NMC did not report much data at $x \gtrsim 0.4$. In any case, the discrepancy between the NMC data and the quark parameterisations at large x is unrelated to the failure of the Gottfried sum rule, and is therefore not our primary concern.

The Pauli correction is largest in the small-x region, for $0.01 \lesssim x \lesssim 0.1$. By reducing the absolute value of $F_{2p} - F_{2n}$ at small x the Pauli correction brings the parameterisation (with $\bar{d} = \bar{u}$) into better agreement with the data in that region. However, for larger x ($0.1 \lesssim x \lesssim 0.3$) the peak in the distributions is still too large to be consistent with the NMC data. On the other hand, when combined with a small mesonic correction (for $\Lambda = 700$ MeV), a very good fit is possible with $\mathcal{P}_N \approx 0.1$.

Integrating the structure function difference between x and 1, we plot in Fig.21 the function $S_G(x, 1)$ including both meson and Pauli effects. The parameterisation is clearly too large for $x \lesssim 0.1$. With the addition of the meson correction, the fit is clearly improved, but still overestimates the NMC data at very small x. This is partly remedied when a small Pauli blocking correction is added. In particular, the apparent saturation of the sum rule below $x \approx 0.01$ is better fitted by including the Pauli term. (In a more recent experiment, the E665 Collaboration at Fermilab reported an even more dramatic saturation of the Gottfried sum rule for $x \lesssim 0.125$, $\int_{0.001}^{0.125} dx (F_{2p} - F_{2n})/x = -0.07 \pm 0.07$ [128].) In the intermediate-x region ($x \gtrsim 0.3$) the meson-corrected curves appear to underestimate the NMC data. This can be understood from the shape of the original $F_{2p} - F_{2n}$ distributions in Fig.20, where for x above ~ 0.3 the curves tend to lie beneath the NMC data points.

For the Gottfried sum, from equs.(136) and (119) we obtain

$$S_G = S_G^{MB} \, (1 - 2\,\mathcal{P}_N) + \frac{10}{9} (\mathcal{P}_N - \mathcal{P}_\Delta) \left(\langle n \rangle_{\Delta\pi} + \langle n \rangle_{\Delta\rho} \right) + \frac{2}{9} (3\mathcal{P}_N + \mathcal{P}_K - 4\mathcal{P}_\Sigma) \langle n \rangle_{\Sigma K} + \frac{2}{3} (\mathcal{P}_N - \mathcal{P}_K) \langle n \rangle_{\Lambda K} \tag{137}$$

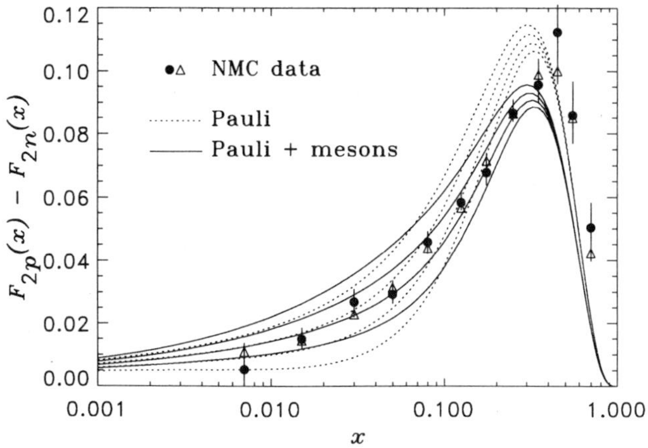

Fig. 20. Effect of the Pauli exclusion principle on the proton – neutron structure function difference, as a function of x. The dotted (without meson corrections) and solid (with Λ = 700 MeV meson corrections) curves are for $\mathcal{P}$ = 0 (largest curves) 0.05, 0.1, and 0.15 (smallest curves).

where

$$S_G^{MB} = \frac{1}{3}\left(1 - \frac{4\langle n\rangle_{N\pi}}{3} - \frac{4\langle n\rangle_{N\rho}}{3} + \frac{2\langle n\rangle_{\Delta\pi}}{3} + \frac{2\langle n\rangle_{\Delta\rho}}{3}\right). \tag{138}$$

is the sum rule with meson/baryon corrections only. Dropping the negligible strange contributions, and assuming that the difference between the Pauli blocking in the nucleon and Δ is not large, only the first term in equ.(137) remains. In Fig.22 we show the variation of S_G with both Λ and $\mathcal{P}_N$. It is clear that the net effect of the virtual meson–baryon states is to decrease S_G. The πN state alone can reproduce the quoted value of S_G for $\Lambda \sim$ 1.3 GeV. The addition of $\pi\Delta$ components would require a slightly larger cut-off (since this produces an excess of $\bar{u}$ over $\bar{d}$, which cancels some of the $\bar{d}$ excess generated by the πN states). Including the ρN state however restores, and actually enhances, the $\bar{d}$ excess, although some of this is again cancelled by the $\rho\Delta$ states. With all components included, the NMC value of S_G can be reproduced with $\Lambda \approx 1.1-1.2$ GeV. For $\Lambda \lesssim$ 700 MeV, as suggested by the antiquark data in the previous section, mesons can generate only about half of the asymmetry required to satisfy the experimental sum rule. For $\Lambda \approx 0.7$ GeV, the experimental sum rule can be obtained with $\mathcal{P}_N \approx 0.1$. Thus various phenomenological constraints seem to imply the need for both mechanisms.

Before finishing this discussion, we should mention some alternative explanations for the Gottfried sum rule violation. It was suggested by Martin, Stirling and Roberts [114] that there may not be any violation of the quark-parton model S_G prediction at all, if large contributions to the Gottfried integral come from the unmeasured, $x < 0.004$, region. By parameterising their valence quark distributions to be more singular at small x than what would otherwise be expected from Regge theory (namely, $q_V \sim x^{-0.5}$), and also compared with what the NMC used in their $x \to 0$ extrapolation, it was shown in

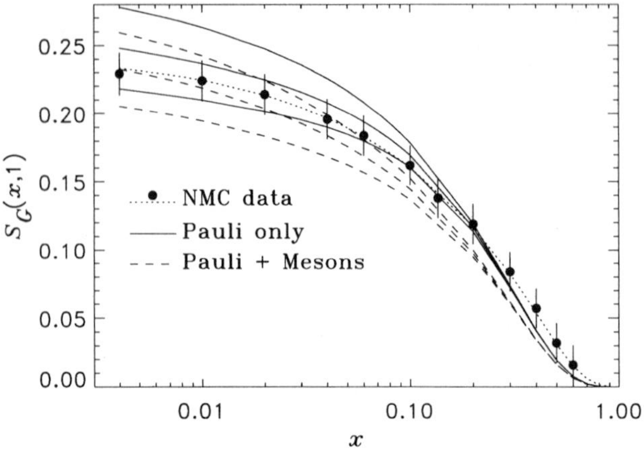

Fig. 21. Effect of Pauli blocking on the Gottfried sum integrand with no mesons (solid curves), and with mesons for Λ = 700 MeV (dotted curves). The three curves in each case are for $\mathcal{P}_N = 0$ (largest curves), 0.05 and 0.10 (smallest curves).

ref.[114] that a value of 1/3 could be recovered. Although this more singular behaviour seems rather artificial, without data at such small x it remains a possibility. However, one problem with this hypothesis of late onset (in the sense of decreasing x) of Regge behaviour is the data from the E665 Collaboration [128], which suggests early saturation of the Gottfried sum rule, and would therefore tend to rule out this option.

It was also suggested by Kaptari and Umnikov [129] that nuclear effects in deuterium may introduce errors in the extraction of the neutron structure function from the deuteron DIS data. In particular, it was claimed that meson exchange currents in the deuteron could lead to substantial antishadowing corrections, so that F_{2n} extracted in a naive manner would be overestimated. With this correction taken into account, it was argued that a value roughly consistent with 1/3 could again be recovered. Furthermore, although expected to be small, genuine nuclear shadowing in deuterium could also introduce corrections to the naively-extracted neutron structure function [75, 130]. In ref.[131] the nuclear effects in deuterium were investigated for a variety of deuteron models. It was found that the combined effects of shadowing resulted in an increase for F_{2n} of $\lesssim 1-2\%$ for $x \sim 0.004$. Consequently, including shadowing corrections would mean that the experimental value for S_G should be lowered from 0.24 to ≈ 0.22 when the "true" neutron structure function is used. Within the above model such a decrease can easily be accommodated by increasing the Pauli blocking correction from $\mathcal{P}_N \approx 0.1$ to ≈ 0.15, if the meson—baryon form factor cut-off is kept at the same value ($\Lambda \approx 0.7$ GeV). Of course a larger Λ could also produce a smaller S_G, but, as we saw in the preceding sections, increasing Λ would also produce a depletion in $F_{2p} - F_{2n}$ at intermediate x, together with an increase at small x. This would be contrary to the behaviour of the shadowing-corrected proton—neutron structure function difference seen in Fig.20. A reduction of $F_{2p} - F_{2n}$ at $x \lesssim 0.3$ can only be explained by a larger Pauli

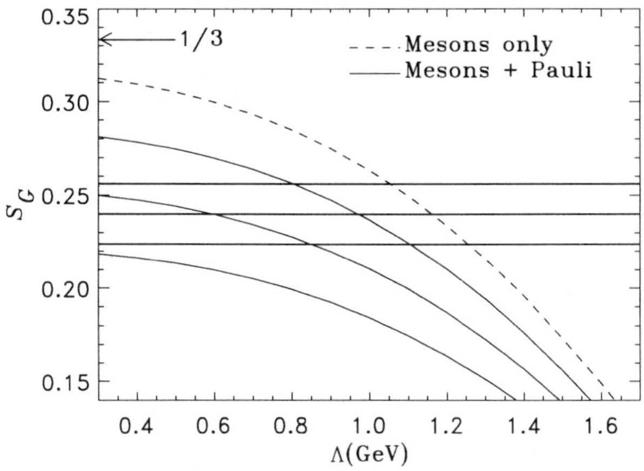

Fig. 22. Gottfried sum rule with mesonic and Pauli corrections. The solid curves represent Pauli normalisation of $\mathcal{P}_N = 0.05$ (largest), 0.1 and 0.15 (smallest).

blocking correction, such as the one required to reproduce the corrected S_G.

6.3 The EMC Spin Effect

In the past few years a great deal of attention has focussed on the QCD improved parton model as a result of the EMC spin effect (or proton "spin crisis"). The EMC [32] extended the earlier SLAC measurement [132] of the structure function $g_{1p}(x, Q^2)$ of the polarised proton to smaller x and hence improved the accuracy with which the first moment was determined. In the naive parton model g_1 is written as:

$$g_{1p}(x) = \frac{1}{2} \sum_q e_q^2 \Delta q(x) \tag{139}$$

where

$$\Delta q(x) = (q^\uparrow + \overline{q}^\uparrow)(x) - (q^\downarrow + \overline{q}^\downarrow)(x) \tag{140}$$

is the polarised quark distribution. It is helpful to rewrite $g_{1p}(x)$ in terms of the SU(3) flavor combinations:

$$\Delta u(x) - \Delta d(x), \qquad \Delta u(x) + \Delta d(x) - 2\Delta s(x) \tag{141}$$

and

$$\Delta u(x) + \Delta d(x) + \Delta s(x). \tag{142}$$

Then the first moment of the flavor singlet piece is related to the fraction of the proton's spin which is carried by its quarks. After a smooth Regge ($x \to 0$) extrapolation of their data ($g_{1p} \sim x^{-0.12}$) EMC determined this quantity to be [32]

$$\Delta u + \Delta d + \Delta s = 0.120 \pm 0.094(stat.) \pm 0.138(syst.) \tag{143}$$

which is consistent with zero and two standard deviations from the Ellis-Jaffe hypothesis, which says that strange quarks should not play a significant role. Our present discussion closely follows that of Bass and Thomas [39], in which a detailed review of the EMC spin experiment and its theoretical interpretation was given.

This result is a violation of Zweig's rule in the flavor singlet channel [133]. As this is the only one of the three SU(3) flavor combinations which can involve the U(1) axial anomaly it seems highly likely that this might be the source of the spin effect. If this is the case, and we strongly suggest that it is, it is clearly an important experimental problem to map out the x dependence of the axial anomaly in the inclusive DIS cross-section. We wish to discuss how important information may be obtained about this in current and future experiments, in particular we examine the (C-odd) spin dependent structure function $g_{3p}(x)$ and also the polarised deuteron structure function $g_{1d}(x)$.

As discussed in sect.3, in QCD the inclusive deep-inelastic process is described by the operator product expansion (OPE) and the renormalisation group. The interesting physics of g_{1p} is in the flavor singlet part, which receives contributions from both quark and gluon partons, viz.

$$g_{1p}(x,Q^2)\big|_S = \frac{1}{3}\sqrt{\frac{2}{3}}\int_x^1 \frac{dz}{z}\left[\Delta q_0(z,Q^2)C_S^q(\frac{x}{z},\alpha_s(Q^2)) + \frac{1}{\sqrt{6}}\Delta g(z,Q^2)C_S^g(\frac{x}{z},\alpha_s(Q^2))\right] \quad (144)$$

The C-even, spin dependent quark $\Delta q_0(x,Q^2)$ and gluon $\Delta g(x,Q^2)$ distributions are defined with respect to the operator product expansion. Their odd moments project out the target matrix elements of the renormalised, spin odd, composite operators

$$2Ms_+(p_+)^{2n}\int_0^1 dx\, x^{2n}\Delta q_k(x,Q^2) = \langle p,s|[\overline{q}(0)\gamma_+\gamma_5(iD_+)^{2n}\frac{\lambda^k}{2}q(0)]^{GI}_{Q^2}|p,s\rangle_c \quad (145)$$

and

$$2Ms_+(p_+)^{2n}\int_0^1 dx\, x^{2n}\Delta g(x,Q^2) = \langle p,s|[\mathrm{Tr}\, G_{+\alpha}(0)(iD_+)^{2n-1}\widetilde{G}^{\alpha}_{\ +}(0)]^{GI}_{Q^2}|p,s\rangle_c \quad (n\geq 1). \quad (146)$$

>From unpolarised DIS experiments we know that the gluon distribution is concentrated at small x. In polarised DIS the hard photon scatters from a gluon via a quark-antiquark pair, described in $C^g(x,\alpha_s)$. This dissipates the gluon's already small momentum so that $\Delta g(x,Q^2)$ is relevant to g_{1p} only at small x ($x \leq 0.03$) [134]. It makes a negligible contribution to the measured sum rule between $x = 0.01$ and 1, where the three constituent quarks are expected to dominate.

The clue to understanding the spin effect lies in the identification of the axial-vector current (and the higher spin axial tensors in equ.(145)) with spin. Classically the axial vector current looks like a gauge invariant operator, with the quark field operator transforming as

$$q(x) \to U(x)q(x) \quad (147)$$

and

$$\overline{q}(x)\gamma_\mu\gamma_5 \rightarrow \overline{q}(x)\gamma_\mu\gamma_5 U^\dagger(x) \tag{148}$$

under a given gauge transformation U. On the other hand, in quantum field theory the axial vector current operator is not just $\overline{q}(0)\gamma_\mu\gamma_5$ multiplied by $q(0)$. It is a composite operator which has to be renormalised and there are extra divergences which are intrinsic to the operator itself. It turns out that one cannot renormalise the axial tensor operators in a gauge invariant way so that they describe spin at the same time. In general, for a given choice of renormalisation prescription R, the renormalised axial tensor operator differs from the gauge invariant operator by a multiple of a gauge-dependent, gluonic counterterm $k_{\mu\mu_1...\mu_{2n}}$, viz.

$$\left[\overline{q}(0)\gamma_\mu\gamma_5 D_{\mu_1}...D_{\mu_{2n}}q(0)\right]^R_{Q^2} = \left[\overline{q}(0)\gamma_\mu\gamma_5 D_{\mu_1}...D_{\mu_{2n}}q(0)\right]^{GI}_{Q^2} + \lambda_{R,n}\left[k_{\mu\mu_1...\mu_{2n}}\right]_{Q^2} \tag{149}$$

where the coefficients $\lambda_{R,n}$ are fixed by the choice of renormalisation prescription. This axial anomaly was discovered for the axial vector current in QED [135, 136].

Not only does the axial anomaly lead to a difference between the renormalised axial currents which preserve gauge invariance and chiral symmetry, but in addition the gauge invariant axial current is scale dependent (in this case the scale is Q^2). (The anomalous dimension of the first moment, Δq_0, was first calculated in QCD by Kodaira [137].) This means that one cannot derive the generators of the spin algebra SU(2) from it. It follows [138] that the gauge invariant axial-vector current and the higher spin operators which appear in equ.(145) do not describe a distribution of quark spin in the proton. One can construct a distribution which does measure spin. It differs from the physical distribution $\Delta q(x, Q^2)$ which is measured in deep-inelastic scattering by by a gauge dependent gluonic term related to the $k_{\mu\mu_1...\mu_{2n}}$ in equ.(149) — the anomaly. In other words, one can say that the gauge symmetry screens the spin of the quarks.

We now compare g_1 with the other structure functions measured in deep-inelastic scattering. The axial anomaly is not relevant to the unpolarised quark distributions, which are described in OPE language by the operators $\overline{q}(0)\gamma_+(iD_+)^n q(0)$. Nor is it relevant to g_3, which is the polarised version of F_3. Since g_3 is odd under charge conjugation, and gluons are C-even, it can have no anomalous gluonic contribution. This means that it does make sense to talk about F_1, F_3 and g_3 in terms of quarks with explicit spin degrees of freedom — the clash of symmetry between gauge invariance and spin does not manifest itself in these structure functions.

It is clearly an important problem to map out the x dependence of the anomaly in g_{1p}, and a comparison between g_{1p} and g_{3p} would be the ideal way to do it. Unfortunately, the cross-section for DIS with a neutrino beam and proton target is very small — enough to make direct measurements of g_{3p} impracticable at the present time. However, if one assumes that the quark fragmentation functions are spin independent it may be possible to extract the C-odd distribution from the g_{1p} measurements by detecting fast pions from among the final state hadrons [139]. This experiment is planned by the HERMES collaboration at HERA [140].

Important information about the x dependence of the axial anomaly in polarised deep-inelastic scattering will also come from measurements of $g_{1n}(x,Q^2)$. The axial anomaly occurs only in the flavor singlet part of g_1 and therefore it will be present equally in g_{1p} and g_{1n} as a function of x. If the anomaly acts to screen the quark spin at large x in g_{1p} it follows that the same should be true in g_{1n}. The combination appearing in the Bjorken sum rule $(g_{1p}-g_{1n})$ has no flavor singlet component and is anomaly free. On the other hand, the flavor singlet component is enhanced in the deuteron structure function $g_{1d} = (g_{1p}+g_{1n})/2$, which has no isotriplet piece $\Delta q_3(x,Q^2)$. Thus the deuteron structure function g_{1d} is an ideal place to test model predictions about how the anomaly should contribute in the nucleon structure function $g_1(x,Q^2)$.

As we saw in sect.4.1 the usual quark model calculations, which do not include the anomaly, suggest that g_{1n} will change sign and become small and positive at large x [20, 22, 141]. To the extent that these models do not include any OZI violation, a large-x anomaly would tend to render g_{1n} negative at large x. As a specific example we consider the quark model calculation of the structure functions described in sect.4.1. As we have seen these calculations provide reasonable agreement with the unpolarised structure function data. However the bag model has not yet been extended to satisfy the U(1) chiral Ward identity. That is, it does not include an OZI violation induced by the anomaly. On the other hand it does seem reasonable that these model calculations for $g_1(x)$ might describe $g_3(x)$ at large x — i.e. correspond to a world without the OZI violations due to the anomaly.

As we saw in Fig.6 the bag model prediction for $g_{1p}(x)$ overestimates the data throughout all of the large-x region. Hence, it is tempting to associate the difference between the model results and the data as that associated with the anomaly. In this case, the same flavor singlet correction should therefore be applied to the neutron. In recent work Bass and Thomas [41] have added a purely phenomenological correction to the bag model results to fit the proton data. Adding the same correction to the prediction for the neutron and combining them to get the deuteron structure function they obtained the result shown in Fig.23. (For the present purposes we make the simple approximation that $g_{1d} = (g_{1p}+g_{1n})/2$, thus ignoring corrections due to shadowing, Fermi-motion and the D-state probability of the deuteron. These are expected to be important at the few-percent level [130, 131, 142] - — well below the present experimental accuracy.) The corrected curve is in good agreement with the recent SMC measurement of the deuteron spin structure function $xg_{1d}(x,Q^2)$ [143].

To summarise the results of this section, we have discussed how one could map out the x dependence of the axial anomaly in g_1. If the anomaly is a large-x effect then it can be isolated as a finite difference between g_1 and g_3 in the large-x bins. If it is purely a small-x effect the anomaly would be lost among the sea and gluon distributions which dominate the data at small x (say ≤ 0.1). We stress that the comparison with g_3 is the only definitive experimental test of whether the anomaly is a large or a small x effect. Certainly, it is an intrinsic part of the spin dependent quark distribution and there is no good theoretical reason to believe that it is confined to small x. We strongly urge that consideration be given to the challenging experimental problem of how to measure g_3. In the interim it would be very useful to obtain more data (with reduced errors) on the deuteron spin structure function g_1^d. This deuteron data will help constrain theoretical models of the structure functions.

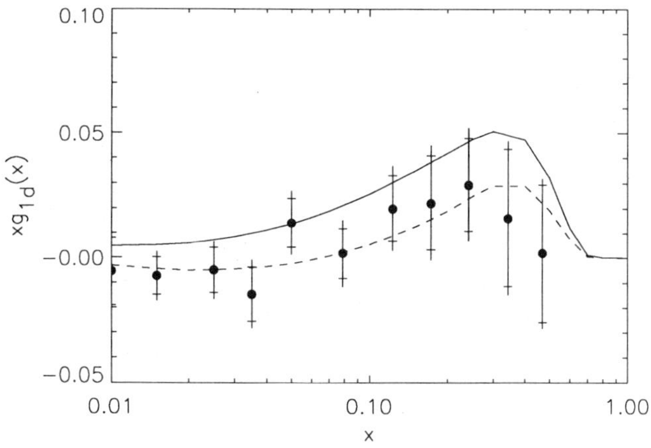

Fig. 23. Prediction for the structure function of the deuteron [41] based on the bag model (with R = 0.8 fm) – solid curve. The dashed line shows the effect of adding a phenomenological term representing the effect of the axial anomaly on both the proton and the neutron. The data is from SMC [143].

6.4 Nuclear structure functions

Since the discovery by the EMC that the structure functions of nuclei did not all have the same shape [144-146] there has been considerable further investigation. On the experimental side some features of the data, like the apparently dramatic increase in the sea with mass number, have become much less distinct. Other new features such as shadowing at small x have become apparent. However the outstanding feature of the data, namely the softening of the valence quark distribution below x = 0.7 (at which point Fermi motion takes over) has not changed much [32, 147, 148]. In our discussion of the nuclear EMC effect in this section we will closely follow Saito et al. [149, 150].

While one would eventually like a unified theoretical treatment of all the features of the EMC data, we shall concentrate on the softening of the valence quark distribution, which is possibly its most surprising feature. Early attempts to understand this aspect of the data were based upon conventional ideas like nucleon binding, calculated in impulse approximation (IA) [151-154] -— for a review see Bickerstaff and Thomas [155]. Other ideas included a possible enhancement of the cloud of virtual pions around a nucleon in a nucleus [68, 156, 157]. More exotic proposals included the possibility of multi-quark clusters [158-160] and quark percolation through the nucleus [161]. Extensive work has also been put into the idea that the nucleon may swell in the nucleus (dynamical rescaling) [162]. Our approach has been to extend to nuclei the same technique that has been successfully used to calculate free nucleon structure functions for the MIT bag model — see sect.4.1.

An earlier investigation by Thomas et al. [163] used the Guichon model [164], in which nuclear matter consists of non-overlapping bags bound in mean-field-approximation (MFA) by the self-consistent exchange of scalar (σ) and vector (ω) mesons. The results emphasised that the EMC effect provides information on the momentum *and* energy distribution of *quarks* in nuclei. In particular the usual impulse approximation based on nuclear binding was shown to significantly overestimate the suppression of the nuclear valence quarks because of the neglect of the binding of the quarks that are spectators to the hard collision. This conclusion remains valid in the more recent work which is significantly more sophisticated [149, 150]. Moreover, in that treatment it is possible to understand the experimental data quantitatively for finite nuclei.

Because it is the energy *and* momentum distribution of quarks that matters one needs a quark model for nuclear matter. At the present stage of development such models are necessarily quite crude. We have used both the Guichon model *and* a further development along the lines of the Boguta model [165] — which enables us to fit not only the binding energy and saturation density of nuclear matter, but also its surface energy and thickness. In order to apply these models to finite nuclei we have used the local density approximation

$$q_A^{(2)}(x,\mu^2) = \int d\vec{r}\,\rho(r) \int_x dy\; f(y,\rho)\, q_N^{(2)}(\frac{x}{y},\rho,\mu^2). \tag{150}$$

Here ρ is the nuclear density distribution, $f(y,\rho)$ accounts for Fermi motion and $q_N^{(2)}$ is the twist-2 quark distribution of the bound nucleon. Note that the recent discovery that convolution fails in a more sophisticated treatment of the off-shell behaviour of the nucleon structure function — c.f. sect.4.2 — has not yet been incorporated in any calculation of the EMC effect.

In Fig.24 the solid curve is the complete calculation for the ratio of the structure functions of Fe and D (at 10 GeV^2), for the case where the free bag radius, R_0, is 0.6 fm. (There is in fact little sensitivity to R_0 [150].) For comparison we also show the most recent data from EMC, BCDMS and SLAC [32, 147, 148]. Clearly the calculation provides a semi-quantitative description of the data. Notice that one cannot use the local density approximation for deuterium. Instead a conventional convolution model which allowed for 8 MeV binding and recoil was used. The nucleon momentum distribution in the deuteron was given by the Paris potential. This particular calculation of the nuclear structure functions included no nucleon momenta higher than about 1.3 fm^{-1}. Therefore a cut-off (p_c) at 1.3 fm^{-1} was also imposed on the deuteron momentum wave function. In fact, this only affects the EMC ratio at large x [150].

The Guichon model [164] predicts the ratio shown by the dotted curve in Fig.24 which is not quite as good. From the phenomenological point of view it is important to know whether our earlier conclusion about the inaccuracy of the impulse approximation remains true. The dashed curve in Fig.24 shows the effect of ignoring the interaction of the residual two-quark bag with the nucleus. Clearly it dramatically overestimates the EMC effect. On physical grounds this makes good sense. Deep-inelastic scattering measures the energy-momentum distribution of the struck *quark*, not the struck nucleon.

Let us now briefly summarise our conclusions and comment on what remains to be done. It is very satisfying that a quark level description of nuclear matter, together with the local density approximation and a microscopic method of calculating twist-two structure functions does give a reasonable fit to the EMC data on Fe. Although it is disappointing in some ways that the impulse approximation fails, we have seen that there is a sound physical reason for this failure, namely that it effectively assigns the

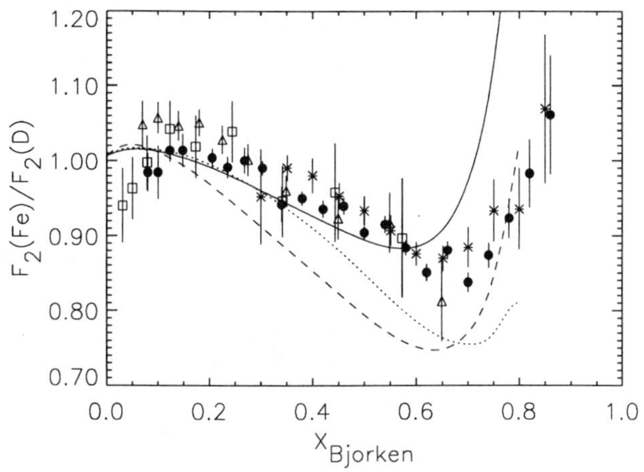

Fig. 24. Local density model for the ratio of the structure functions of Fe and D. See the text for details – from ref.[150].

binding of an entire nucleon to a single struck quark. On the positive side we can be sure that nuclear deep-inelastic scattering does tell us about the binding of quarks in nuclei.

Of course, having achieved this much one necessarily wants much more. One would like to go beyond the local density approximation which would require a quark model for the structure of finite nuclei. It is also unsatisfactory to be limited to a mean-field approximation for non-overlapping bags. Even though σ- and ω-exchange may be viewed at least partly as a macroscopic treatment of more complicated short distance processes (perhaps involving quark and gluon exchange) one would like to do better.

The MIT bag model is also a fairly crude representation of nucleon structure. One should explore the consequences of using more sophisticated models — e.g. with a more reasonable surface, a better treatment of c.m. motion and perhaps a pion cloud. (The latter might eventually help us understand why there is no evidence for an enhancement in the pion field of the nucleus in recent Drell-Yan data [166].) As we have explained elsewhere [167], we are limited to leading order QCD unless the model used has a well defined connection to QCD.

In view of our findings in sect.4.2 [43] it will be crucial to go beyond the convolution model in the treatment of nucleon Fermi motion. Finally, our analysis has not yet had anything new to say about the nuclear sea or shadowing at small x. Eventually one may hope to understand all the features of the data within a single unified theory.

7 Conclusion

In these lectures we have covered an enormous amount of ground, beginning with elementary kinematics, and finishing with some of the exciting problems that remain unsolved. Along the way we have seen how we can make a clear connection between familiar low-energy quark models and the parton distributions measured in high-energy deep-inelastic scattering. This is a very exciting and enouraging step along the way to meeting the ultimate challenge of nuclear physics — understanding the relationship between QCD and low-energy nuclear physics.

We have seen in detail that pions and possibly heavier mesons are needed if we are to make sense of some of the recent deep-inelastic scattering data. Certainly it is quite likely that a meson cloud around the proton is at least partially responsible for the observed flavor asymmetry in the proton sea. In addition, a meson cloud may also play a role in the spin structure of the proton, since some of the spin (as well as flavor) quantum numbers may actually be carried by the cloud.

While still not a completely understood subject, it is clear that a fully consistent description of the spin structure of the nucleon requires a thorough re-examination of the very foundations of the parton model. To shed some light on this fundamental question we eagerly anticipate new data on the polarised structure functions of protons and neutrons with which we can compare some of the model predictions.

Finally, we have touched briefly on the important question of how the nuclear medium affects the properties of quarks inside nucleons. In particular, we now understand that microscopic structure is indeed crucial to any quantitative description of nuclear properties as seen by high-energy probes. Furthermore, improvements in calculating nuclear structure functions by incorporating relativistic effects also indicate the necessity of including both quark and nuclear degrees of freedom within a fully consistent treatment.

If we have succeeded in passing on on some of the excitement we feel about this field, then the effort of preparing the lectures has been worthwhile. With an energy upgrade underway at SLAC, new experiments planned at CERN and Fermilab, CEBAF about to come on line and ELFE (in Europe) a real possibility, this is a field that is just beginning to flower.

Acknowledgements

We would like to thank the organisers of the Shimoda meeting, particularly Professors T.Yamazaki, S.Homma, Y.Akaishi and K.Yazaki for their generous hospitality. Financial support from the JSPS, INS and the Inoue Foundation is greatfully acknowledged. We would also like to thank S.Bass, K.Saito, A.W.Schreiber, A.I.Signal and F.M.Steffens for their contributions to our understanding of the issues discussed here. This work was supported by the Australian Research Council.

References

1. A.W.Thomas, in *Progress in Nuclear and Particle Physics*, ed. D. Wilkinson, Vol.11 (1984) 325-378.
2. A.W.Thomas, in *Progress in Particle and Nuclear Physics*, ed. A. Faessler, Vol.20 (1988) 21-46.
3. F.E.Close, *At Introduction to Quarks and Partons*, Academic Press, New York (1979).

4. E.Leader and E.Predazzi, *An Introduction to Gauge Theories and the New Physics*, Cambridge (1982).
5. T.Muta, *Foundations of Quantum Chromodynamics*, World Scientific, Singapore (1987).
6. C.T.Sachrajda, in *Gauge Theories in High Energy Physics*, North Holland, Amsterdam (1983).
7. F.Yndurain, *Quantum Chromodynamics*, Springer, Berlin (1983).
8. J.L.Alonso and R.Tarrach (eds.), *Quantum Chromodynamics*, Springer, Berlin (1980).
9. G.Altarelli, *Phys.Rep.* **81** (1982) 1.
10. N.S.Craigie et al., *Phys.Rep.* **99** (1983) 69.
11. J.D.Jackson, G.G. Ross and R.G. Roberts, *Phys.Lett.* **B226** (1989) 159.
12. S.Dasu et al., *Phys.Rev.Lett.* **64** (1988) 2591.
13. J.G.Morfin, talk given at New Directions in Neutrino Physics at Fermilab, Illinois, Sep.1988, published in *Fermilab Neutrino Phys.* **13** (1988).
14. H.Abramowicz et al. (CDHS Collaboration), *Z.Phys* **C15** (1982) 19; **C17** (1983) 1283.
15. M.Lissia, M.-C.Chu, J.W.Negele and J.M.Grandy, *Nucl.Phys.* **A555** (1993) 272.
16. C.T.Sachrajda, plenary talk given at Annual Lattice Conf., Lattice '88, Batavia, IL, Sep. 1988, published in *Fermilab Lattice* **121** (1988).
17. R.L.Jaffe, *Nucl.Phys.* **B229** (1983) 205.
18. C.H. Llewellyn Smith, in *Proc. 1988 CAP-NSERC Summer Institute* (Singapore, World Scientific).
19. A.I.Signal and A.W.Thomas, *Phys.Rev.* **D40** (1989) 2832.
20. F.E.Close and A.W.Thomas, *Phys.Lett.* **B212** (1988) 227.
21. A.W.Schreiber, A.W.Thomas and J.T.Londergan, *Phys.Lett.* **B237** (1989) 120.
22. A.W.Schreiber, A.W.Thomas and J.T.Londergan, *Phys.Rev.* **D42** (1990) 2226.
23. D.W.Duke and J.T.Owens, *Phys.Rev.* **D30** (1984) 49.
24. A.D.Martin, R.G.Roberts and W.J.Stirling, *Phys.Rev.* **D37** (1988) 1161.
25. E.Eichten, I.Hinchliffe, K.Lane and C.Quigg, *Rev.Mod.Phys.* **56** (1984) 579.
26. M.Diemoz, F.Ferroni, E.Longo and G.Martinelli, *Z.Phys.* **C39** (1988) 21.
27. M.Glück, E.Reya and W.Vogelsang, *Nucl.Phys.* **B329** (1990) 347.
28. A.W.Thomas, *Nucl.Phys.* **A518** (1990) 186.
29. A.W.Schreiber, P.J.Mulders, A.I.Signal and A.W.Thomas, *Phys.Rev.* **D45** (1992) 3069.
30. A.Schafer, *Phys.Lett.* **B208** (1988) 175.
31. G.Farrar and D.Jackson, *Phys.Rev.Lett.* **35** (1975) 1416.
32. J. Ashman et al., *Phys.Lett.* **B206** (1988) 364; *Nucl.Phys.* **B328** (1990) 1.
33. J.Ellis and R.L.Jaffe, *Phys.Rev.* **D9** (1974) 1444.
34. A.W.Schreiber and A.W.Thomas, *Phys.Lett.* **B215** (1988) 141.
35. H.Hogaasen and F.Myhrer, *Phys.Lett.* **B214** (1988) 123.
36. F.Myhrer and A.W.Thomas, *Phys.Rev.* **D38** (1988) 1633.
37. G.Altarelli and G.G.Ross, *Phys.Lett.* **B212** (1988) 391.
38. R.D.Carlitz et al., *Phys.Lett.* **B214** (1988) 229.
39. S.D.Bass and A.W.Thomas, *J.Phys.* **G 19** (1993) 925.
40. R.L.Jaffe and A.V.Manohar, *Nucl.Phys.* **B337** (1990) 509; A.V.Manohar, *Phys.Rev.Lett.* **66** (1991) 289; S.D.Bass, B.L.Ioffe, N.N.Nikolaev and A.W.Thomas, *J.Moscow Phys.Soc.* **1** (1991) 317.
41. S.D.Bass and A.W.Thomas *Phys. Lett.* **B312** (1993) 345.
42. C.J.Benesh and G.A.Miller, *Phys.Lett.* **B222** (1989) 476.
43. W.Melnitchouk, A.W.Schreiber and A.W.Thomas, Adelaide/PSI preprint ADP-93-210/T128, PSI-PR-93-13 (1993).
44. P.J.Mulders, A.W.Schreiber and H.Meyer, *Nucl.Phys.* **A549** (1992) 498.
45. V.M.Belayev and B.L.Ioffe, *Nucl.Phys.* **B310** (1988) 548.

46. G.E.Brown and M.Rho, *Phys.Lett.* **82B** (1979) 177.
47. A.W.Thomas, *Adv.Nucl.Phys.* **13** (1984) 1.
48. S.Theberge, G.A.Miller and A.W.Thomas, *Can.J.Phys.* **60** (1982) 59.
49. R.F.Alvarez-Estrada and A.W.Thomas, *J.Phys.* **G9** (1983) 161.
50. J.D.Sullivan, *Phys.Rev.* **D5** (1972) 1732.
51. A.W.Thomas, *Phys.Lett.* **126B** (1983) 97.
52. A.I.Signal, A.W.Schreiber and A.W.Thomas, *Mod.Phys.Lett.* **A6** (1991) 271.
53. E.M.Henley and G.A.Miller, *Phys.Lett.* **B251** (1990) 497.
54. S.Kumano and J.T.Londergan, *Phys.Rev.* **D44** (1991) 717.
55. W.-Y.P.Hwang, J.Speth and G.E.Brown, *Z.Phys.* **A339** (1991) 383; A.Szczurek and J.Speth, *Nucl.Phys.* **A555** (1993) 249.
56. W.Melnitchouk, A.W.Thomas and A.I.Signal, *Z.Phys.* **A340** (1991) 85.
57. K.Gottfried, *Phys.Rev.Lett.* **18** (1967) 1174.
58. P.Amaudruz et al. (New Muon Collaboration), *Phys.Rev.Lett.* **66** (1991) 2712.
59. M.Lusignoli and Y.Srivastava, *Nucl.Phys.* **B138** (1978) 151.
60. G.G.Arakelyan, K.G.Boreskov and A.B.Kaidalov, *Sov.J.Nucl.Phys.* **33** (1981) 247.
61. M.T.Jeong and Il-T.Cheon, *Phys.Rev.* **D43** (1991) 3725.
62. A.W.Thomas, *Prog.Theor.Phys.Supp.* **91** (1987) 204.
63. L.L.Frankfurt, L.Mankiewicz and M.I.Strikman, *Z.Phys.* **A334** (1989) 343.
64. S.Kumano, *Phys.Rev.* **D43** (1991) 59; 3067.
65. W.Rarita and J.Schwinger, *Phys.Rev.* **60** (1941) 61.
66. M.Benmerrouche, R.M.Davidson and N.C.Mukhopadhyay, *Phys.Rev.* **C39** (1989) 2339.
67. V.Dmitrašinović and R.Tegen, *Phys. Rev.* **C46** (1992) 1108.
68. E.L.Berger, F.Coester and R.B.Wiringa, *Phys.Rev.* **D29** (1984) 398.
69. R.L.Jaffe, in *Relativistic Dynamics and Quark-Nuclear Physics*, ed. M.B.Johnson and A.Pickleseimer (Wiley, New York, 1985).
70. T.Hippchen, J.Haidenbauer, K Holinde and V.Mull, *Phys.Rev.* **C44** (1991) 1323.
71. B.Kämpfer, A.I.Titov and E.L.Bratkovskaya, *Phys.Lett.* **B301** (1993) 123.
72. A.Bodek and J.L.Ritchie, *Phys.Rev.* **D23** (1981) 1070.
73. G.Dunne and A.W.Thomas, *Phys.Rev.* **D33** (1986) 2061.
74. F.Güttner, G.Chanfray, H.J.Pirner and B.Povh, *Nucl.Phys.* **A429** (1984) 389.
75. V.R.Zoller, *Z.Phys.* **C54** (1992) 425.
76. S.Weinberg, *Phys.Rev.* **150** (1966) 1313.
77. S.D.Drell, D.J.Levy and T.M.Yan, *Phys.Rev.* **D1** (1970) 1035.
78. W.Melnitchouk and A.W.Thomas, *Phys.Rev.* **D47** (1993) 3794.
79. R.Machleidt, K.Holinde and Ch.Elster, *Phys.Rep.* **149** (1987) 1.
80. G.N.Khachatryan and Yu.G.Shakhnazaryan, *Sov.J.Nucl.Phys.* **29** (1979) 73; G.I.Gakh and A.P.Korzh, *Ukr.J.Phys.* **26** (1981) 480.
81. R.V.Reid, Jr., *Ann.Phys.* **50** (1968) 411.
82. M.M.Nagels, Th.A.Rijken and J.J.de Swart, *Phys.Rev.* **12** (1975) 744.
83. M.Lacombe et al., *Phys.Rev.* **D12** (1975) 1495; *Phys.Rev.* **C21** (1980) 861; *Phys.Lett.* **101B** (1981) 139.
84. A.D.Jackson, D.O.Riska and B.Verwest, *Nucl.Phys.* **A249** (1975) 397.
85. G.E.Brown and A.D.Jackson, *The Nucleon-Nucleon Interaction* (North-Holland, Amsterdam, 1976).
86. S.Deister, M.F.Gari, W.Krumpelmann and M.Mahlke, *Few Body Systems* **10** (1991) 1.
87. A.W.Thomas and K.Holinde, *Phys.Rev.Lett.* **63** (1989) 2025.
88. K.Holinde and A.W.Thomas, *Phys.Rev.* **C42** (1990) R1195.
89. B.G.Yu and Il-T.Cheon, *J.Korean Phys.Soc.* **25** (1992) 571.

90. U.G.Meissner, N.Kaiser and W.Weise, *Nucl.Phys.* **A466** (1987) 685; C.Schuren, E.Ruiz Arriola and K.Goeke, Ruhr University preprint RUB-TPII-01-92 (1992).
91. We thank K.Holinde for a discussion relating to this point.
92. T.Uematsu and T.F.Walsh, *Nucl.Phys.* **B199** (1982) 93; P.Hoodbhoy, R.L.Jaffe and A.Manohar, *Nucl.Phys.* **B312** (1989) 571; F.E.Close and S.Kumano, *Phys.Rev.* **D42** (1990) 2377.
93. D.Allasia et al. (BEBC Collaboration), *Z.Phys.* **C28** (1985) 321.
94. A.I.Signal and A.W.Thomas, *Phys.Lett.* **B191** (1987) 205.
95. B.Betev et al. (NA10 Collaboration), *Z.Phys.* **C28** (1985) 15.
96. M.Glück, E.Reya and A.Vogt, *Z.Phys.* **C53** (1992) 651.
97. P.J.Sutton, A.D.Martin, R.G.Roberts and W.J.Stirling, *Phys.Rev.* **D45** (1992) 2349.
98. S.J.Brodsky and R.Blankenbecler, *Phys.Rev.* **D10** (1974) 2973; S.J.Brodsky and G.Farrar, *Phys.Rev.Lett.* **31** (1975) 1193.
99. P.D.B.Collins, *An Introduction to Regge Theory and High Energy Physics* (Cambridge University Press, 1977).
100. J.Badier et al., *Phys.Lett.* **93B** (1980) 354.
101. J.R.Bergervoet et al., *Phys.Rev.* **C41** (1990) 1435; R.Machleidt and F.Sammarruca, *Phys.Rev.Lett.* **66** (1991) 564; R.L.Workman, R.A.Arndt and M.M.Pavan, *Phys.Rev.Lett.* **68** (1992) 1653; 2712.
102. R.Koch and E.Pietarinen, *Nucl.Phys.* **A336** (1980) 331.
103. G.Hohler and E.Pietarinen, *Nucl.Phys.* **B95** (1975) 210.
104. W.Grein and P.Kroll, *Nucl.Phys.* **A338** (1980) 332.
105. O.Dumbrajs et al., *Nucl.Phys.* **B216** (1983) 277.
106. P.M.M.Maessen, Th.A.Rijken and J.J.de Swart, *Phys.Rev.* **C40** (1989) 2226; B.Holzenkamp, K.Holinde and J.Speth, *Nucl.Phys.* **A500** (1989) 485.
107. G.E.Brown and W.Weise, *Phys.Rep.* **C22** (1975) 281.
108. J.F.Owens, *Phys.Lett.* **B266** (1991) 126.
109. J.G.Morfin and W.K.Tung, *Z.Phys.* **C52** (1991) 13.
110. D.A.Ross and C.T.Sachrajda, *Nucl.Phys.* **B149** (1979) 497.
111. R.D.Field and R.P.Feynman, *Phys.Rev.* **D15** (1977) 2590.
112. J.F.Donoghue and E.Golowich, *Phys.Rev.* **D15** (1977) 3421.
113. A.D.Martin, W.J.Stirling and R.G.Roberts, *Phys.Rev.* **D47** (1993) 867.
114. A.D.Martin, W.J.Stirling and R.G.Roberts, *Phys.Lett.* **B252** (1990) 653.
115. S.Leone et al. (CDF Collaboration), presented at Rencontres de Moriond, Les Arcs, 1991, in *Moriond 1991: Electroweak*, p.13.
116. A.S.Ito et al., *Phys.Rev.* **D23** (1981) 604.
117. G.G.Arakelyan and K.G.Boreskov, *Sov.J.Nucl.Phys.* **41** (1985) 267.
118. M.Ericson and A.W.Thomas, *Phys.Lett.* **148B** (1984) 191.
119. E.J.Eichten, I.Hinchliffe and C.Quigg, *Phys.Rev.* **D47** (1993) R747.
120. S.D.Ellis and W.J.Stirling, *Phys.Lett.* **B256** (1991) 258.
121. S.Kumano and J.T.Londergan, *Phys.Rev.* **D46** (1992) 457.
122. G.T.Garvey et al., *Measurement of $\bar{d}(x)/\bar{u}(x)$ in the Proton*, Fermilab Proposal P866.
123. J.Levelt, P.J.Mulders and A.W.Schreiber, *Phys.Lett.* **B263** (1991) 498.
124. M.Gronau, F.Ravndal and Y.Zarmi, *Nucl.Phys.* **B51** (1973) 611.
125. M.Arneodo et al. (EM Collaboration) *Nucl.Phys.* **B321** (1989) 541.
126. G.Preparata, P.G.Ratcliffe and J.Soffer, *Phys.Rev.Lett.* **66** (1991) 687.
127. M.Arneodo et al. (NM Collaboration), CERN-PPE/93-117 (July 1993).
128. M.R.Adams et al. (E665 Collaboration), Fermilab preprint (Sep. 1991).
129. L.P.Kaptari and A.Yu.Umnikov, *Phys.Lett.* **B272** (1991) 359.

130. B.Badelek and J.Kwiecinski, *Nucl.Phys.* **B370** (1992) 278.
131. W.Melnitchouk and A.W.Thomas, *Phys.Rev.* **D47** (1993) 3783.
132. G.Baum et al., *Phys.Rev.Lett.* **51** (1983) 1135.
133. G.Veneziano, Okubofest lecture, CERN preprint TH-5840/90 (1990).
134. S.D.Bass and A.W.Thomas, *Nucl.Phys.* **A527** (1991) 519c; J.Ellis, M.Karliner and C.Sachrajda, *Phys.Lett.* **B231** (1989) 497; S.D.Bass, N.N.Nikolaev and A.W.Thomas, Adelaide preprint ADP-90-133/T80 (1990).
135. J.S.Bell and R.Jackiw, *Nuov.Cim.* **51A** (1969) 47.
136. S.L.Adler, *Phys.Rev.* **177** (1969) 2426.
137. J.Kodaira, *Nucl.Phys.* **B165** (1980) 129.
138. S.D.Bass, Cavendish preprints HEP 92/11 (1992) and 93/2 (1993).
139. L.L.Frankfurt et al., *Phys.Lett.* **B320** (1989) 141.
140. HERMES Proposal, K.Coulter et al., DESY/PRC 90-1 (1990); M.Veltri et al., *Proc.Physics at HERA*, Vol. 1, 447 (1991).
141. R.Carlitz and J.Kaur, *Phys.Rev.Lett.* **38** (1977) 673; A.Schafer, *Phys.Lett.* **B208** (1988) 175.
142. R.M.Woloshyn, *Nucl.Phys.* **A496** (1989) 749.
143. The SMC Collaboration, *Phys.Lett.* **B302** (1993) 533.
144. J.J.Aubert et al. (EM Collaboration), *Phys.Lett.* **B123** (1983) 275.
145. J.J.Aubert et al. (EM Collaboration), *Nucl.Phys.* **B293** (1983) 740.
146. R.G.Arnold et al., *Phys.Rev.Lett.* **52** (1984) 727.
147. A.C.Benvenuti et al. (BCDMS), *Phys.Lett.* **B189** (1987) 483.
148. S.Dasu et al., *Phys.Rev.Lett.* **64** (1988) 2591.
149. A.W.Thomas, K.Saito and A.Michels, *Aust.J.Phys.* **46** (1993) 3.
150. K.Saito, A.Michels and A.W.Thomas, *Phys.Rev.* **C46** (1992) R2149.
151. S.V.Akulinichev, S.Shlomo, S.A.Kulagin and G.M.Vagradov *Phys.Rev.Lett.* **55** (1985) 2239.
152. S.V.Akulinichev and S.Shlomo, *Phys.Rev.* **C33** (1986) 1551.
153. G.Dunne and A.W.Thomas, *Nucl.Phys.* **A455** (1986) 701.
154. T.Uchiyama and K.Saito, *Phys.Rev.* **C38** (1988) 2245.
155. R.P.Bickerstaff and A.W.Thomas, *J.Phys.* **G15** (1989) 1523.
156. C.H.Llewellyn Smith, *Phys.Lett.* **128B** (1983) 107.
157. M.Ericson and A.W.Thomas, *Phys.Lett.* **B128** (1983) 112.
158. A.Krzywicki, *Phys.Rev.* **D14** (1976) 152.
159. H.J.Pirner and J.P.Vary, *Phys.Rev.Lett.* **46** (1981) 1376.
160. K.Saito and T.Uchiyama, *Z.Phys.* **A322** (1985) 299.
161. O.Nachtmann and H.J.Pirner, *Z.Phys.* **C21** (1984) 277.
162. F.E.Close, R.L.Jaffe, R.G.Roberts and G.G.Ross, *Phys.Rev.* **D31** (1985) 1004.
163. A.W.Thomas, *Nucl.Phys.* **A532** (1989) 177.
164. P.A.M.Guichon, *Phys.Lett.* **B200** (1988) 235.
165. J.Boguta, *Phys.Lett.* **B106** (1981) 255.
166. G.T.Garvey, *Nucl.Phys.* **A532** (1991) 119c.
167. A.W.Thomas, *Nucl.Phys.* **A532** (1991) 177c.

Lecture given at the JSPS-INS International Spring School, March 1993 at Shimoda

Current Topics in Nuclear High-Spin Phenomena — Nuclear Shell-Structure and Superdeformation —

Ikuko Hamamoto

Department of Mathematical Physics
LTH, University of Lund
P.O.Box 118, S-221 00 LUND, SWEDEN

Abstract Some aspects of superdeformation and related phenomena are reviewed, which provide striking evidence for nuclear shell-structure with much larger deformations than those encountered in the nuclear ground states.

List of contents

1 Introduction

As current topics in nuclear high-spin phenomena I choose to talk about superdeformation (SD) and the related phenomena. Nuclear SD shape, which is a prolate spheroid with a ratio of axes of 2:1, was found in the actinide region as fission isomers (at low spins), already in 1962 [1]. The discovery of the strongly deformed shape provides striking evidence for the shell structure in nuclear potentials with much larger deformations [2] than those encountered in the ground states of heavy nuclei. It has been possible to discover the SD shape in the actinide region thanks to the spontaneous fission decay mode. However, the very presence of the fission decay channels in the actinide region makes it impossible to have a long-lived high-spin SD state.

Already in the middle of the seventies it was predicted [3, 4] that, though the ground states in the A = 150 region is not superdeformed, the possible SD band may become yrast states for higher spins due to the large moments of inertia. First in 1986 the high spin members of an SD band were observed in $^{152}_{66}\mathrm{Dy}_{86}$ [5]; Nineteen gamma-rays with an almost constant spacing of $\Delta E_\gamma = (E(I+2) - E(I)) - (E(I) - E(I-2)) \approx 47$ keV, which connect the members of the band with B(E2) values of about 2600 W.u. [6]. The yrast states are sketched in Fig. 1. The presence of a number of the similar kind of SD bands is at the moment known both in the A = 150 region (a centered nucleus is ^{152}Dy) and in the $A = 190$ region (a centered nucleus is ^{192}Hg) [7]. No definite spin-parity assignments have been made in any of the SD bands. Having no spin-assignments, a quantity, socalled "dynamical moment of inertia" [8], which is defined by

$$\hbar^{-2}\mathcal{J}^{(2)} \equiv \left(\hbar\frac{d\omega}{dI}\right)^{-1} = \left(\frac{d^2E}{dI^2}\right)^{-1} \approx \frac{4}{\Delta E_\gamma} \tag{1}$$

is extracted from the data.

In sect. 2.1. the presence of nuclear shell structure and the resulting strongly deformed nuclear shape are illustrated mainly by using an axially symmetric deformed oscillator model. In sect. 2.2 the observed facts on fission isomers are briefly summarized, while in sect. 2.3 an example of the nuclear spectroscopy using observed dynamical moments of inertia is presented. The presence of "identical bands", which was first reported by examining the rotational spectra of some SD bands, is discussed also in sect. 2.3. In sect. 3 the properties of collective modes at SD are discussed, which are classified into those coming from the quantum-mechanical shell-structure unique at SD and those from exceptionally large quadrupole deformation.

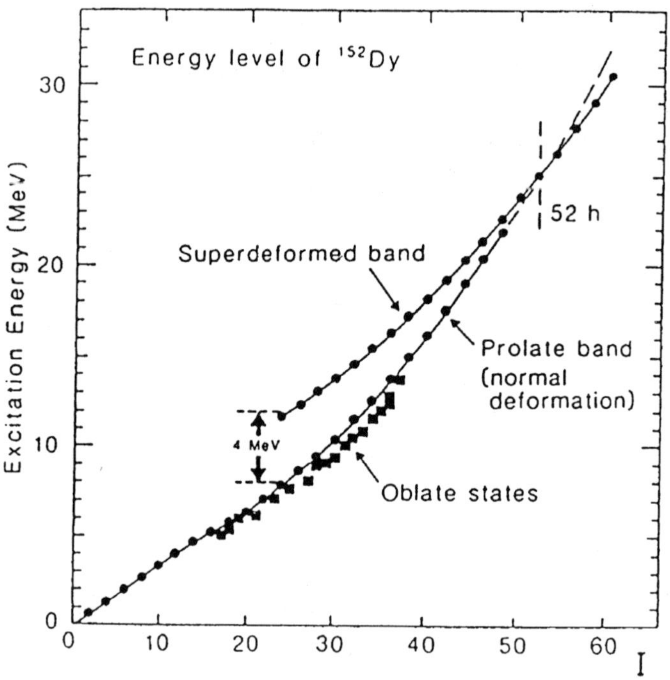

Fig. 1. Yrast states of $^{152}_{66}$Dy$_{86}$. The angular momentum at which the SD band becomes yrast is not experimentally known but is placed at $I = 52$. The figure is taken from ref. [31].

2 Strongly Deformed Shape Resulting from Shell Structure

2.1 Axially-symmetric deformed oscillator model

The Hamiltonian of axially symmetric deformed oscillator potential is written as

$$H = T + \frac{M}{2}\left(\omega_3^2 x_3^2 + \omega_\perp^2\left(x_1^2 + x_2^2\right)\right) \tag{2}$$

The one-particle eigenstate of the Hamiltonian (2) is characterized by the quantum numbers $(n_3, n_\perp)$ with the eigenvalue

$$\begin{aligned}\varepsilon(n_3, n_\perp) &= \hbar\omega_\perp(n_\perp + 1) + h\omega_3\left(n_3 + \frac{1}{2}\right)\\ &= \hbar\bar{\omega}\left(N + \frac{3}{2} - \frac{1}{3}(3n_3 - N)\,\delta\right)\end{aligned} \tag{3}$$

where

$$\begin{aligned}N &= n_3 + n_\perp = n_3 + n_1 + n_2\\ \bar{\omega} &= \frac{1}{3}(2\omega_\perp + \omega_3)\end{aligned}$$

We note that the deformation parameter [9] can be expressed as

$$\delta = 3\frac{\omega_\perp - \omega_3}{2\omega_\perp + \omega_3} \approx \frac{R_3 - R_\perp}{\bar{R}} \tag{4}$$

The one-particle spectrum (3) is shown in Fig. 2, in which some frequency ratios $(\omega_\perp : \omega_3)$ are indicated. Note that the ratio of the radius $(R_3 : R_\perp)$ is equal to $(\omega_\perp : \omega_3)$. Since

$$n_\perp = n_1 + n_2 \qquad \text{where} \qquad n_1 = 0, 1, \ldots, n_\perp \tag{5}$$

the state with $(n_3, n_\perp)$ has a degeneracy of $2(n_\perp + 1)$ when the instrinsic spin degree of freedom of nucleons is included.

The one-particle energy eigenvalues (3) can be rewritten [9] as

$$\varepsilon_\nu = \hbar\omega_{SH}\left(N_{SH} + a + \frac{1}{2}b\right) \tag{6}$$

where

$$N_{SH} = an_\perp + bn_3 \tag{7}$$

which may be interpreted as a "shell quantum number" in the case of a simple integer ratio of (a:b) , and

$$\hbar\omega_{SH} = \frac{\hbar\omega_\perp}{a} = \frac{\hbar\omega_3}{b} = \hbar\,\overset{0}{\omega_0}\left(a^2 b\right)^{-1/3} \tag{8}$$

From Fig. 2 it is seen that if the ratio of $(\omega_\perp : \omega_3)$ is a simple integer ratio, the levels are assembled into degenerate major shells and the degeneracy of each major shell is larger. For $\delta = 0$ (namely, the (1:1) ratio) the spectrum exhibits the degeneracies of the spherical harmonic oscillator, but those degeneracies are removed for $\delta = 0.2 - 0.3$, which is a typical quadrupole deformation of the ground states of well-deformed nuclei.

However, for $\delta = 0.6$ (namely, the (2:1) ratio) a striking feature of shell-closure comes back with a different set of magic numbers from those for a spherical shape. It is seen that at the prolate deformation with the (3:1) ratio another prominent level bunching appears though with less degeneracy than at the (2:1) deformation. The deformation corresponding to the (3:1) ratio is sometimes called as hyperdeformation (HD), and some evidence for its presence was also reported.

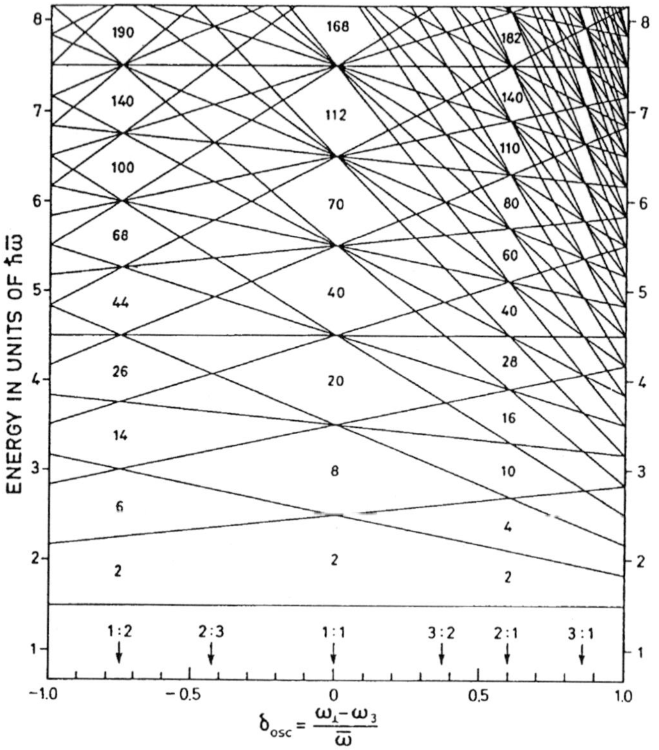

Fig. 2. Single-particle spectrum for axially-symmetric oscillator potentials. The arrows mark the deformations corresponding to the indicated rational ratios of ($\omega_\perp : \omega_3$). The figure is taken from ref.[9].

The nuclear potential differs from the oscillator potential partly through a rather well-defined surface and partly through the presence of the spin-orbit coupling. The modification of the oscillator potential by including these two elements rearranges the one-particle spectrum, however, a relatively strong shell-structure turns out to appear around $\delta = 0.6$ with a set of magic numbers approximately equal to 66, 86, 116 and 148. See, for example, ref. [9].

2.2 Fission isomers — superdeformation at low spins

In 1962 the first observation of a fission isomer in ${}^{242}_{95}\mathrm{Am}_{147}$ was reported [1], which had a half-life of 14 m sec. The excitation energy measured from the ground state was not known, but it was supposed to be 2-3 MeV. The gamma-decay of the isomer to the ground state was observed to be strongly delayed, but it had a spontaneous fission probability, which was much larger than that of the ground state.

Examining the presently known fission isomers, the related magic neutron-number is indicated to be around 146-148. The only measured excitation energies of the lowest-lying fission isomers are 2.56 MeV in ^{238}U and 2.77 MeV for ^{236}U. Measured electric quadrupole moments of fission isomers are, for example, 37 b for the 37 ps isomer in ^{236}Pu [10] and 36 b for the 8 μs isomer in ^{239}Pu [11]. These large quadrupole moments are consistent with a prolate spheroid with a ratio of axes of (2:1). Though the sign of the quadrupole moment is not known experimentally, an oblate shape is very unlikely since the liquid-drop deformation energy required to produce the shape with observed magnitudes of quadrupole moment is over 50 MeV. It should be noted that the observation of fission isomers (namely, the construction of the local energy minimum along the path to the fission) was possible, since the (2:1) deformation corresponds approximately to the saddle point of the fission barrier as calculated from the liquid-drop model. (See the illustration in Fig. 3.) The fission barrier comes out from a large surface energy balanced by a large Coulomb energy.

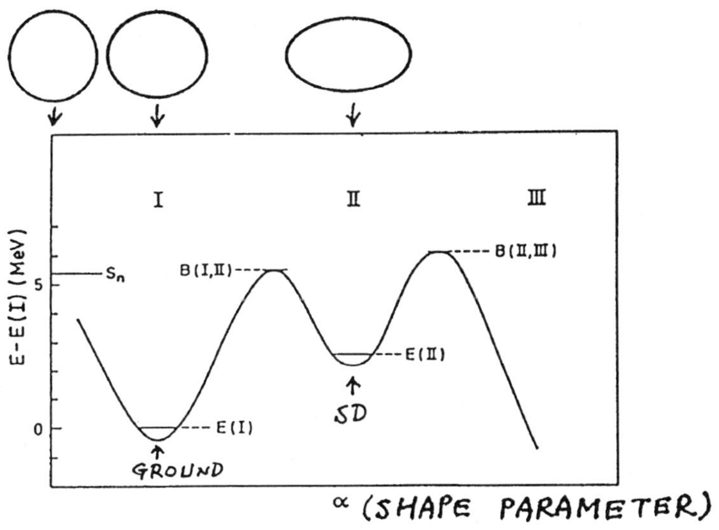

Fig. 3. Schematic illustration of potential energy function as a function of nulcear shape parameter, exhibiting a second minimum. The major part of the figure is taken from ref.[9].

The experimental determination of magnetic moments of fission isomers specifies the intrinsic configurations involved. For example, the reported relatively large positive value of the g-factor of the 110 ns isomer in ^{237}Pu [12] requires the $N = 143$ orbit, in which the spin and the orbital angular momentum are anti-parallel to each other. Since there are few such neutron-orbits expected in the neighbourhood of the closed shell with $N = 148$, the requirement gives a good measure of the strength of the nuclear spin-orbit potential at the deformation of (2:1) axis ratio [13].

A rotational band structure (0+, 2+, ..., 8+) based on a fission isomer is seen [14] in ^{240}Pu, in which the observed moment of inertia is 75-80% of the rigid moment of inertia estimated at the corresponding shape. However, due to the fission instability one cannot expect to observe high-spin states in the actinide region.

2.3 High-spin rotational bands — superdeformation at high spins

Already in 1975 the presence of higher-spin SD states in the $A = 150$ region was predicted [3,4]. At high spins the centrifugal force creates a saddle point at the large deformation on which the (2:1) shell structure could show its presence. It is shown that the (2:1) shell structure remains for high spins, if it is already prominent at $I = 0$. In Fig. 4 an example of one-particle Routhians for neutrons ($N = 86$) as a function of cranking frequency, which are calculated using a cranked Woods-Saxon potential. Namely, the cranking Hamiltonian is written as

$$H_{CR} = H - \hbar\omega_{CR} J_1 \tag{9}$$

where H expresses a one-particle Hamiltonian with deformed Woods-Saxon potential and the expectation value of the operator J_1

$$\langle J_1 \rangle = \sum_{n=1}^{A} \langle n | j_1 | n \rangle \tag{10}$$

is identified as the total angular momentum of the nuclear system. One-particle eigenvalues, E'_n, obtained from the diagonalization of H_{CR} in (9) express one-particle energies in the rotating coordinate system and are called as Routhians [16]. In Fig. 4 it is seen that the gap (or low level-density) at $N = 86$, which is present at zero cranking frequency, remains until very high cranking frequencies. This is because the majority of one-particle orbits except a few "high-j" orbits are so strongly aligned to the large quadrupole deformation that the Coriolis force can hardly change the nature of the orbits. Namely, those orbits obtain little alignment and, thus, those levels are almost parallel to the abscissa in Fig. 4. (In reality, at a large quadrupole deformation such as SD, "high-j" orbits are no more pure high-j. Nevertheless, the unique properties of those orbits different from the remaining orbitals are very clear. Thus, in the present lecture we use the terminology, "high-j", to spectify those unique orbits.)

The recent development of the gamma-ray multi-detector system made it possible to observe a number of high-spin SD bands both in the $A = 150$ region and in the $A = 190$ region. Reported quadrupole moments obtained from measured life-times (in other words, measured B(E2) values) are consistent with the (2:1) shape of those SD bands. Examples of observed dynamical moments of inertia, (1), of the SD bands in the $A = 150$ region are shown in Fig. 5. The values of dynamical moments of inertia are determined by the local structure of the band and are known to be very sensitive

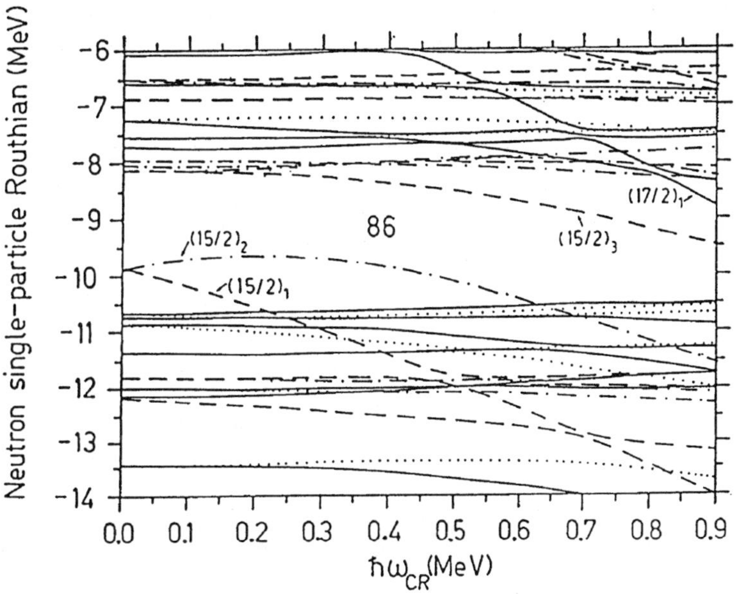

Fig. 4. One-particle Routhians for neutrons(N=86) as a function of cranking frequency, which are calculated from a cranked Woods-Saxon potential. The deformation parameters used, β_2 = 0.62 and β_4 = 0.12, are characteristic for the SD band in ^{152}Dy. The "high-j" orbits are marked by the j quantum-number indexed by an ordering number. Defining the signature quantum-number, α, by $\{j_1 = \alpha \, mod \; 2\}$, the orbits with ($\alpha$ = 1/2,π = +),(α = $-1/2$,π = +),(α = 1/2,π = $-$) and (α = $-1/2$,π = $-$) are denoted by solid lines, dotted lines, dot-dashed lines, and long-dashed lines, respectively. The figure is provided by the authors of ref.[15]

to a small change in spin-alignments, shape- and pairing-parameters as well as band-crossings [8].

Without knowing spin-parity, it is still possible to obtain some spectroscopic information, if one obtains the frequency-dependence of dynamical moments of inertia as well as their nucleon-number dependence. Namely, in the cranking approximation the total angular momentum is a sum of the alignment of all occupied particle orbits, n. Thus, one obtains

$$\hbar^{-2}\mathcal{J}^{(2)} = \frac{1}{\hbar}\sum_{n} \frac{d}{d\omega_{CR}} \langle n \,|j_1|\, n\rangle = -\hbar^{-2}\sum_{n} \frac{d^2 E'_n}{d\omega_{CR}^2} \tag{11}$$

which is a sum of the derivative of the (negative) slopes of Routhians in Fig. 4. From Fig. 4 it is seen that any appreciable variation of dynamical moments of inertia as cranking frequency increases comes exclusively from the particles in high-j orbits. Furthermoe, the variation depends sensitively on the occupation numbers of the high-j orbits. Thus, if we can assume that the possible change and fluctuation of the shape- and pairing-parameter contribute negligibly to the observed variation of dynamical moments of inertia as rotational frequency changes (this assumption can presumably be made for the SD bands in the A = 150 region), it is possible to extract the occupation numbers of high-j orbits from the variation seen in Fig. 5 [19]. From the occupation numbers thus obtained one can estimate, for example, the strength of the spin-orbit potential at SD.

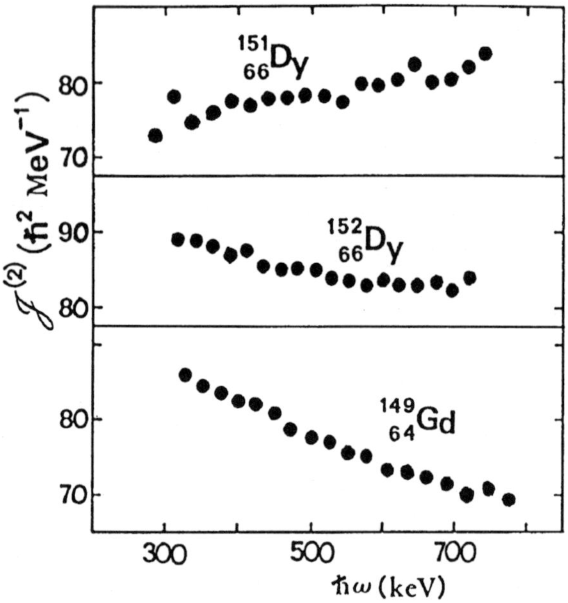

Fig. 5. Examples of measured dynamic moments of inertia of SD bands. Data are taken from refs. [5, 17, 18] for ^{151}Dy, ^{152}Dy and ^{149}Gd, respectively. The figure is taken from ref. [32].

As a last subject of the present section I mention the observation of "identical" SD bands. Having several observed SD bands in a given nucleus, an unexpected discovery [18, 20] is that some pairs of SD bands in neighbouring nuclei have almost identical gamma-ray transition energies. Some of the early reported such SD pairs are ($^{151}_{65}\mathrm{Tb}^{*}_{86}$, $^{152}_{66}\mathrm{Dy}_{86}$), ($^{150}_{64}\mathrm{Gd}^{*}_{86}$, $^{151}_{65}\mathrm{Tb}_{86}$) and ($^{150}_{65}\mathrm{Tb}_{85}$, $^{149}_{64}\mathrm{Gd}^{*}_{85}$), in which the asterisk denotes an excited SD band. (Since the relative positions of observed SD bands in a given nucleus are usually unknown, the SD band with the strongest population intensity is customarily identified as the lowest-lying one.) The differences between observed "identical" transition energies are shown in Fig. 6 as a function of the transition energy. In those pairs of bands the transition energies were found to be equal to within 1-3 keV over a span of 14 transitions. Namely, the transition energies are equal to better than 3 parts in 1000. Similar examples are found also in pairs of the SD bands in the A = 190 region (see ref. [21] and also [7]). An estimate of the differences between the transition energies in those adjacent nuclei can be made using the fact that the transition energies should be scaled with moments of inertia. If we use the rigid moments of inertia which is proportional to $A^{5/3}$, the transition energies of the A and the $(A-1)$ nuclei for A = 150 would differ by about 12 keV, which is an order of magnitude larger than 1-3 keV shown in Fig. 6. Furthermore, the angular momenta of corresponding transitions in each pair necessarily differ by $(1/2)\hbar$, since one is an even-A nucleus, while the other is odd-A.

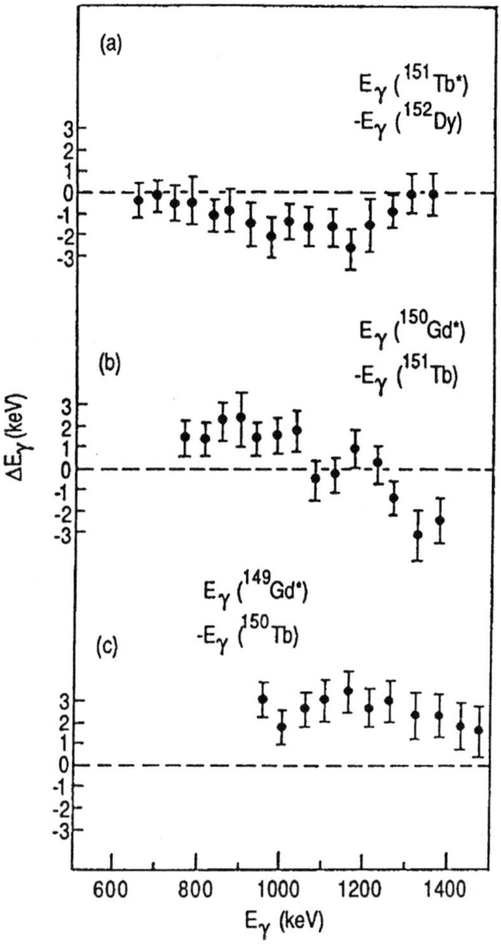

Fig. 6. Differences in the gamma-ray energies ΔE_γ between the SD bands. The figure is taken from ref. [7].

Using the particle-rotor model [9] in the first-order Coriolis coupling, one can evaluate the contribution of the last-odd particle to transition energies and, thereby, one may study the dependence of the possible occurrence of "identical" transition energies on the orbitals occupied by the additional odd-particle. The Hamiltonian of an even-even nucleus is written as

$$\begin{aligned} H_{ROT} &= \frac{\hbar^2}{2\mathcal{J}}\left(\vec{R}\right)^2 \\ &\rightarrow \frac{\hbar^2}{2\mathcal{J}} R(R+1) \end{aligned} \tag{12}$$

where the angular momentum of the system is expressed by $\vec{R}$ and R is an even integer. Denoting the total angular momentum of the adjacent $(A+1)$ nucleus by $\vec{I}$,

$$\vec{I} = \vec{R} + \vec{j} \tag{13}$$

and the total Hamiltonian by

$$H = H_{particle} + H_{ROT} \tag{14}$$

where

$$\begin{aligned} H_{ROT} &= \frac{\hbar^2}{2\mathcal{J}} \left(\vec{R}\right)^2 \\ &= \frac{\hbar^2}{2\mathcal{J}} \left[\vec{I}^2 + \vec{j}^2 - 2\vec{I}\cdot\vec{j}\right] \end{aligned} \tag{15}$$

Assuming an axially-symmetric nuclear shape of the A nucleus and including only the first-order Coriolis coupling, the energy of the $(A+1)$ system is written as

$$E_K(I) = \varepsilon_K + \frac{\hbar^2}{2\mathcal{J}} \left[I(I+1) - K^2 + a\left(I+\frac{1}{2}\right)(-1)^{I+\frac{1}{2}}\delta\left(K, \frac{1}{2}\right)\right] \tag{16}$$

where

$$H_{particle}\Phi_K = \varepsilon_K \Phi_K \tag{17}$$

and

$$a = -\left\langle \Phi_{K=1/2} \left| j_+ \right| \Phi_{K=1/2} \right\rangle \tag{18}$$

is called as "decoupling parameter". From (12) and (16) one finds three cases [22] in which there is a simple relation between the transition energies in the A and the $(A+1)$ system. Namely, writing $E_\gamma(R)_A$ for the gamma-ray energies for the transition from angular momentum R to $(R-2)$ in the A system, etc;

(i) $[a = 0]$ case: (for example, $K > 1/2$ case)

$$\frac{1}{2}\left\{E_\gamma\left(R+\frac{1}{2}\right) + E_\gamma\left(R-\frac{1}{2}\right)\right\}_{(A+1)} \Leftrightarrow E_\gamma(R)_A$$

(ii) $[a = 1]$ case: (for example, $j = 1/2$ odd-particle)

$$E_\gamma\left(R+\frac{1}{2}\right)_{(A+1)} = E_\gamma\left(R-\frac{1}{2}\right)_{(A+1)} \Leftrightarrow E_\gamma(R)_A$$

where R is an even integer.

(iii) $[a = -1]$ case:

$$E_\gamma\left(\bar{R}+\frac{1}{2}\right)_{(A+1)} = E_\gamma\left(\bar{R}-\frac{1}{2}\right)_{(A+1)} \Leftrightarrow E_\gamma\left(\bar{R}\right)_A$$

where $\bar{R}$ is an odd integer.

The above case (i) is known to correspond, for example, to the observed "idential" SD pair of ($^{153}_{66}\mathrm{Dy}^*_{87}$, $^{152}_{66}\mathrm{Dy}_{86}$). The two excited SD bands in ^{153}Dy have been interpreted as a structure of ($^{152}\mathrm{Dy} \otimes n[514\,9/2]$) with no signature splitting. The case (ii) with the odd-proton orbit [301 1/2] provides a natural explanation [23] for the three "idential SD pairs shown in Fig. 6.

Now, since the additional contribution to moments of inertia of the $(A + 1)$ system from the presence of an extra odd-particle comes from the second-order Coriolis coupling, one has to include the second order effect when the physical quantities corresponding to moments of inertia are discussed. Since moments of inertia depend on mass, deformation, polarization, alignment and pair correlation, one may write the contribution of the extra odd-particle (or odd-hole) as

$$\delta\mathcal{J} = \frac{\partial\mathcal{J}}{\partial A}dA + \frac{\partial\mathcal{J}}{\partial\beta}d\beta + \frac{\partial\mathcal{J}}{\partial\omega}d\omega + \frac{\partial\mathcal{J}}{\partial\Delta}d\Delta + \ldots \tag{19}$$

As was already described, the first term on r.h.s. of (19) for an $A = 150$ nucleus leads to about 12 keV contribution to the transition energy, if we use the rigid-rotor expression. Therefore, in order to obtain "identical" bands we must have a striking cancellation between various terms on r.h.s. of (19). If one uses a deformed oscillator model in the absence of pair correlation, one can show that for some orbitals with particular quantum-numbers (for example, $n_z = 0$ orbitals) there is a cancellation [24] between the first three terms in (19). However, more realistic calculations have been only very partially successful in explaining the observed striking cancellation. Whether this striking cancellation may mean a presence of some fundamental symmetry or not has been certainly explored, but at the moment there seems no widely-accepted fundamental explanation.

3 Collective Modes at Superdeformation

Up till now the properties of collective modes at SD are experimentally very little explored. Nevertheless, in the present section I try to describe our expectation of the properties, since this is a very interesting subject, partly because the quantum-mechanical shell-structure is very unique at SD and partly because the exceptionally large quadrupole deformation gives rise to specific features of collective modes. Except at some part of the description of scissors mode (in sect. 3.2), I use exclusively the deformed oscillator model, since the essence of the characteristic feature of collective modes at SD is described in a very simple way using the oscillator model.

3.1 Modes from quantum-mechanical shell-structure unique at superdeformation

When there are many particles outside of the closed shell in the spherical oscillator potential, it is known that we obtain a lower energy of the system if it gets quadrupole deformed. The deformation is favoured by using a large degeneracy of one-particle energies in the spherical potential (Jahn-Teller effect), where many pairs of the degenerate one-particle orbitals have large quadrupole matrix elements [25]. In a quite similar way it can be seen that a Y_{31} deformation is favoured in the system in which there are many particles outside of SD closed shells, while a Y_{41} deformation is favoured in the system of many particles outside of hyperdeformation (HD, namely, spheroid with 3:1 ratio of axes) closed shells.

In tables 1 and 2 possible sets of quantum-numbers for $N_{SH} = 8$ and 9 at SD and for $N_{SH} = 11$ and 12 at HD are shown. Note that

$$N_{SH} = n_3 + 2n_\perp \qquad \text{at } SD$$
$$N_{SH} = n_3 + 3n_\perp \qquad \text{at HD}$$

In the deformed oscillator model all one-particle orbitals for a given value of N_{SH} are degenerate, and two protons (or neutrons) with spin up and down, respectively, can occupy an orbital with a given set of the quantum numbers, $(n_\perp, n_3, \pi, \Lambda)$. From tables 1 and 2 we see that one-particle orbitals with different parities are degenerate at SD, while only those with a given parity are present in a given major shell (namely, a given value of N_{SH}) at HD.

Table 1. Possible sets of quantum numbers for N_{SH}=9 and 8 at superdeformation. Parity is denoted by π, while Λ expresses the component of orbital angular momentum along the 3-axis.

$N_{SH} = 9$				$N_{SH} = 8$			
$n_\perp$	n_3	π	$\lvert\Lambda\rvert$	$n_\perp$	n_3	π	$\lvert\Lambda\rvert$
0	9	−	0	0	8	+	0
1	7	+	1	1	6	−	1
2	5	−	2,0	2	4	+	2,0
3	3	+	3,1	3	2	−	3,1
4	1	−	4,2,0	4	0	+	4,2,0

Table 2. Possible sets of quantum numbers for N_{SH}=12 and 11 at hyperdeformation. See the caption to Table 1.

$N_{SH} = 12$				$N_{SH} = 11$			
$n_\perp$	n_3	π	$\lvert\Lambda\rvert$	$n_\perp$	n_3	π	$\lvert\Lambda\rvert$
0	12	+	0	0	11	−	0
1	9	+	1	1	8	−	1
2	6	+	2,0	2	5	−	2,0
3	3	+	3,1	3	2	−	3,1
4	0	+	4,2,0				

Examining the selection rules of oscillator quantum numbers, within a given N_{SH} shell at SD there are many pairs of orbitals which can be connected by Y_{31} operator, but no pairs by Y_{30} or Y_{32} or Y_{33}. Within a given N_{SH} shell at HD there are many pairs of orbitals which can be connected by Y_{41} operator, but none by other types of $Y_{4\mu}$ operators. Therefore, it is expected that due to Jahn-Teller effect the system at SD (HD) could get softer towards a deformation of Y_{31} (Y_{41}) type when there are many particles outside of respective closed shells. One may obtain a banana-like shape when a small amplitude of Y_{31} deformation is superposed on an SD shape. Thus, this mode is sometimes called as a "banana" mode.

Using tables 1 and 2 one can also study which type of collective modes comes down in excitation energy, when we have a closed-shell configuration. Among Y_3-type of operators one finds that at SD Y_{30} and Y_{32} have non-vanishing matrix elements between the orbitals belonging to an N_{SH} shell and those to an $(N_{SH}+1)$ shell. A more detailed examination shows that the Y_{30} type mode (namely, $K\pi$ = 0− mode) is expected to be the most collective one for an SD closed-shell configuration. A further study of table 1 leads to the fact that the properties of the low-lying collective $K\pi$ = 0− mode depend on whether the last filled shell has an even or an odd value of N_{SH}. In Fig. 7 we show the excitation energy of the lowest $K\pi$ = 0− mode, which is calculated at SD closed-shell configurations using RPA in the deformed oscillator model, as a function of the last filled N_{SH} value [26]. One indeed sees the presence of the effect of odd or even N_{SH} values on the calculated RPA frequency.

When the nuclear system begins to rotate, the quantum shell-structure directly obtained from tables 1 and 2 starts to be disturbed. However, since most one-particle orbitals are strongly aligned to large quadrupole deformation at SD, the characteristic feature of the shell-structure present at a non-rotating system remains until appreciably high rotational frequency (see, for example, ref. [27]).

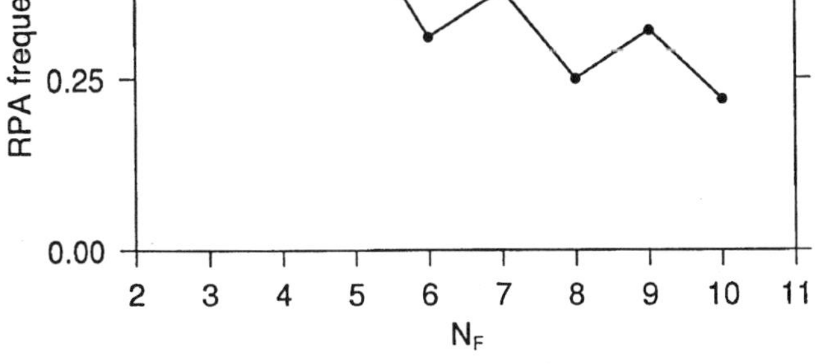

Fig. 7. RPA frequency of the lowest-lying $K\pi$ = 0− mode, which is calculated at SD closed-shell configurations using the deformed oscillator model, as a function of the last filled N_{SH} value (= N_F). The calculated result is taken from ref. [26].

3.2 Modes from exceptionally large quadrupole deformation

As a first example of the characteristic properties of collective modes coming from exceptionally large quadrupole deformation, I mention the expected large energy-splitting of higher-lying giant resonances related to shape oscillations. In spherical nuclei it is known that observed frequencies of giant resonances are [9]

$$E = 58A^{-1/3}\ \text{MeV} \qquad \text{for isoscalar } (T = 0) \text{ quadrupole mode}$$
$$E = 80A^{-1/3}\ \text{MeV} \qquad \text{for isovector } (T = 1) \text{ dipole mode}$$

where A expresses the mass number. In Fig. 8 the frequencies of isoscalar quadrupole, isovector dipole, and (high-lying parts of) isoscalar octupole giant resonances, which are calculated by using the deformed oscillator model, are plotted as a function of axially-symmetric quadrupole deformation. In the presence of the quadrupole deformation the frequencies depend on the absolute value of K quantum-number, which is the angular momentum component along the symmetry axis. For a prolate shape the frequencies with smaller K values become lower.

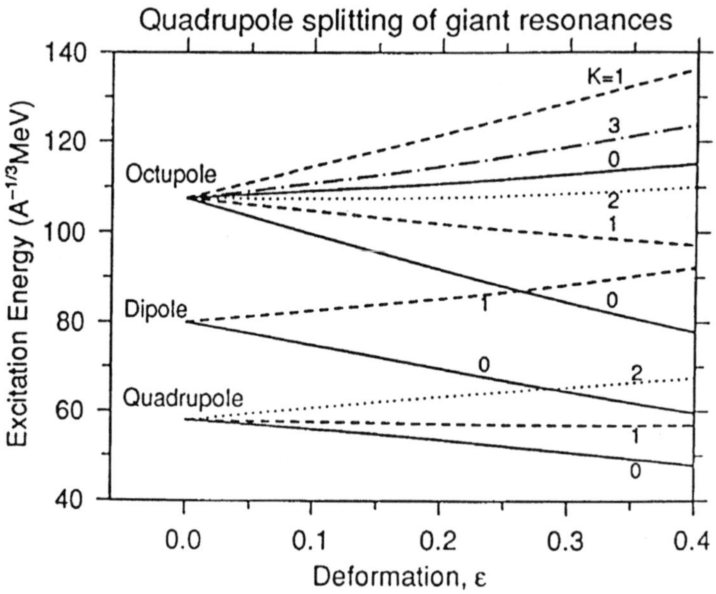

Fig. 8. Deformation dependence of giant quadrupole resonance ($T = 0$), giant dipole resonance ($T = 1$) and the high-lying parts of giant octupole resonance ($T = 0$). Solid lines mark $K = 0$, dashed $K = 1$, dotted $K = 2$ and dot-dashed $K = 3$, respectively. The figure is taken from ref. [27].

In the case of giant dipole resonance one may intuitively expect that the frequency is proportional to the inverse of the radius in the corresponding direction. Therefore, for example, at SD one obtains the relation $(E(K = 1)/E(K = 0)) = 2$. As a numerical example, in the following I compare the frequencies of giant resonances for spherical shape with those at SD, which are calculated by using the deformed oscillator model for an $A = 160$ nucleus (for which $\hbar\omega_0 = 7.6$ MeV)

	$\delta = 0$	$SD\,(\delta = 0.6)$	
$(T = 1)$ dipole	15 MeV	18 MeV	$(K = \pm 1)$
		9 MeV	$(K = 0)$
$(T = 0)$ quadrupole	10.5 MeV	16 MeV	$(K = 0)$
		13 MeV	$(K = \pm 2)$
		10.2 MeV	$(K = \pm 1)$
		7 MeV	$(K = 0)$
$(T = 0)$ monopole	13.5 MeV	16 MeV	$(K = 0)$
		7 MeV	$(K = 0)$

The second example is magnetic dipole 1+ modes ("scissors mode"?) at SD [28]. Using the deformed oscillator model (in which the spin contribution to the M1 strength is neglected and high-j orbitals are absent), in the absence of pair correlations, it is known that in the limit of small quadrupole deformation the B(M1) values of the M1 transition from the lowest RPA state to the ground state in even-even nuclei are proportional to δ (for example, see ref. [29, 30]). The frequencies of the RPA isovector modes of the oscillator model are given by

$$\Omega_{\pm}^{T=1} = \left[\left(\omega_{\perp}^2 + \omega_3^2\right)\left(1 - \frac{a}{2}\right) \pm \sqrt{\left(\omega_{\perp}^2 + \omega_3^2\right)^2 \left(1 - \frac{a}{2}\right)^2 - \left(\omega_{\perp}^2 - \omega_3^2\right)^2 (1 - a)} \right]^{1/2} \tag{20}$$

The mode with $\Omega_{-}^{T=1}$ represents the low-frequency $K\pi = 1+$ isovector quadrupole vibration, while the energy of isovector $K\pi = 1+$ giant quadrupole resonance is given by $\Omega_{+}^{T=1}$. The associated $B(\mathrm{M1})_{\pm}$ values can be expressed as

$$B(\mathrm{M1})_{\pm} \propto \frac{B(\mathrm{E2})_{\pm}}{\Omega_{\pm}^2} \left(\omega_{\perp}^2 - \omega_3^2\right)^2 (1 - a)^2 \tag{21}$$

where the $B(\mathrm{E2})_{\pm}$ values are given by

$$B(\mathrm{E2})_{\pm} \propto \frac{\Omega_{\pm}}{\omega_3} \left(1 - a - \frac{1}{2} \frac{\left(\omega_{\perp}^2 + \omega_3^2\right) \Omega_{\pm}^2 a^2}{\left(\omega_{\perp}^2 + \omega_3^2 - \Omega_{\pm}^2\right)^2 - 4\omega_{\perp}^2 \omega_3^2} \right)^{-1} \tag{22}$$

The excitation energies in (20) and the B(M1) values in (21) are plotted in Fig. 9 as a function of deformation parameter δ. In Fig. 9 it is seen that the calculated B(M1) value for the low-frequency mode increases approximately linearly up till SD deformation, while the one for the high-frequency (giant resonance) mode is a strongly increasing function of deformation. In a more realistic calculation the presence of nuclear surface, pair correlation, spin-orbit potential and the spin contribution to the M1 strength changes the above estimate for B(M1) values based on the oscillator model.

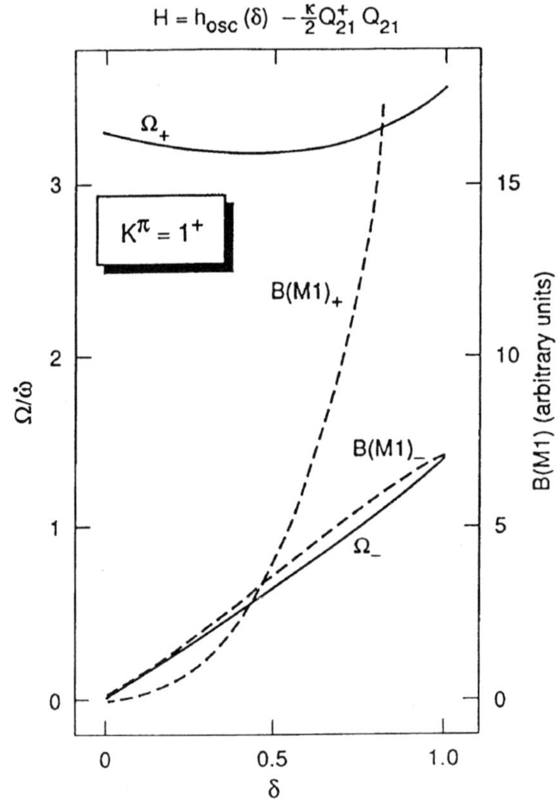

Fig. 9. Excitation energies (solid lines, in units of $\hbar\overset{0}{\omega} = 41/A^{1/3}$) and B(M1) values (dashed lines, arbitrary units) as a function of deformation δ for the two RPA $K\pi$ = 1+ isovector modes of the oscillator model, see eqs. (20) and (21). The figure is taken from ref. [28].

In Fig. 10 we show the $B(\text{M1}; g.s., K\pi = 0+, I\pi = 0+ \rightarrow K\pi = 1+, I\pi = 1+)$ strength carried by the $K\pi$ = 1+ QRPA (quasiparticle random phase approximation) excitation modes obtained from a realistic calculation, which was done based on a deformed Woods-Saxon potential with a monopole-pairing field, an isoscalar as well as isovector long-ranged residual interaction and spin-spin residual interactions. The spurious component in the $K\pi$ = 1+ QRPA solutions with non-zero frequencies was strictly taken away. For comparison, we show the M1 strength in ^{152}Dy and ^{192}Hg calculated for both the SD and the ground-state (GS) deformation. In order to show the reason why the M1 strength at SD shapes is so large compared with the one at GS shapes, the M1 strength arising from only the proton convection current (i.e. $g_s = 0$) and the M1 strength from only the spin part (i.e. $g_\ell = 0$) are also plotted in Fig. 10. It is seen that the enhancement in the M1 strength at large deformation arises from the increased orbital contribution. However, the coherent (or destructive) interference between the orbital and the spin contribution plays an important role in producing the obtained M1 strength even at very low excitation energies.

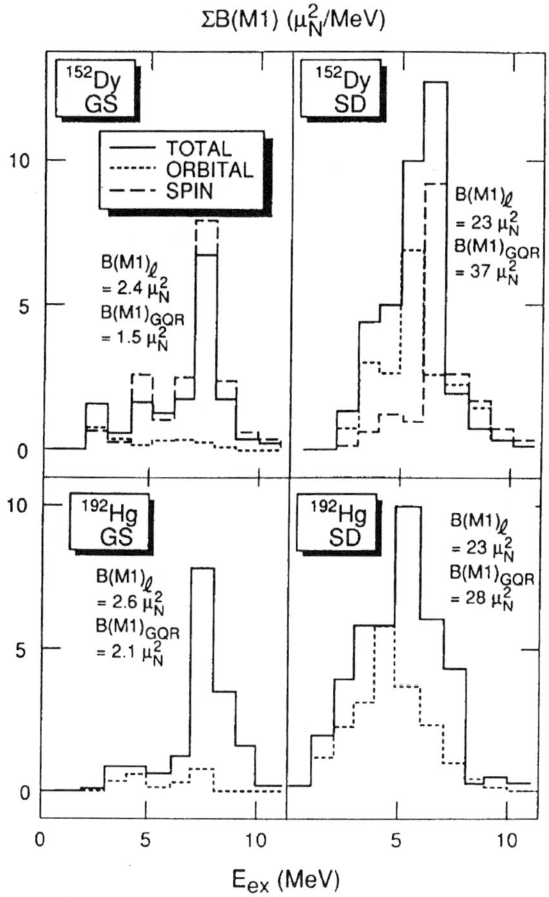

Fig. 10. $B(\mathrm{M1}; g.s., K\pi = 0+, I\pi = 0+ \rightarrow K\pi = 1+, I\pi = 1+)$ values calculated in QRPA as a function of the excitation energies of 1+ states. The summed values per 1 MeV energy bin are plotted as a histogram. For reference, the B(M1) values associated with spin part only ($g_\ell = 0$) or orbital part only ($g_s = 0$) are also shown. $B(\mathrm{M1})_\ell$ denotes the total B(M1) orbital strength in the energy region below 20 MeV. This can be compared with the value for the GQR, $B(\mathrm{M1})_{GQR}$, which consists almost exclusively of the orbital contribution. The g-factors used are $g_\ell = g_\ell^{free}$ and $g_s = (0.85)g_s^{free}$. The figure is taken from ref. [28].

From Fig. 10 it is seen that in the low-energy region (say, $E_{ex} < 10$ MeV) the M1 strength distribution at SD resembles a large resonance with a peak around 5-6 MeV, while at GS shapes the low-lying M1 strength is much more scattered and has one peak at about 7 MeV, mainly attributed to the spin-flip component (i.e. the spin contribution to the M1 strength).

We remark that at SD the $K\pi$ = 1+ isovector quadrupole giant resonance lying, for example, around E_{ex} = 28 MeV in ^{152}Dy, carries a very large M1 strength coming exclusively from the orbital contribution. The M1 strength is in fact as large as the one summed over the region of $E_{ex} < 10$ MeV, and the giant resonance contains a major component of the "scissors mode" [28].

Before finishing the present lecture, I briefly mention the deformation dependence of the effective g-factors of one-particle M1 operators in the static limit (namely, in the limit that the collective vibration frequency, which induces a polarization, is much larger than the relevant transition energy of the one-particle). We may write the M1 operator for a free nucleon as

$$\vec{\mathrm{M1}} \sim g_s^{free} \vec{s} + g_\ell^{free} \vec{\ell} \tag{23}$$

The effective M1 operator for a nucleon in spherical nuclei as [9]

$$(\mathrm{M1})_\mu \sim g_s^{eff} s_\mu + g_\ell^{eff} \ell_\mu + g_s'(r)\,(Y_2 s)_{1\mu} \tag{24}$$

where

$$g_\ell^{eff} \approx g_\ell^{free} \qquad \text{and} \qquad g_s^{eff} = (0.6 - 0.7) g_s^{free}$$

The effective M1 operator in axially symmetric deformed nuclei may be written as [9]

$$(\mathrm{M1})_{\nu=0} \sim g_{\ell 0} \ell_3 + g_{s0} s_3 + g'_{s0}\,(Y_{20} s_3) + g''_{s0} \frac{1}{\sqrt{2}} \left(Y_{2-1} s_{+1} + Y_{2+1} s_{-1}\right)$$

$$(\mathrm{M1})_{\nu=1} \sim g_{\ell 1} \ell_{+1} + g_{s1} s_{+1} + g'_{s1}\,(Y_{20} s_{+1}) + g''_{s1}\,(Y_{21} s_3) + R - conj.$$

Now, what we know or what we have found about the dependence of the values of $\{g_{\ell\nu}, g_{s\nu}, g'_{s\nu}, g''_{s\nu}\}$ on quadrupole deformation is summarized as follows;

$$\left.\begin{array}{l} g_{\ell 0} \rightarrow g_\ell^{free} \\ g_{s0} \rightarrow g_s^{free} \end{array}\right\} \qquad \text{for} \quad \delta \rightarrow \text{large}$$

$g'_{s\nu}$ and $g''_{s\nu}$	change a lot for $\delta = 0.0 \rightarrow 0.3$, while change little for $\delta = 0.3 \rightarrow 0.6$.
g_{s1}	has no definite dependence on deformation [13].
$g_{\ell 1}$	will not have a strong deformation dependence, if only the $Q - Q$ type residual interaction is effective and, for example, an appreciable amount of $\vec{\ell} \cdot \vec{\ell}$ type residual interaction is not present.

References

1. S.M. Polikanov et al., J. Exptl. Theoret. Phys. (USSR) 42 (1962) 1464.
2. V.M. Strutinsky, Nucl. Phys. **A95** (1967) 420; **A122** (1968) 1.
3. R. Bengtsson et al., Phys. Lett. **57B** (1975) 301.
4. K. Neergaard and V.V. Pashkevich, Phys. Lett. **59B** (1975) 218.
5. P. Twin et al., Phys. Rev. Lett. **57** (1986) 811.
6. M.A. Bentley et al., Phys. Rev. Lett. **59** (1987) 2141.
7. For a recent review of this field, see, for example; R.V.F. Janssens and T.L. Khoo, Ann. Rev. Nucl. Part. Sci. **41** (1991) 321; The presence of SD bands in the A=190 region was first predicted by R.R. Chasman, Phys. Lett. **219B** (1989) 227.
8. A. Bohr and B.R. Mottelson, Physica Scripta, **24** (1981) 71.
9. A. Bohr and B.R. Mottelson, NUCLEAR STRUCTURE, vol. 2, Benjamin, New York, 1975.
10. V. Metag and G. Sletten, Nucl. Phys. **A282** (1977) 77.
11. D. Habs et al., Phys. Rev. Lett. **38** (1977) 387; H. Backe et al, Phys. Rev. Lett. **42** (1979) 490.
12. R. Kalish et al., Phys. Rev. Lett. **32** (1974) 1009; D. Shackleton et al., Proc. Int. Conf. on Hyperfine Interaction, Uppsala, 1974.
13. I. Hamamoto and W. Ogle, Nucl. Phys. **A240** (1975) 54.
14. H.J. Specht et al., Phys. Lett. **41B** (1972) 43.
15. W. Nazarewicz, R. Wyss and A. Johnson, Nucl. Phys. **A503** (1989) 285.
16. A. Bohr and B.R. Mottelson, Proc. Int. Conf. Nuclear Structure, Tokyo, 1977 (published as; Jpn. J. Phys., Suppl.44 (1978) 157).
17. G.E. Rathke et al., Phys. Lett. **209B** (1988) 177.
18. B. Haas et al., Phys. Rev. Lett. **60** (1988) 503.
19. T. Bengtsson, S. Aaberg and I. Ragnarsson, Phys. Lett. **208B** (1988) 39.
20. T. Byrsky et al., Phys. Rev. Lett. **64** (1990) 1650.
21. F.S. Stephens, Nucl. Phys. **A520** (1990) 91c.
22. W. Nazarewicz, Recent Advances in Nuclear Structure, eds. D. Bucurescu, G. Catadanil and N.V. Zamfir, (World scientific Publ., 1991), p. 175.
23. W. Nazarewicz et al., Phys. Rev. Lett. **64** (1990) 1654.
24. I. Ragnarsson, Nucl. Phys. **A520** (1990) 74c.
25. See, for example, ref. [9].
26. T. Nakatsukasa, S. Mizutori and K. Matsuyanagi, Progr. Theor. Phys. **87** (1992) 607.
27. R. Nazmitdinov and S. Aaberg, Phys. Lett. **B289** (1992) 238.
28. I. Hamamoto and W. Nazarewicz, Phys. Lett. **B297** (1992) 25.
29. D.R. Bes and R.A. Broglia, Phys. Lett. **B137** (1984) 141.
30. I. Hamamoto and S. Aaberg, Phys. Lett. **B145** (1984) 163; Phys. Scr. **34** (1986) 697.
31. P.J. Twin, Proc. Conf. on Contemporary Topics in Nuclear Structure Physics, Cocoyoc, 1988, World Scientific.
32. I. Hamamoto and B. Mottelson, Nucl. Phys. **A507** (1990) 65c.

Lecture given at the JSPS-INS International Spring School, March 1993 at Shimoda

Nuclear Physics with Strangeness

Kenichi Imai

Department of Physics,
Kyoto University,
Kyoto, Japan

1 Introduction

The frontier of the experimental nuclear physics is to discover unknown nuclei and find a new physics. We have already a long history of study about the ordinary nuclei which exist around us and their low lying excited states. In today's nuclear physics, therefore, new nuclei or nuclear matter have been searched for along new axes such as isospin, spin, temperature and new flavor, namely strangeness. In the strangeness nuclear physics, we produce new nuclei which has strangeness quantum number and study its structure and "nuclear force with strangeness".

What happens if nuclei have many strange (s) quarks? From the Pauli principle at the quark level, it is natural to conclude that nuclear matter with s-quarks would be more stable than ordinary matter. It should be true if the mass of s-quark is same as u and d-quark. In reality, the mass of the s-quark is heavier than the others and SU(3) flavor symmetry is broken. The answer to the above question is not trivial. Many theoretical works have been made about the stability of quark (nuclear) matter with strangeness from 6-quark system (H-dibaryon) to the large system such as a star. At the present stage, it would be fair to say that the possibility of a phase transition from baryonic matter to quark matter still exists at a certain number of s-quarks in nuclear matter. Apart from the extreme case such as a phase transition, there exists a possibility to observe a quark degree of freedom explicitly in nuclei with strangeness. This is one of the most interesting aspect of the strangeness nuclear physics.

Secondly, the nuclear physics with strangeness is regarded as a physics with impurity. For example, the four nucleon system with Z=1 (^{4}H) becomes stable if one neutron is replaced by a Λ hyperon. Another example is ^{8}Be which becomes also stable by a Λ hyperon. Thus, one Λ hyperon, an impurity, can change the state of matter. This is often observed in the solid state physics.

A hyperon can penetrate into deep inside of nuclear matter, since hyperons have no Pauli exclusion from nucleons. The hyperon is, therefore, a good probe to study deep inside of nuclei. One can study nuclear matter by the energy levels of hyperon in nuclei and its decay. In the same time, the property of hyperon inside nuclei is also an interesting subject.

One of the basic subject of strangeness nuclear physics is to study hyperon-nucleon and hyperon-hyperon interaction. We have a long history of the systematic study of the nucleon-nucleon scattering, and the nucleon-nucleon interaction is well known phenomenologically. Recently, theoretical efforts to understand the nuclear force based on QCD have been made extensively, and theoretical models such as the quark cluster model, bag model and Skyrme model are proposed. The hyperon-nucleon interaction is an important place to check these models. The hyperon-nucleon interaction can be studied by hyperon-nucleon scattering experiments and also the spectroscopy of hypernuclei. It is also very interesting to study SU(3) flavor symmetry of the generalized nuclear force. It may provide the information about the origin of isospin symmetry and SU(3) symmetry of the strong interaction.

The objective of the nuclear physics is to understand nuclear structure and nuclear force consistently. For the nuclear structure and nuclear force with strangeness, can we understand them with the present nuclear theory? If it does not work, it would be a "new physics" in the nuclear physics.

We have already a long history of strangeness nuclear physics since the first hypernucleus was discovered in a nuclear emulsion as a fragment from the nuclear reaction by cosmic ray in 1952 [1]. However, the recent active researches both theoretical and experimental, especially at KEK and BNL, have opened the second phase of strangeness nuclear physics. Here I will describe a brief review of Λ-hypernuclear physics and discuss a new strangeness nuclear physics, namely nuclei with S=−2.

For more complete review for Λ and Σ hypernuclei, one should refer the review paper by H. Bando, T. Motoba and J. Zofka [2]. The most recent status of this field, in general, can be learned by a recent conference proceedings [3].

2 Λ hypernuclei

2.1 Hypernuclei studied by nuclear emulsion

In 1950's and 60's, several tens thousands hypernuclei were identified by scanning the nuclear emulsion exposed to K^- beams. The results of the emulsion experiments in this period were well summarized in the ref. [4]. In the nuclear emulsion, a hypernucleus is observed as a nuclearfragment from a nucleus which absorbed K^- meson. The probability of the K^- absorption by nuclei is very large if the K^- stops in the emulsion. In the elementary process of the stopped K^- reaction, Σ is more likely produced than Λ. The ratio of $K^-+p \rightarrow \Lambda+\pi$ to $K^-+p \rightarrow \Sigma+\pi$ for stopped K^- is 0.07. In the nuclei, however, Σ turns to Λ by the conversion reaction, Σ+N$\rightarrow$ Λ+N. In either direct or indirect Λ production process, the Q-value of the Λ production from stopped K^- is about 180 MeV. After absorbing K^-, a "hot" compound nucleus with S=−1 is formed, and emits a hyperfragment occasionally. The observed rate of the hyperfragment in the stopped K^- in the emulsion is as high as 6.5%. The hyperfragment is identified as a track which has a decay vertex. As elementary decay processes, there are mesonic decay modes, $\Lambda \rightarrow p\pi^-$ (Q=38 MeV) and $n\pi^0$, and nonmesonic decay modes, $\Lambda p \rightarrow np$ (Q=176 MeV) and $\Lambda n \rightarrow nn$. From the range of decay particles from the hyperfragment, one can obtain the mass of the hyperfragment, by using the energy and momentum conservation. If two neutral particles are emitted, the mass of the fragment can not be uniquely reconstructed. Practically, the mass can be obtained only when the fragment shows the π^- mesonic decay. Even in the emulsion the tracks of heavy hyperfragment are too short to identify, and the branching ratio of π^- mesonic decay of heavy hypernuclei is very small. The mass of the light hypernuclei ($<^{15}$N) was, therefore, obtained by emulsion experiments [5, 6]. The binding energies (=separation energies) of Λ (B_Λ) for light hypernuclei were shown in Fig. 1 [7]. The binding energy increases monotonically as the mass number

A increases at the rate of about 1 MeV/nucleon. It shows marked difference from the neutron separation energies shown in the figure, mainly due to the Pauli principle. There is no bound state for A=2, Λ n or Λ p. The A = 3 hypernucleus is $^3H_\Lambda$ which is just barely bound (B_Λ=0.13±0.05 MeV) and expected to be an extended object compared with ordinary nuclei.

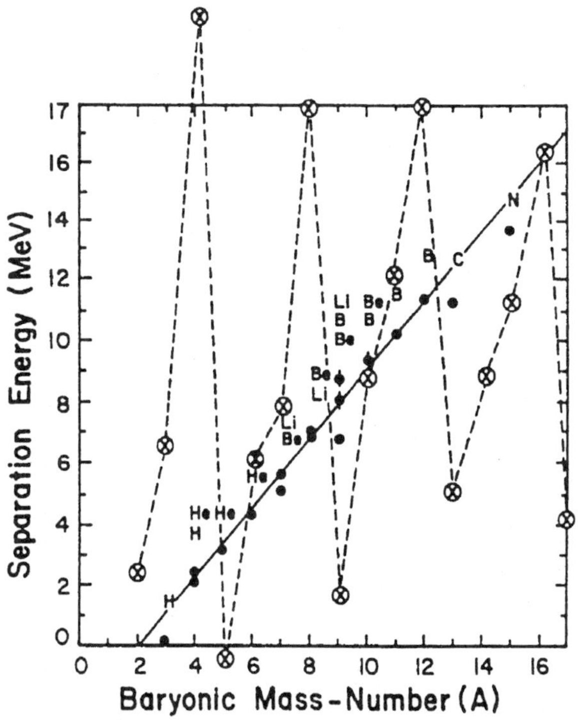

Fig. 1. A plot of Λ hyperon (solid circle) and neutron (cross) separation energies as a function of mass number. The lines are guide for eye. Taken from ref.[7].

2.2 Production of hypernuclei (counter experiment)

Typical reactions to produce hypernuclei are (K^-, π^-), (π^+, K^+) and (stopped K^-,π) reactions from nuclear targets. The heavy ion reaction have been tried at Dubna [8]. The (γ,K) reaction is planned at CEBAF [9]. The production cross section for hypernuclei depends on the cross section for the elementary process to produce a hyperon and the sticking probability of the hyperon to a nucleus. The distortion of K and π in a nucleus should be also taken into account for the calculation of the production cross sections. The laboratory differential cross sections for the elementary process to produce hyperons at θ=0° are shown in Fig. 2. The incident momentum dependence of the cross sections is complicate because of the effect of various hyperon resonances. The distortion effect due to a nucleus can be calculated with eikonal approximation. The effective nucleon (proton or neutron) number is often used to describe the distortion effect. The cross section for the hyperon production from a nucleus can be written by the effective nucleon number, $N_{\rm eff}$, and the cross section of the elementary process, $d\sigma/d\Omega$(a+N→b+Y),

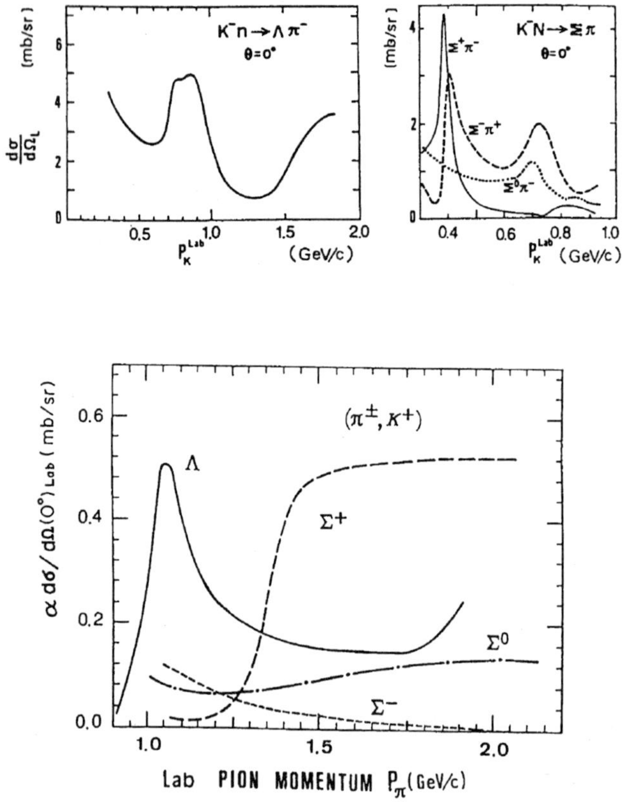

Fig. 2. Differential cross sections (Lab.) at $\theta_{\pi K}$=0° for K^- + N → π +Y and π + N →K^++Y reactions are shown as a function of incident momentum. Taken from ref.[2].

$$
\begin{aligned}
d\sigma/d\Omega &= \alpha_{\rm kin} N_{\rm eff} d\sigma/d\Omega(\mathrm{a} + N \to \mathrm{b} + Y) \qquad (1)\\
&= \int \rho_N(\mathbf{r}) \exp\{-\langle\sigma_{\rm aN}\rangle \int^z \rho(x,y,z')dz' - \langle\sigma_{\rm bN}\rangle \int_z \rho(x,y,z')dz'\} d\mathbf{r},
\end{aligned}
$$

where $\alpha_{\rm kin}$ is a kinematical factor to convert from the nucleon target to the nuclear target. The effective nucleon number, N_{eff}, is calculated with nuclear density, $\rho(\mathbf{r})$, and total cross sections for a+N and b+N reactions. The sticking probability of hyperons depends on the momentum matching and can be described as following for the harmonic oscillator model,

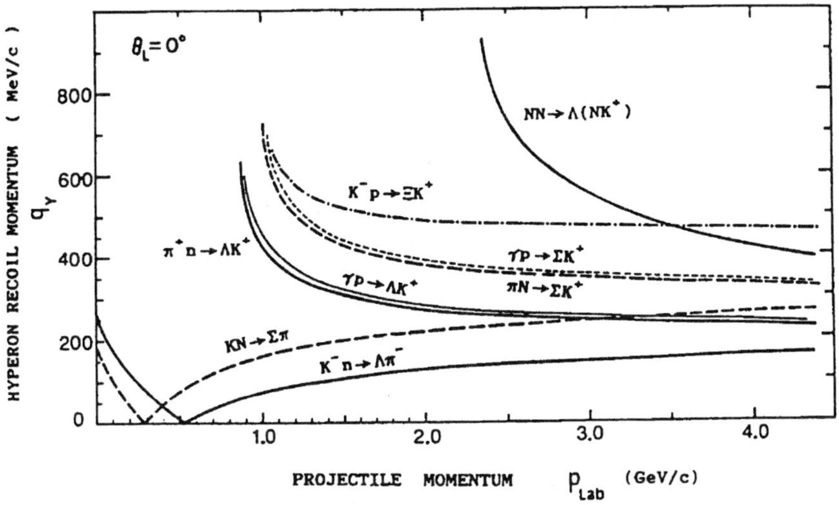

Fig. 3. The momentum transfer (q_Y) to the hyperon (Y) as a function of the projectile momentum ($P_{\rm Lab}$) in the reaction, a+N →Y+b, at θ_b=0°.

$$S_k(q_Y; n_N l_N, n_Y l_Y) = |\langle \phi_{n_Y l_Y}(r)| jk(qr) |\phi_{n_N l_N}(r)\rangle|^2, \tag{2}$$

where q_Y is the hyperon momentum, $n_N l_N$ is the quantum number of the state of the nucleon and $n_Y l_Y$ is the state where the hyperon is captured. If q_Y is small, the hyperon is most likely captured to the same state where the initial nucleon is populated. This is called the substitutional state. As the q_Y increases, the larger angular momentum transfer $l_N \rightarrow l_Y$ becomes feasible, and, for example, the production of the ground state is enhanced compared to the substitutional states. The selection of q_Y is important to study the specific hypernuclear states, and the state can be identified from the q_Y-dependence of the cross section. The incident particle momentum where q_Y becomes zero in the elementary process is called the magic momentum, where the production cross section for the substitutional states is expected to be maximum. In order to produce hypernuclei, the cross section for the elementary hyperon production process and the q_Y are the important parameters. For the Λ hypernuclei, the (K$^-$,π^-) reaction is usually done at P_{K^-}=800 MeV/c to study substitutional states and the (π^+,K$^+$) reaction at about P_{π^+}=1.1 GeV/c to study the states which are populated by the large angular momentum transfer.

2.3 Shell structure of Λ hypernuclei

The (K$^-$, π^-) reaction was successfully used for the spectroscopic study of the light hypernuclei. The excitation spectra obtained from the (K$^-$, π^-) reaction on ^{12}C and ^{16}O are shown in Fig. 4 [10]. If the two states (peaks in the spectrum) of $^{12}C_\Lambda$ are interpreted as a neutron hole in $P_{3/2}$ and the Λ hyperon in $P_{3/2}$ and $S_{1/2}$ states, a; ($P_{3/2^{-1}}$,$P_{3/2^\Lambda}$)

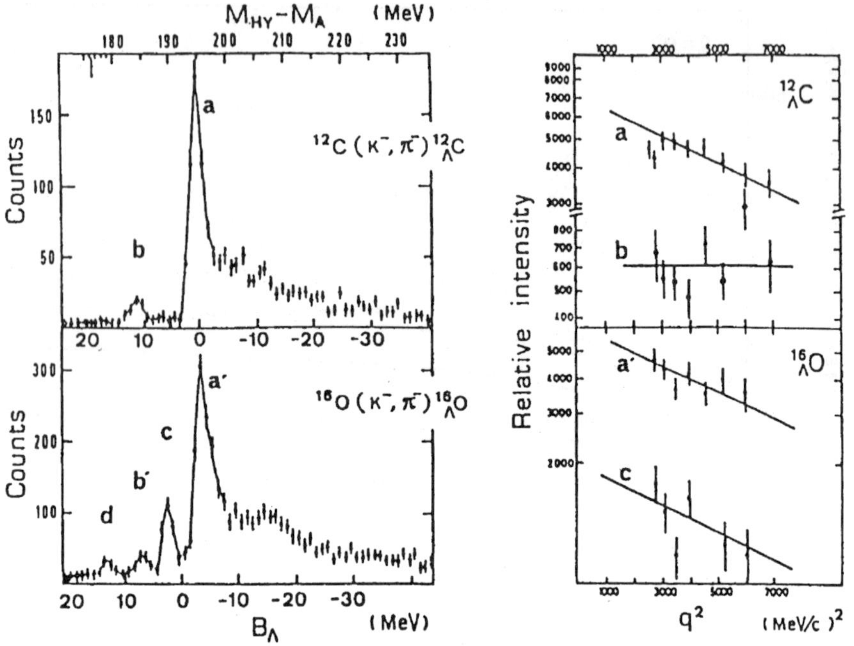

Fig. 4. Excitation spectra obtained from the (K^-, π^-) reaction on ^{12}C and ^{16}O at P_K=715 MeV/c are plotted as a function of the Λ-hyperon binding energy. The q^2-dependence (angular distribution) of the peak intensities is shown in the right half [10].

and b; $(P_{3/2^{-1}}, S_{1/2^\Lambda})$, the four peaks for $^{16}O_\Lambda$ can be consistently understood. Namely, the two states, a$'$ and b$'$, have the same configuration as a and b, and other two states, c and d, can be interpreted as c; $(P_{1/2^{-1}}, P_{1/2^\Lambda})$ and d; $(P_{1/2^{-1}}, S_{1/2^\Lambda})$. This interpretation is supported by the fact that i) the energy difference between a$'$ and c peaks is consistent with the known energy difference of $P_{1/2}$ and $P_{3/2}$ nucleon shell and ii) the q-dependence of the cross sections for a, a$'$ and c states show its maximum at q=0 which is characteristic for the substitutional states. From this data, existence of s and p-shell of Λ in the light hypernuclei was established. Since the P state is, however, degenerated, we conclude that the spin-orbit splitting of Λ p-shell ($p_{1/2}$ and $p_{3/2}$), which is originated from the LS force of Λ-N interaction, is very small. It is estimated to be less than 0.3 MeV which can be compared with 6 MeV for the nucleon.

It is difficult to produce the ground state of heavy hypernuclei with the (K^-,π^-) reaction directly. The (π^+,K^+) reaction was therefore proposed [11] and measured at both BNL and KEK [12, 13]. The excitation energy spectrum for $^{89}Y_\Lambda$ obtained by the (π^+,K^+) reaction is shown in Fig. 5 [12]. The several states up to the ground state are seen due to the large momentum transfer of the (π^+,K^+) reaction. From theoretical consideration based on the spectra of light hypernuclei, these peaks are expected to be due to s, p, d, f and g shell of Λ hyperon. This data proved that the (π^+,K^+) reaction is a fruitful method to do the spectroscopic study of hypernuclei.

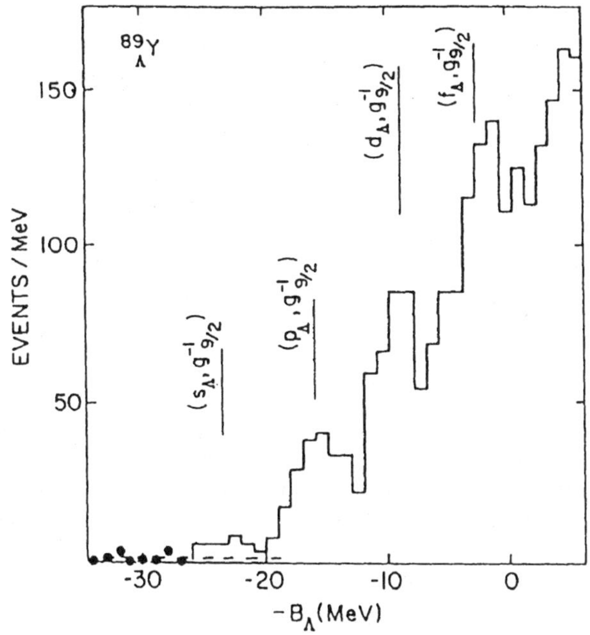

Fig. 5. The excitation spectrum for the hypernucleus $^{89}Y_\Lambda$ obtained by (π^+, K^+) reaction [12].

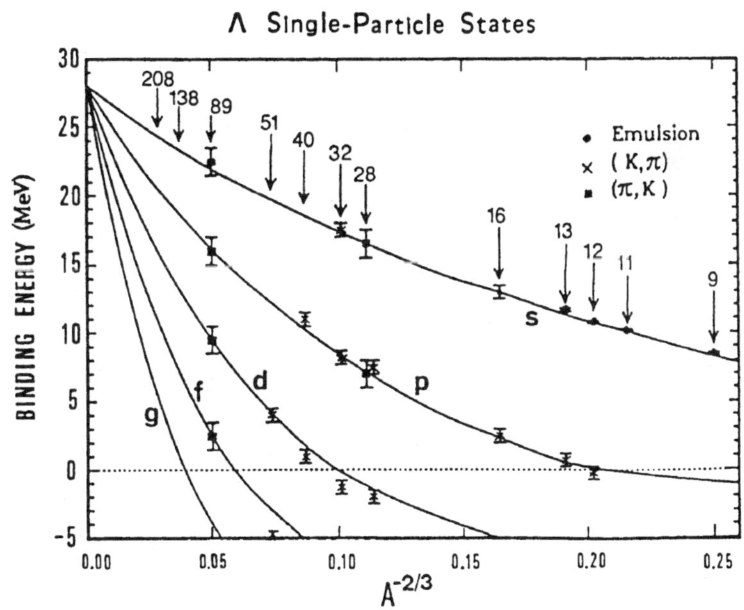

Fig. 6. Binding energy of Λ single-particle states as a function of $A^{2/3}$. Solid lines are calculations with a Woods-Saxon type potential [14].

The binding energy of Λ single-particle states obtained from various experiments discussed here is summarized in Fig. 6 [14]. All the experimental values are surprisingly well reproduced by the calculation with the Woods-Saxon type potential [14]. It indicates that binding energy of Λ in hypernuclei is understood with the shell model picture. In other words, the single particle picture persists up to $A = 90$ hypernuclei, fairly deep inside the nuclear medium. However, the assignment of the peaks of $^{89}Y_\Lambda$ was not made experimentally and the energy resolution is not good enough. The measurement of angular dependence is necessary to determine the shell of the Λ-hyperon. At KEK, a superconducting kaon spectrometer (SKS) has been recently built to measure the (π^+,K^+) reaction with a better energy resolution ($\Delta E \leq 2$ MeV FWHM) and higher statistics for various nuclei up to Pb [15]. The new systematic data are awaited.

2.4 Decay of hypernuclei

For the spectroscopy of ordinary nuclei, γ spectroscopy has been the most fruitful tool. However, there has been only a few experiment which measure γ-rays from excited hypernuclei. One of the best measurement is shown in Fig. 7. The γ-rays from the excited states of $^4H_\Lambda$ were measured with NaI detectors [16]. The peak is identified as due to the γ-ray transition between the first excited state of $^4H_\Lambda(1^+)$ and its ground state (0^+). The energy difference (1.1 MeV) provides the strength of the Λ-N spin-spin interaction. The energy resolution and statistics are, however, very poor. The advanced detector which has a higher energy resolution, 4π acceptance and fast response will be necessary to extend the γ-ray spectroscopy of hypernuclei.

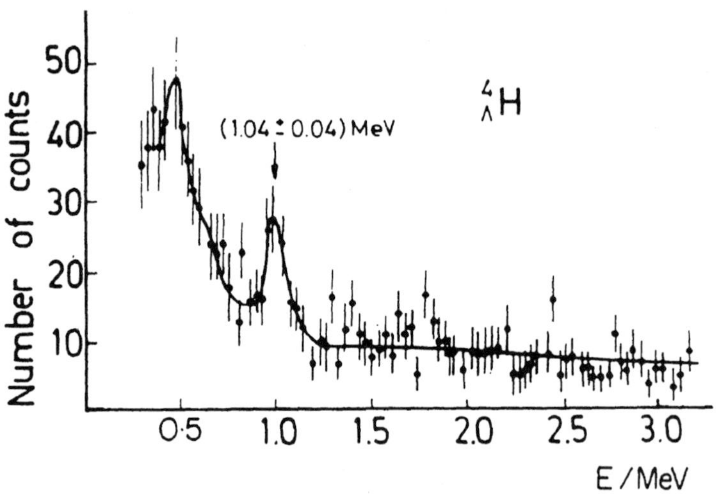

Fig. 7. The γ-ray spectrum from $^4H^*_\Lambda$. A peak due to the $1^+ \rightarrow 0^+$ transition can be seen.

The weak decay of hypernuclei can be used as a spectroscopic tool to study hypernuclei by detecting π^- or π^0. But few efforts have been made except the emulsion

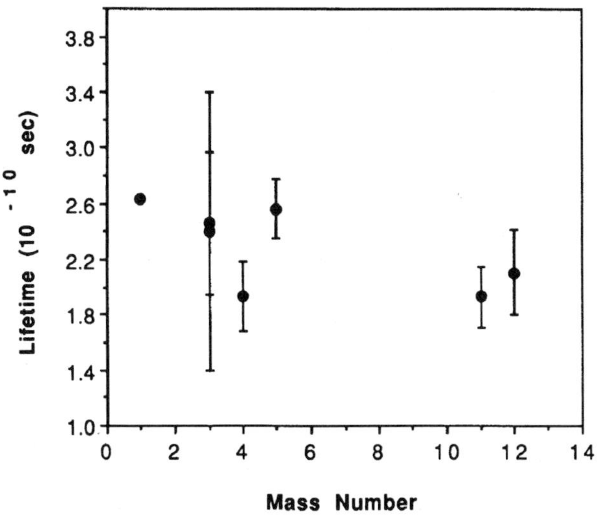

Fig. 8. Experimental data of lifetime of Λ-hypernuclei.

experiments. The emphasis on the study of the weak decay of hypernuclei has been to study the hyperon and the weak interaction inside nuclei. The lifetime is one of the most important observable of hypernuclei, and has been measured with various methods. The lifetime of H_Λ and He_Λ hypernuclei was first measured by their flight length in emulsion and bubble chamber produced by K^--beam [17] and recently in the streamer chamber produced by the relativistic heavy ion beam [18]. Recently, precise values have been obtained by the direct time measurement with counters for $^4H_\Lambda$, $^5He_\Lambda$, $^9Be_\Lambda$, $^{11}B_\Lambda$ and $^{12}C_\Lambda$ [19]. For the heavy nuclei, the lifetime of Bi_Λ and U_Λ was measured by detecting the delayed fission following at-rest anti-proton capture [20]. The data obtained so far are summarized in Fig. 8. The observed lifetimes of light hypernuclei are not very different from that of the free Λ hyperon. The measurement of the lifetime for heavy hypernuclei is ,however, important to study the property of the Λ hyperon deep inside nuclear matter.

In the case of hypernuclei, not only mesonic decay but nonmesonic decay is possible. The mesonic decay is expected to be suppressed especially in heavier nuclei because of the smaller Q-value and Pauli blocking in the final state. This tendency is supported by experiments. The branching of the π^- decay is as low as about 10% at $A = 10$ and 1% at $A = 100$. The systematic experimental study of the weak decay of hypernuclei is not sufficient to discuss the systematics of each branching ratio, π^- and π^0 mesonic decay, $\Lambda\, n \to nn$ and $\Lambda\, p \to np$ decay, although a few experiments provide invaluable data [21]. Theoretically, systematic calculations for these decay branching have been made, taking π distortion in nuclei into account [22]. However two body amplitude of the nonmesonic decay itself is not well understood, apart from the nuclear medium effect. Understanding the nonmesonic weak interaction is one of the goal of studying the weak decay of hypernuclei.

Recently, polarized hypernuclei were produced by the (π^+, K$^+$) reaction from C target at KEK and the asymmetry of nonmesonic decay of $^{12}C_\Lambda$ and $^{11}B_\Lambda$ were measured for the first time [23]. It will elucidate the nonmesonic decay mechanism.

The (π, γ) spectroscopy of hypernuclei will greatly contribute to understand the strangeness nuclear physics just as the (β, γ) spectroscopy did contribute to the progress of the nuclear physics.

3 Hypernuclei with S=−2

3.1 H-dibaryon and QCD

The H-dibaryon was predicted as a stable 6-quark system of SU(3) flavor singlet state (ssuudd, I=J=0) by R. Jaffe [24]. It is expected to be stable because the QCD color magnetic interaction is attractive for this state. It is well known that the mass difference between octet and decuplet baryon is explained by the QCD hyperfine interaction, namely color magnetic interaction [25]. The interaction energy due to the color magnetic interaction can be written by the following operator

$$\Delta_c = -\alpha \sum_{ij} \sigma_i \sigma_j \lambda_i \lambda_j, \tag{3}$$

where σ_i and λ_i are Pauli and SU(3) operators, respectively. Under the assumption of SU(3) flavor symmetry, one can easily calculate the expectation values of this energy for baryons and dibaryons. For example, the values for a nucleon and Δ_{33} are,

$$\Delta_c(N) = -8\alpha, \Delta_c(\Delta_{33}) = +8\alpha. \tag{4}$$

From the mass difference of nucleon and Δ_{33}, one can obtain the parameter α=18.8 MeV. The expectation value is -24α for the H-dibaryon, which is minimum among all 6-quark states. In the case of proton-proton in the $^1S_0$ state it is +8α , which is consistent with the existence of a short range repulsive core in p-p system. The "binding energy" of a dibaryon system due to this interaction can be calculated as following,

$$\Delta M = \Delta_c(\text{dibaryon}) - \Delta_c(\text{baryon}) \times 2, \tag{5}$$

$$\Delta M(H) = \quad -8\alpha \quad = -150 \quad \text{MeV}, \tag{6}$$

$$\Delta M(NN\,^1S_0) = \quad 24\alpha \quad = +450 \quad \text{MeV}. \tag{7}$$

In this simple calculation, the mass of the H-dibaryon is 150 MeV below the two Λ mass and it is stable against the strong decay. This argument is too simple. The SU(3) flavor symmetry is broken and s-quark has larger mass than u and d-quark. Since the H-dibaryon is a symbolic problem of the low energy QCD, the calculations of the H-dibaryon mass with all the effective models based on QCD, such as a bag model [26], Skyrme model [27] and quark-cluster model [28], and also lattice QCD [29] have been made. The calculated mass value ranges from as light as a deuteron to an unbound state. It should be determined experimentally. The search for the H-dibaryon has, therefore, great importance. There has been a few experimental works on the H-dibaryon but no positive result on its existence has been confirmed so far [30].

3.2 H-dibaryon and nuclei with S=−2

The existence of the stable H-dibaryon is related with stable nuclei with $S = -2$ which are denoted as double hypernuclei hereafter. In the conventional theory, it is expected to be double Λ hypernuclei. If the mass of the H-dibaryon is much smaller than that of two Λ , a double Λ hypernucleus can decay into the H-dibaryon and a residual nucleus by the strong interaction, ${}^{A}Z_{\Lambda\Lambda} \to$ H+${}^{A-2}Z$. In such a case, the branching ratio for the weak decay of double hypernuclei is expected to be smaller than for the strong H-decay by many orders of magnitude. The weak decay, therefore, can not be observed practically. The observation of the weak decay of a double hypernucleus excludes the existence of the light H-dibaryon. The lower bound of the mass of the H-dibaryon is determined by the mass of such a double Λ hypernucleus as,

$$M_H > M({}^{A}Z_{\Lambda\Lambda}) - M({}^{A-2}Z) = 2M_\Lambda - B_{\Lambda\Lambda}, \tag{8}$$

where $B_{\Lambda\Lambda}$ is the binding energy of two Λ in the nucleus. This conclusion is not changed if it is not a double Λ hypernucleus but an H-nucleus or any kind of a stable nucleus with $S = -2$. It is very interesting question what is the ground state of nuclei with S=−2. It may be a mixed state of two Λ, Ξ and H, which depends on the interactions of these particles and nuclei which are not known yet.

More than 20 years ago, observation of the weak decay of a double Λ hypernucleus was reported by two groups [31]. They were found in the general scanning of nuclear emulsion exposed to K$^-$-beam. According to the papers, they were produced following a Ξ^- capture at rest by emulsion nuclei. The probability of such an observation is, however, estimated to be very small. Their observations are too lucky, if true, and no one could repeat this type of experiment. Recently, new techniques such as an emulsion-counter hybrid method and a scintillating fiber visual detector have been applied to study nuclei with S=−2 and the H-dibaryon at KEK, which are discussed in sec.3-4 and 3-5.

3.3 (K$^-$, K$^+$) reaction and Ξ hyperon

There are two possible ways to produce nuclei with S=−2 or the H-dibaryon with a K-beam. One is coherent or incoherent production by the (K$^-$, K$^+$) reaction from nuclei. The H-dibaryon can be directly produced through the reaction from a proton pair in nuclei, K$^-$+(pp)→K$^+$+H. The cross section was estimated for ^{3}He target [32]. The double hypernuclei can be produced through either secondary interaction of Ξ^- in nuclei or two step (K$^-$, K$^+$) reaction as shown in Fig. 9a [33, 34]. From the K$^+$ momentum spectrum from the (K$^-$, K$^+$) reaction from nuclei, one can study ground and excited states of nuclei with $S = -2$ as in the same way as those of Λ hypernuclei by (K$^-$, π^-) or (π^+, K$^+$) reactions. The cross sections to produce possible Ξ^- hypernuclear states and ground state of double Λ hypernuclei in the coherent process were calculated for several target nuclei. However, due to a large momentum transfer of the (K$^-$,K$^+$) reaction as shown in Fig. 3, the cross section is as small as an order of or less than μb for the Ξ-hypernuclei [35] and much less (nb) for the double Λ hypernuclei [36].

The other way is the production through Ξ^- capture reaction either in flight or at rest, where the Ξ^- is also produced by the (K$^-$, K$^+$) reaction. In particular, the at-rest capture reaction is a very efficient way to produce the double strangeness system. The stopped Ξ^- is first captured in an atomic orbit and then absorbed by the nucleus from low-lying orbits, and forms a "compound nucleus with S=−2" [37]. Particles such as Λ, H and sometimes double hyperfragment can be emitted from the "compound nucleus" due to the following Ξ^- induced reactions,

$$1)\ \Xi^- + \mathrm{p} \rightarrow \Lambda + \Lambda + Q\ (28\quad \mathrm{MeV}), \tag{9}$$

$$2)\ \Xi^- + \mathrm{p} \rightarrow \Xi^o + \mathrm{n} + Q\ (5\quad \mathrm{MeV}), \tag{10}$$

$$3)\ \Xi^- + \mathrm{p} \rightarrow H + Q\ (?). \tag{11}$$

This process is illustrated in Fig. 9b. Since the expected yield for the double hyperfragment or the H-dibaryon is not very large, a visual detector such as a nuclear emulsion and scintillating fiber visual detector, which can identify those event by event, is very powerful. We need, otherwise, a high intensity K$^-$ beam which will be available at KAON.

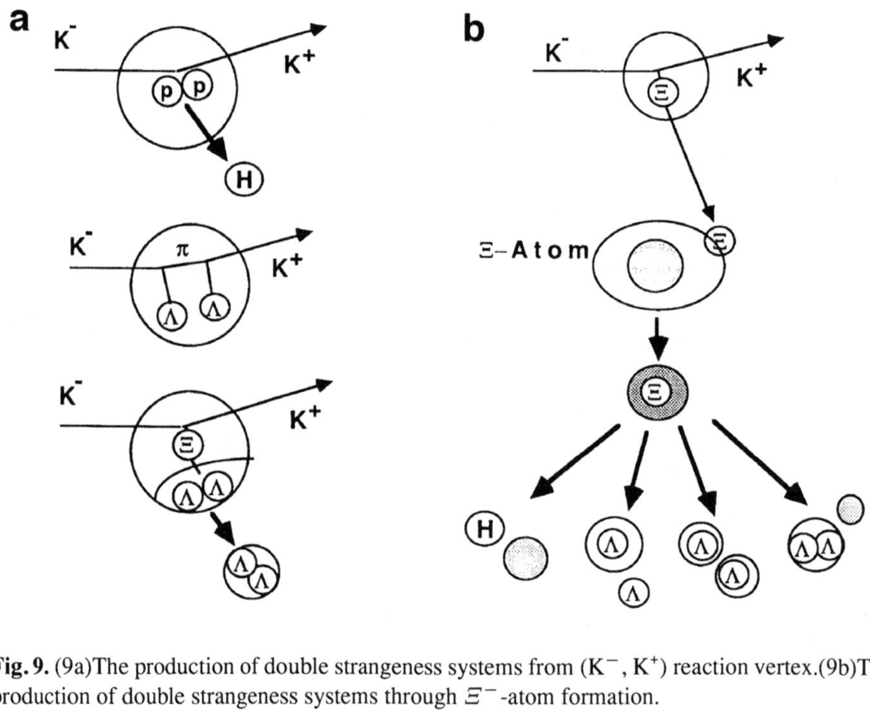

Fig. 9. (9a)The production of double strangeness systems from (K$^-$, K$^+$) reaction vertex.(9b)The production of double strangeness systems through Ξ^--atom formation.

The forward-angle laboratory cross section of K$^-$+p→K$^+$+Ξ^- reaction has maximum at around P_{K^-}=1.8 GeV/c. Recently the (K$^-$,K$^+$) reaction from several nuclei has been studied at P_{K^-}=1.66 GeV/c [38]. The obtained momentum spectra of K$^+$ shown in Fig. 10 are characterized by a peak in the high-momentum region, which is due to the quasi-free process, K$^-$+p→K$^+$+Ξ^-. This part of the spectra is well reproduced by DWIA (eikonal approximation) calculation (solid curves) discussed in sec.2-2. In the low-momentum region, two step processes are, however, expected to be dominant from the A-dependence of the cross section [39].

In the following sections, three experiments; 1) emulsion-counter hybrid experiment (KEK-E176), 2) H-search experiment with a scintillating fiber visual detector (KEK-E224) and 3) H-search through Ξ-d atom (BNL-813), are described.

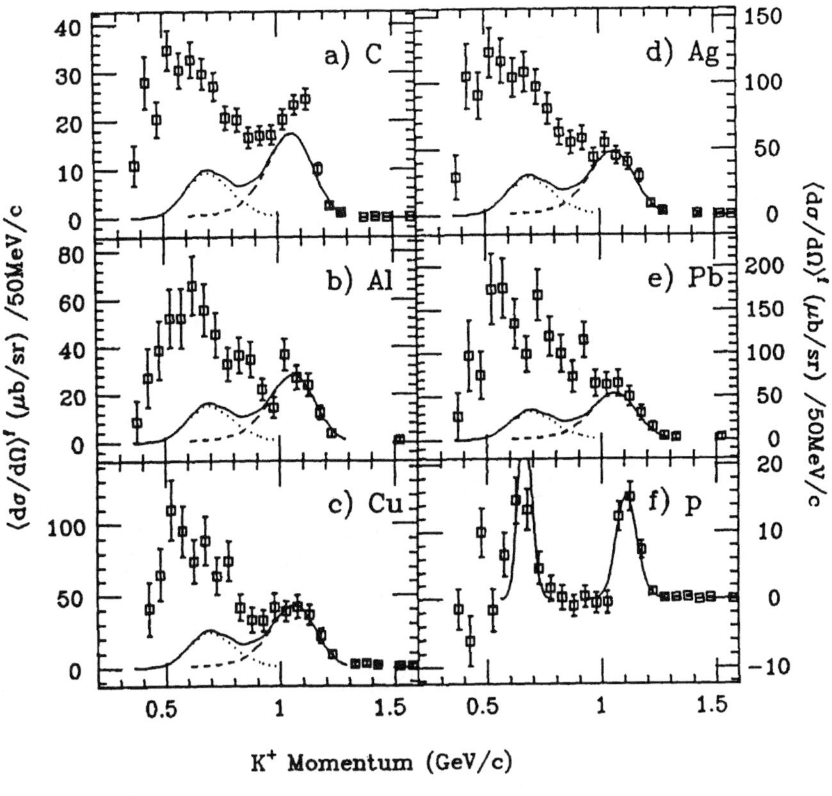

Fig. 10. Momentum spectra of K^+ from the (K^-, K^+) reaction at P_{K^-}=1.66 GeV/c for various target nuclei; a) C, b) Al, c) Cu, d) Ag, e) Pb, f) proton. The solid curves are calculated sum spectra for $K^-+(p) \rightarrow K^++\Xi^-$ and $K^-+(p) \rightarrow K^++ \Xi^{-*}(1530)$ under the assumption of quasi-free reaction.

3.4 Emulsion-counter hybrid experiment (KEK-E176)

The emulsion-counter hybrid method has an advantage for the detection of rather rare events under large backgrounds such as charm and beauty particle search in high energy reactions [40]. This is the first experiment of this type applied at low energy. The experimental set up, together with a schematic drawing of the target region, of the emulsion-counter hybrid experiment (KEK-E176) is shown in Fig. 11. The emulsion stack is sandwiched between the silicon microstrip detectors (BSSD and VSSD) which have a good spatial resolution (δx=16 μm), enabling us to find a tagged K$^+$ track in the emulsion. A semi-automated emulsion scanning system was employed to find out the K$^+$ track in the vicinity of the position in the last emulsion sheet predicted by the silicon microstrip detectors and trace it back to a (K$^-$, K$^+$) reaction vertex through the emulsion sheets. If a Ξ^- candidate is emitted, we follow the track to find its stopping point where the Ξ^- is absorbed by a nucleus. These vertices are studied in detail under a microscope. Details concerning the experimental aspects are described elsewhere [41]. In this method, the emulsion can be exposed to the K$^-$-beam of which integrated number is higher than the old emulsion experiments by four orders of magnitude. Therefore, the (K$^-$, K$^+$) reaction which is only a very small fraction of all the K$^-$ induced reactions can be studied with high statistics. It is another great advantage that the (K$^-$, K$^+$) reaction is clearly identified in this method. It was not possible in the old emulsion experiments.

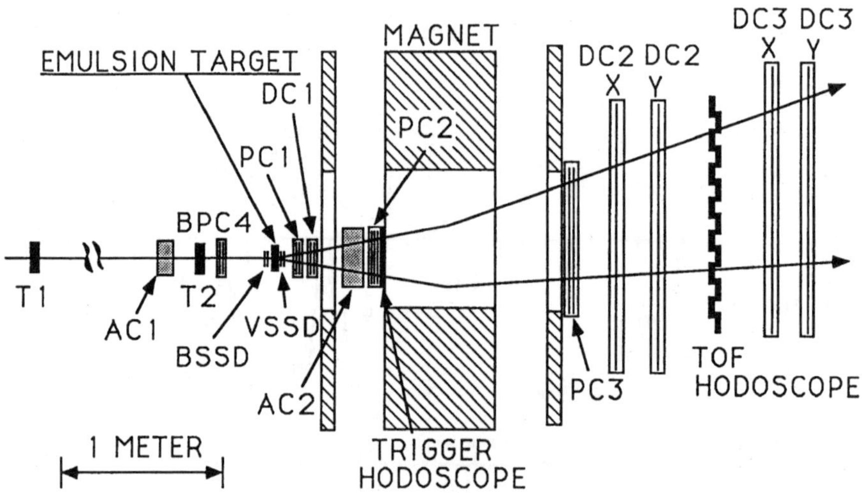

Fig. 11. Experimental setup for the emulsion-counter hybrid experiment (KEK-E176). The incident and scattered K-meson are identified with aerogel Cherenkov counters (AC1 and AC2) and time-of-flight measurements (T2-T1 and TOF hodoscope -T2).

We have studied about 2000 (K$^-$, K$^+$) events in the emulsion. Here we summarize the results of this experiment briefly, since they have been already reported in several papers [42–45].

1. Direct observation of cascade weak decay of double hypernuclei.
   Characteristics of double hypernuclei in the emulsion is the topology of the cascade weak decay of the hyperfragment. A double hypernuceus decays into a Λ-hypernucleus and other particles and then the Λ-hypernucleus decays into several particles. If the fragments are light, they can be identified as tracks in the emulsion. We have observed 4 candidates from the Ξ^- stopping points and 3 candidates from the (K$^-$,K$^+$) reaction vertices. Among them, one event (Fig. 12) from a stopping

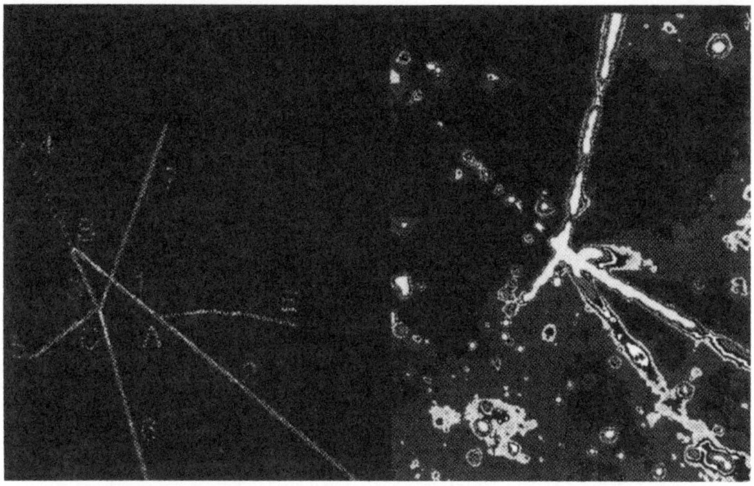

Fig. 12. A schematic drawing of the event. A Ξ^- hyperon produced by the reaction K^-+(p) $\rightarrow$ K^++Ξ^- is captured by an emulsion nucleus at point A. Two particles (track #1,2) are emitted from the stopping point of the Ξ^- in the back-to-back direction. Then a hyperfragment (track #1), with a very short range of 3.9 μm, decays into a daughter hyperfragment (track #3) and a π^- meson in the back to back direction at point B. The daughter hyperfragment, having a very short track, decays into three charged particles at point C. This event can be interpreted as the following two reactions; 1) Ξ^- +^{12}C$\rightarrow$^{3}H+^{10}Be$_{\Lambda\Lambda}$, ^{10}Be$_{\Lambda\Lambda}$ $\rightarrow$ π^-+^{10}B$_\Lambda$, ^{10}B$_\Lambda$ $\rightarrow$(H, He, He, n, n etc.) 2) Ξ^- +^{14}N$\rightarrow$^{1}H+n+^{13}B$_{\Lambda\Lambda}$, ^{13}B$_{\Lambda\Lambda}$ $\rightarrow$ π^- +^{13}C$_\Lambda$, ^{13}C$_\Lambda$ $\rightarrow$(He, He, He, n, n etc.)

Ξ^- point showed π^- mesonic decay. From the kinematical reconstruction of this event, the double hyperfragment was identified as either ^{10}Be$_{\Lambda\Lambda}$ or ^{13}B$_{\Lambda\Lambda}$. (A theoretical argument supports the assignment as ^{13}B$_{\Lambda\Lambda}$ [46].) The binding energy of two Λ (B$_{\Lambda\Lambda}$) was estimated to be 8.5±0.7 MeV for ^{10}Be$_{\Lambda\Lambda}$ and 27.6±0.7 MeV for ^{13}B$_{\Lambda\Lambda}$. From these values, the lower limit of the mass of the H-dibaryon was determined to be 2203.7±0.7 MeV [43]. It should be noted that this "double hypernucleus" is not necessarily a "double Λ hypernucleus". The conclusion about the H-dibaryon mass is, however, remains unchanged even if it is an H-nucleus.

2. Evidence of the weak decay of heavy double hypernuclei.
   If heavy hypernuclei are formed after the Ξ^- capture by heavy emulsion nuclei such as Ag and Br, the formation point and the decay point can not be separated in the emulsion, and the tracks of such hypernuclei are not visible. In order to identify the weak decay of such heavy double hypernuclei, the total visible energy-release from the Ξ^- capture was compared with that from Σ^- capture and that from the non-mesonic decay of heavy hyperfragments (Fig. 13). Here the total visible energy-release is defined as the sum of kinetic energies of all charged particles calculated from the range assuming each particle as a proton and binding energies of protons (8 MeV/proton). In the higher energy region, the event rate is clearly enhanced for the Ξ^- capture, even though the Q-value for Σ^-+p$\rightarrow$ Λ +n+Q(80 MeV) reaction is larger than that for Ξ^-+p$\rightarrow$ Λ + Λ + Q(28 MeV) reaction by 52 MeV. Both comparisons indicate weak decays of two units of strangeness occur for the Ξ^--

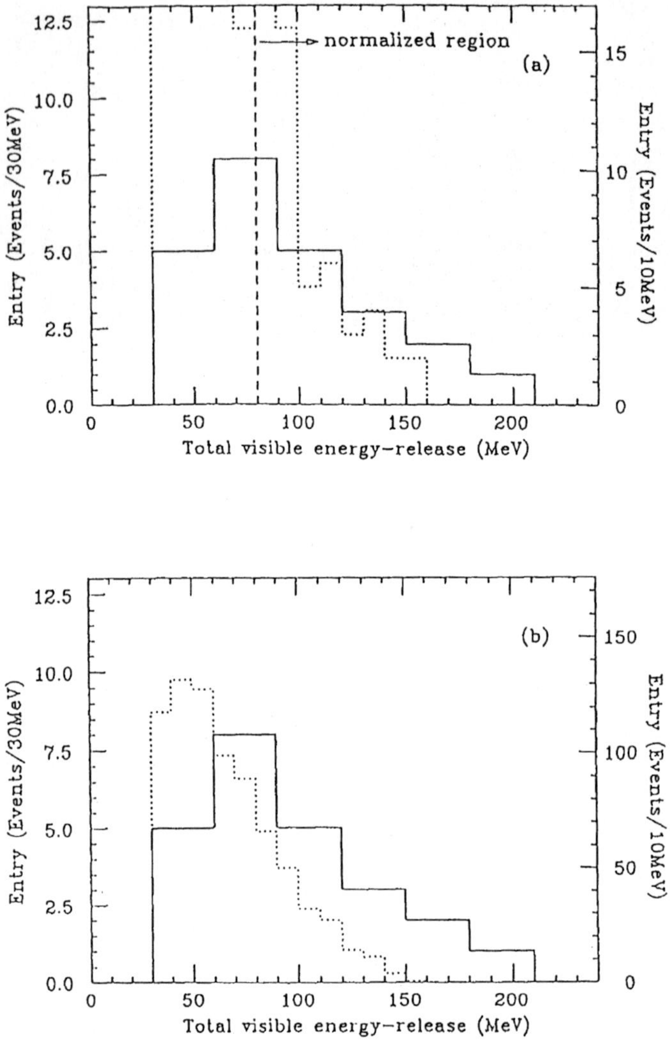

Fig. 13. Distribution of the total visible energy-release for Ξ^--capture stars (solid line). It is compared with the equivalent distribution (dotted line) (a) for Σ^-- capture stars and (b) for heavy hyperfragment decays. The vertical coordinates on the left side are for Ξ^--capture stars, and that on the right side are (a) for Σ^--capture and (b) for heavy hyperfragment decays. The number of events for Σ^--capture is normalized to the present data for a total visible energy release greater than 80 MeV for a comparison.

capture stars. The precise measurement of the vertex points was made for several events which have largest visible energy-release, and we found that two or three fast protons are emitted from one vertex at an accuracy of 0.1 mm in all cases. We, therefore, concluded that the "two weak decays" is considered to be the sequential weak decay of double hypernuclei. Details are described in ref. [44].

3. Observation of Ξ^- bound system
   We have found two events which emit two single hyperfragments both in back to back direction from the Ξ^- capture at rest by light nuclei [45]. In both cases, one of the hyperfragments was clearly identified as $^4H_\Lambda$ due to its characteristic decay into π^- and ^{4}He. From the momentum balance of the two hyperfragments, the kinematical reconstruction was tried to obtain the mass of the parent system (Ξ^- bound system). Although the assignment is not completely unique, the parent system was found to be most probably Ξ^--C system for both events. We have estimated the binding energies of the Ξ^--C system to be 0.53 +0.22 MeV for one event and 3.70$\pm$0.19 MeV for the other event. The former is consistent with a Ξ^--atomic state and the latter is consistent with a Ξ^--C nuclear bound state predicted by Akaishi [47]. These are the first observation of clearly identified Ξ^- bound systems.

In this experiment, the Ξ^--capture at rest in the emulsion was found to be an efficient way to produce the double strangeness systems. Therefore, a second-phase hybrid-emulsion experiment is now planned to obtain 10 times more Ξ^--capture events. For this experiment, a new hybrid method, a combination of a scintillating fiber live target and emulsion, is proposed [48].

3.5 The (K$^-$,K$^+$) reaction with scintillating fiber visual detector (KEK-E224)

In the E224 at KEK, the emulsion target is replaced by a scintillating fiber target for the (K$^-$,K$^+$) reaction. Fig. 14 shows the target surrounded by a cylindrical drift chamber. Since the H-dibaryon was found to be heavier than 2203 MeV/c^2 if it exists, it can decay into H$\rightarrow$ Σ^-+p and Λ +p+π^- which can be easily identified. In the case of the emulsion, we could search for a decay vertex of the H-dibaryon only within 1mm from the production point. In the scintillating fiber target which has a sensitive volume of $10 \times 8 \times 8$ cm^3, the H-dibaryon of which lifetime is up to 10^{-8} sec can be searched. Fig. 15 shows the expected sensitivity of the branching ratio of the H-emission following the Ξ^--capture at rest by C in the E224 and the obtained upper limit in the emulsion experiment (E176). Moreover, the neutral hyperons such as Λ and Ξ^0 can be observed as well as π^- from the decay of double and single hypernuclei.

This is the first experiment to utilize the three dimensional scintillating fiber detector as a visual detector like a bubble chamber, after an extensive R&D of the detector [49]. The scintillating fiber target is made of 500 μm thick plastic scintillating fiber sheets packed in x and y direction alternately, so that a track of a charged particle can be reconstructed in the three dimension. The photon image of the tracks of charged particles are detected with two large-aperture image-intensifier-tubes (IIT) viewed by CCD cameras (Fig. 16), and CCD data is read out through the image digitizing system. The data taking system was also developed, including the second level trigger system with on-line mass calculation to select the (K$^-$,K$^+$) events. The image data can be taken at 20

Hz under a beam rate of 10^5 particles/sec. The detail of the scintillating fiber detector system is described elsewhere [50].

We have taken the image data of about 40 k(K^-, K^+) events which is ten times more than the E176 and old bubble chamber experiments. A preliminary scanning of the image data have been already tried. Fig. 17 shows a typical event of a (K^-, K^+) reaction observed with the scintillating fiber detector system (digitally recorded). Among 1600 preliminary scanned events, we have observed about 900 Ξ^--decay events and many candidates of Ξ^- interaction in flight or at rest. Fig. 18 shows one of these events recorded with a video tape, where the secondary interaction of Ξ^-, $\Xi^- p \rightarrow \Lambda\Lambda$, is clearly seen. The total number of Ξ^- tracks expected to be observed in the fiber is about 20000. Among them, we expect about 300 Ξ^--C atom formation and several hundreds Ξ^--scattering in flight. The cross section for $\Xi^- p$ scattering can be obtained for the first time.

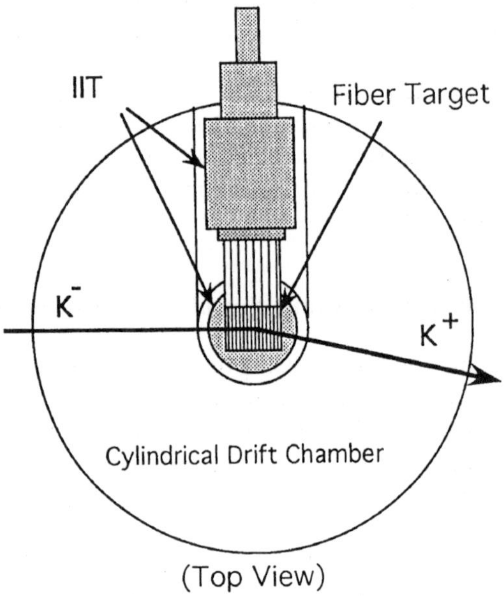

Fig. 14. A schematic view of the scintillating fiber live target.

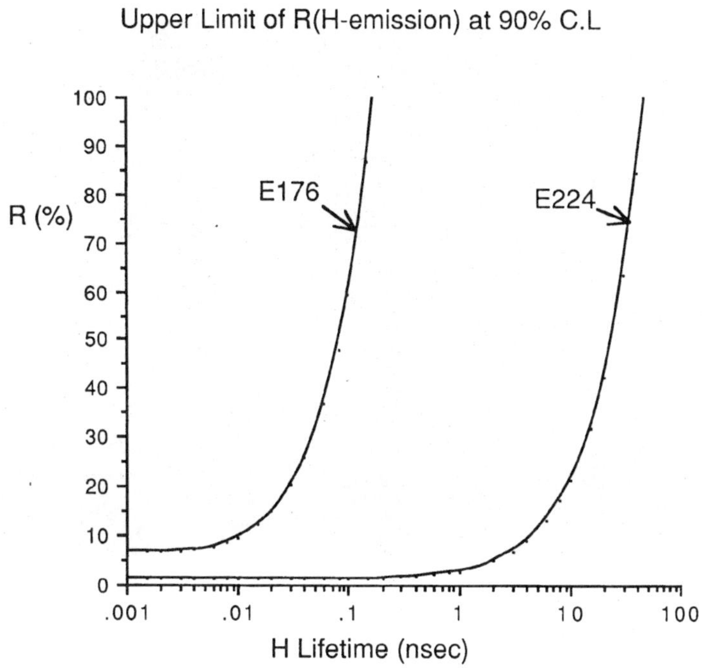

Fig. 15. The upper limit (90% C.L.) for the branching ratio (R) of the H-emission obtained from E176 and expected for E224, as a function of the H-lifetime. The curves are for the H-mass of 2200 MeV/c^2.

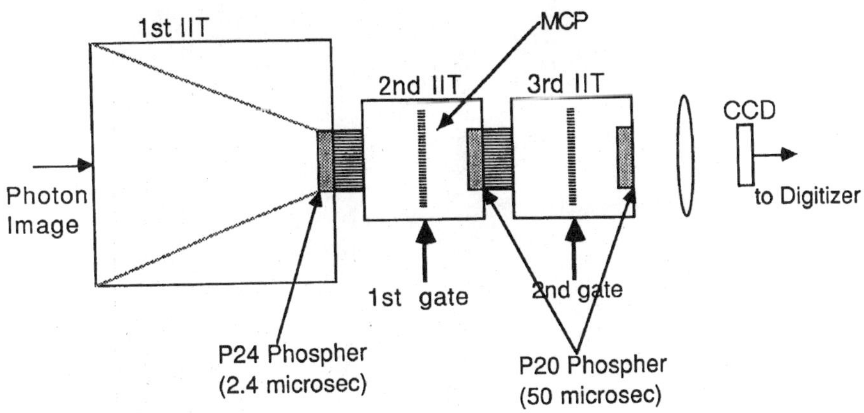

Fig. 16. A schematic view of the image intensifier system used for the scintillating fiber detector (E224).

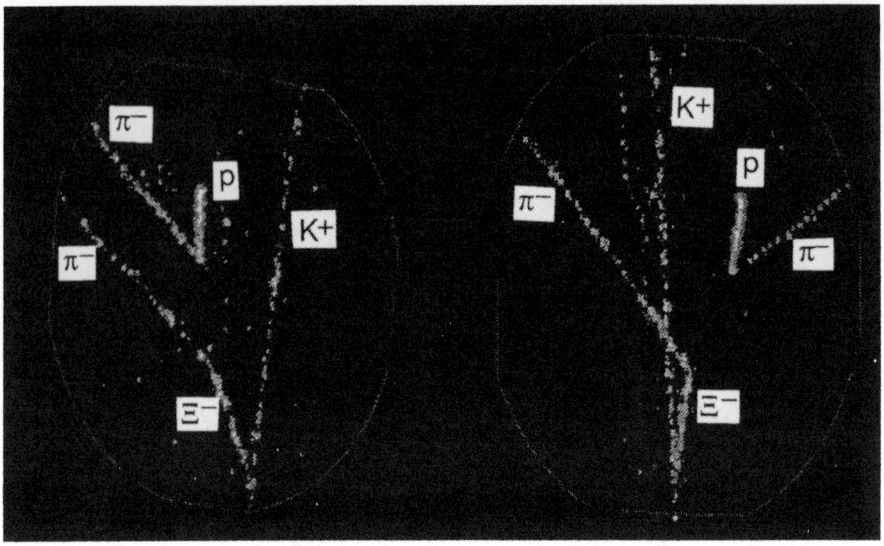

Fig. 17. A typical event of a (K^-, K^+) reaction, $K^-+p \rightarrow K^+ + \Xi^-$, $\Xi^- \rightarrow \pi^- + \Lambda$, $\Lambda \rightarrow \pi^- + p$, seen with the scintillating fiber detector from x and y directions. White circles show fiducials (8×10 cm).

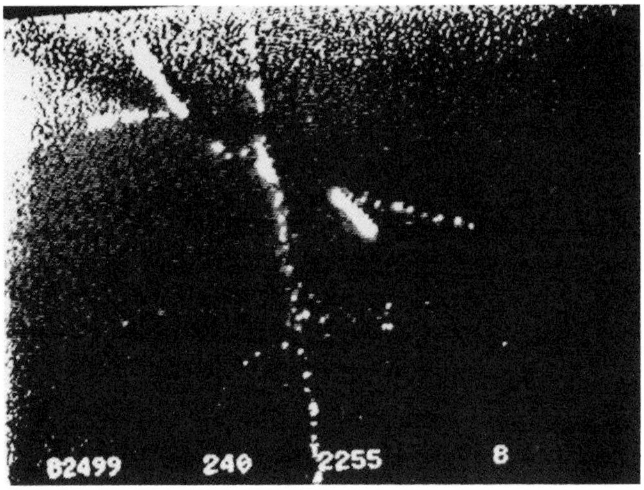

Fig. 18. A typical event of the Ξ^- reaction in flight, $\Xi^- + (p) \rightarrow \Lambda + \Lambda$.

3.6 H-dibaryon search through Ξ^--d atom (BNL-E813)

Recently, a new 2 GeV K^--beam line was built at BNL for the study of double strangeness systems, primarily for E813/836 which are H-search experiments. The beam line is optimized to provide an intense clean beam of 1-2 GeV/c K^- and incorporates two stages of particle separation with the optics corrected to third order. The detail of the beam line is described elsewhere [51]. We have obtained 1.5×10^6 $K^-/5\times10^{12}$ ppp at 1.8 GeV/c and the K^-/π^- ratio of 1/2.5 even with an unfortunate failure of one of the separator power-supply. This success opened new era for the strangeness nuclear physics.

The E813 being done at BNL is the experiment to search for the H-dibaryon through the following two step reactions; $K^-+p\rightarrow K^++\Xi^-$ and $\Xi^-d(\text{atom})\rightarrow n+H$ [52]. The schematic view of the detector system is shown in Fig. 19. The K^+ scattered from a liquid hydrogen target is detected with a K^+-spectrometer. A missing mass spectrum for the reaction shows a peak of Ξ^- production (Fig. 20). The Ξ^- is degraded by a tungsten degrader, tagged with a silicon detector and brought to rest in a liquid deuterium target. In this experiment, a neutron (TOF) is detected in coincidence with K^+ and Ξ^-. The H-dibaryon can be observed as a peak in the neutron TOF spectrum. This experiment has advantage over the previous ones in the sense that it is sensitive to the H even if the H-dibaryon is unbound or does not decay. According to the model calculation [53], the branching ratio (R) for the H-production from Ξ^-d atom is expected to be more than 10% for M_H >2200 MeV/c^2. The first round data taking was made in 1992 and the result will come out soon.

By replacing the liquid deuterium by ^{6}LiH, one can form Ξ^--Li atom and detect $^6He_{\Lambda\Lambda}$ which is similar to an alpha particle in the double strangeness system, called "lampha". The branching ratio, $\Xi^-{}^6$Li atom $\rightarrow$ $^6He_{\Lambda\Lambda}$+n, is recently estimated to be about 3% [54]. This measurement is now being planned by using the same experimental setup [55].

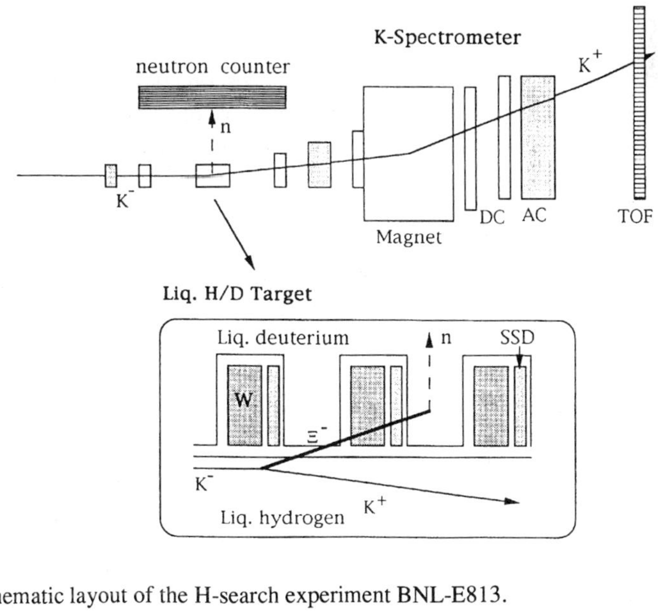

Fig. 19. Schematic layout of the H-search experiment BNL-E813.

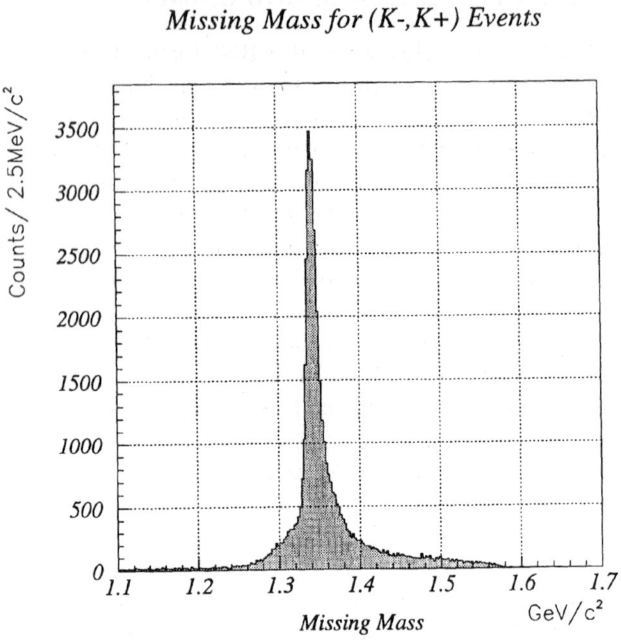

Fig. 20. A missing mass spectrum in the (K^-, K^+) reaction from the hydrogen target. A peak corresponds to the production of Ξ^-.

4 Future Prospect

In the study of $S = -1$ nuclei, more extensive spectroscopy will be performed for heavier nuclei such as Pb with a better energy resolution. we have now the superconducting spectrometer, SKS, at KEK-PS for this purpose. At CEBAF, a high resolution spectroscopy of hypernuclei will be done by (γ, K^+) reaction. The detection of decayed particles such as γ and π from hypernuclei is also important to study the structure and weak interaction of hypernuclei. For the (γ,π) spectroscopy of hypernuclei, large acceptance γ and π detectors are necessary. The Troidal Spectrometer recently built at KEK-PS is suited for this purpose. Further experiments with these new devices will provide us more systematic and precise data of low lying states of hypernuclei and their decays, which are the key to understand the structure of hypernuclei and hyperon-nucleon interaction. The key word is higher resolution and heavier nuclei.

Another future direction is to increase the number of strangeness. The study of $S = -2$ system has just begun. We have studied many things through the (K^-, K^+) reaction with emulsion and scintillating fiber detectors as described before. They are, however, first round experiments. The second phase experiments to find more $S = -2$ nuclei and search for the H-dibaryon with better sensitivity need a high K^--beam intensity. The newly built K^--beam line at BNL is appropriate for this purpose to some extent.

The KAON will be undoubtedly very powerful for the study of $S = -2$ system. For the multi-strangeness system ($S<-2$), relativistic heavy ion collision seems only way to produce, because the average multiplicity of Λ in the central Au-Au collision at 12 GeV/A is expected to be about 10. In the case of the multi-strangeness system, there is a big question whether the ground state is a strange quark matter (strangelet) or multi Λ nuclei. A meta-stable strangelet has been or will be searched in the relativistic heavy ion collision at BNL as well as the H-dibaryon [56]. However, it is very difficult to detect short life multi Λ nuclei in the very high multiplicity environment of heavy ion collisions with use of the present technology.

One of the purpose of the study of hypernuclei is to learn hyperon-nucleon interaction. The hyperon-nucleon scattering experiment is straight forward way to study the hyperon-nucleon interaction. There is a few data from bubble chamber experiments more than 25 years ago. However, the statistics is very poor and energy range of the data is very limited. Recently, the scintillating fiber detector has been applied to detect Σ^++p scattering to measure not only the cross section but also polarization parameter at KEK [57]. The key of the technology is to develop the automatic pattern recognition of the scattering events out of thousands of unscattered events. Future developments will enable to provide hyperon-nucleon scattering data with high statistics to determine the hyperon-nucleon interaction including spin-dependent terms.

References

1. M. Danysz and J. Pniewski, *Phil. Mag.* **44** (1953) 348.
2. H. Bando, T. Motoba and J. Zofka, *Int. J. Mod. Phys.* **5** (1990) 4021-4198.
3. Proceedings of the International Symposium on Hypernuclear and Strange Particle Physics, (Shimoda Japan 1991.12). *Nucl. Phys.* **A547** (1992).
4. D.H. Davis and J. Sacton, *High Energy Phyics*, **vol.2** ed. E.H.S.Burhop (Academic Press, 1967).
5. M. Juric et al., *Nucl. Phys.* **B52** (1973) 1. The binding energies were reported based on some 27000 π^- mesonically-decaying hypernuclei.
6. The most recent values are in; D. Davis and J. Pniewski, *Contemp. Phys.* **27** (1986) 91.
7. D.H. Davis, Proceedings of Hypernuclear Physics Conference (Cracow, 1992).
8. Experiment; S. Avramenko et al., *Nucl. Phys.* **A547** (1992) 95c and Theory; H. Bando, M. Sano, J. Zofka and M. Wakai, *Nucl. Phys.* **A501** (1989) 900.
9. C. Bennhold, *Nucl. Phys.* **A547** (1992) 79c.
10. W. Bruckner et al., *Phys. Lett.* **79B** (1978) 157 and R. Chrien et al., *Phys. Lett.* **B89** (1979) 31.
11. C. Dover, L. Ludeking and G.E. Walker, *Phys. Rev.* **C22** (1980) 2073 and H. Bando and T. Motoba, *Prog. Theor. Phys.* **76** (1986) 1321.
12. P. Pile et al., *Phys. Rev. Lett.* **66** (1991) 2585.
13. M. Akei et al., *Nucl. Phys.* **A534** (1991) 478.
14. D. Millener, C. Dover and A. Gal, *Phys. Rev.* **C38** (1988) 2700.
15. O. Hashimoto et al., *Nuovo Cim.* **102** (1989) 679 and O. Hashimoto, *KEK-E140.*
16. M. Bedjidian et al., *Phys. Lett.* **83B** (1979) 252.
17. G. Keyes et al., *Nucl. Phys.* **B67** (1973) 269 and G. Bohm et al., *Nucl. Phys.* **B23** (1970) 93.
18. J. Szymanski et al., *Phys. Rev.* **C43** (1991) 849 and H. Outa et al., *Nucl. Phys.* **A547** (1992) 109c.

19. S. Avramenko et al., *Nucl. Phys.* **A547** (1992) 95c.
20. S. Polikanov, *Nucl. Phys.* **A478** (1988) 805c.
21. A. Sakaguchi, *Phys. Rev.* **C43** (1991) 73.
22. T. Motoba, *Nucl. Phys.* **A547** (1992) 115c.
23. T. Kishimoto, H. Ejiri and H. Bando, *Phys. Lett.* **B232** (1989) 24, S. Ajimura et al., *Nucl. Phys.* **A547** (1992) 47c.
24. R.L. Jaffe, *Phys. Rev. Lett.* **38** (1977) 195.
25. F. Close, *An Introduction to Quarks and Partons* (Academic Press 1979).
26. P.J.G. Mulders, A.T.M. Aerts and J.J.de Swart, *Phys. Rev.* **D21** (1980) 2953.
27. A.P. Balachandran, F. Lizzi and V.G.J. Rodgers, *Nucl.Phys.* **B256** (1985) 525.
28. M. Oka, K. Shimizu and K. Yazaki, *Nucl. Phys.* **A464** (1987) 700 and U. Straub et al., *Phys. Lett.* **B200** (1988) 241.
29. Y. Iwasaki, T. Yoshie and Y. Tsuboi, *Phys. Rev. Lett.* **60** (1988) 1371.
30. An upper limit of the cross section of $p+p \to K^+ + K^+ + H$ was given by A.S. Carroll et al., *Phys. Rev. Lett.* **41** (1978) 777.
The existence of an H-dibaryon with a mass below 1.875GeV/c^2 is excluded by looking at the decay; $^{74,72}Ge \to {}^{72,70}Ge + H + \gamma$ with use of the double beta decay detector at Kamioka by H. Ejiri et al., *Phys. Lett.* **B228** (1989) 24.
Two candidates for H-decay, $H \to \Sigma^+ + p$, are reported in the p-propane reaction at 10 GeV/c by B.A.Shahbazian, A.O. Kechechyan, A.M. Tarasov and A.S. Martynov, *Z.Phys.* **C39** (1988) 151 and B.A. Shahbazian, V.A. Sashin, A.O. Kechechyan and A.S. Martynov, *Phys. Lett.* **235** (1990) 208. Further confirmation is necessary, especially the strangeness of the neutral particles, to reject other interpretation of these events.
31. M. Danysz et al., *Nucl. Phys.* **49** (1963) 121 (The event was reexamined in the paper by R.H. Dalitz et al., *Proc. R. Soc. Lond.* **A426** (1989) 1.) and D. Powse, *Phys. Rev. Lett.* **17** (1966) 782 (This emulsion event was never reexamined and lost forever.)
32. A.T.M. Aerts and C.B. Dover, *Phys. Rev.* **D28** (1983) 450.
33. Y. Yamamoto, *Nucl. Phys.* **A547** (1992) 233c.
34. Y. Yamamoto, M. Wakai, T. Fukuda and M. Sano, *Prog. Theor. Phys.* **88** (1992) 1163.
35. C.B. Dover and A. Gal, *Ann. Phys.* **146** (1983) 309.
36. A.J. Baltz, C.B. Dover and D.J. Millener, *Phys. Lett.* **123B** (1983) 9.
37. T. Yamazaki, *Trends in nuclear physics* (North Holland, 1989) p.78.
38. T. IIjima et al., *Nucl. Phys.* **A546** (1992) 588.
39. C.B. Dover et al., proposed ϕ-production as the main source of low momentum K^+ in the spectra; private communication.
40. S. Aoki et al., *Nucl. Instr. & Methods* **A274** (1989) 64.
41. S. Aoki et al., to be published in *Nucl. Instr. and Methods.*
42. K. Imai, *Nucl. Phys.* **A547** (1992) 199c and S. Aoki et al., *Phys. Rev. Lett.* **65** (1990) 1729.
43. S. Aoki et al., *Prog. Theor. Phys.* **85** (1991) 1287 and S. Nakanishi, *PhD thesis* (Nagoya Univ. 1991) unpublished.
44. S. Aoki et al., *Prog. Theor. Phys.* **85** (1991) 95 and H. Tajima, *PhD thesis* (Nagoya Univ. 1991) unpublished.
45. S. Aoki et al., *Prog. Theor. Phys.* **89** (1993) 493 and S. Aoki et al., to be published.
46. C. Dover, D. Millener, A. Gal and D. Davis, *Phys. Rev.* **C44** (1991) 1905.
47. Y. Akaishi, *Nucl. Phys.* **A547** (1992) 217c.
48. K. Imai, Proceedings of Seminar on Present and Future Experimental Particle Physics by Hybrid Emulsion-Electronic Detectors (Nagoya 1989) and K. Nakazawa, *KEK-proposal* (1993).
49. A. Konaka et al., *Nucl. Instr. & Methods* **A256** (1987) 70.
50. J.K. Ahn et al., *Nucl. Phys.* **A547** (1992) 211c and Y. Itow et al., to be published.

51. P. Pile et al., *Nucl. Instr & Methods* **A321** (1992) 48.
52. P. Barnes, *Nucl. Phys.* **A547** (1992) 3c.
53. A.T.M. Aerts and C.B. Dover, *Phys. Rev.* **D29** (1984) 433
54. D. Zhu, C. Dover, A. Gal and M. May, *Phys. Rev. Lett.* **67** (1991) 2268.
55. M. May and G. Franklin, *BNL-AGS proposal E889.*
56. K. Imai et al *BNL-AGS E886* and J. Sandwiess et al., *BNL-AGS E864.*
57. M. Ieiri et al., *KEK-PS E251.*

Lecture given at the JSPS-INS International Spring School, March 1993 at Shimoda

Topics on Space-Time Symmetries
— An Experimental Review

Akira Masaike

Department of Physics
Kyoto University
Sakyo-ku, Kyoto, Japan

Abstract Recent topics on fundamental symmetries with respect to P (parity), T (time reversal), CP and CPT transformations are presented. No evidence of violation of CPT-theorem has been found. However, concerning the weak equivalence principle, comparison of matter and anti-matter in a gravitational field must be checked experimentally.

Studies of enhancement of parity-violating processes in hadron-hadron interactions are interesting from the view point of T-violation measurements.

Present status of CP-violation measurements is mentioned briefly. Recent progress of the measurement of electric dipole moments of the neutron and atoms are also presented. In addition, several experiments of the T-invariance test in the β-decays of nuclei and in K-decay are shown.

1 Introduction

The symmetry or invariance of the equations describing a system under an operation is one of the most important concepts in physics. Some of the invariances, however, are still open questions to be tested experimentally with high precision.

Since an evidence of parity non-conservation was found in β-decay in 1956, several experimental approaches to the symmetry tests have been made. The first success of the Weinberg-Salam theory was the discovery of parity non-conservation in high energy electron scattering from deuteron. The source of parity non-conservation can be explained by the weak sector of the electroweak theory.

On the other hand, the origin of the CP-violation is still unknown, even almost 30 years have passed since the discovery of CP-violation in the decays of the neutral kaons. It may be explained by the minimal standard model, or it may be due to a new interaction.

The subject of this report ranges over the basic ideas and experimental tests of symmetries with respect to the spatial reflection (P), time reversal (T), which are related to charge conjugation (C) by the CPT theorem.

First, the CPT-theorem is mentioned briefly. No evidence of violation of the CPT-theorem has been found. However, the gravitational interaction on anti-particles is unknown and it is necessary to check experimentally.

Concerning the symmetries with respect to the spatial reflection, parity-violating processes in hadron-hadron interaction have been studied extensively in recent years. One of the interesting topics is on the large parity-violating effect which has been found in slow neutron radiative capture reactions. The results encourage us to measure T-violation using polarized neutrons and a polarized nuclear target.

Experimental effort over the last 30 years has been devoted to solve the question on CP-violation. Recent results and future projects of CP-violation measurements in kaon and b-meson decays are mentioned.

The measurements of the electric dipole moment of the neutron and atoms have provided constraints on theoretical possibilities on time reversal violation, which is related to the CP-violation by the CPT theorem. β-decays of nuclei can also provide important information on time reversal invariance. Finally, interesting experimental approaches to the test of time reversal invariances are presented.

2 Tests of CPT Symmetry and Weak Equivalence Principle

The CPT theorem states that all interactions are invariant under the succession of the three operations C (charge conjugation), P (parity) and T (time reversal) taken in any order.

The CPT-theorem based on the quantum field theory is a relation between particles and anti-particles. The theorem requires [1]: (1) the field theory is constructed from fields that belong to finite-dimensional representations of the Lorentz group, (2) it has local interactions invariant under the proper Lorentz group, (3) it is described by a Hermitian Lagrangian.

The theorem predicts that a particle and its anti-particle have equal masses, total lifetimes and gyromagnetic ratios. In addition, they have equal and opposite electric charges. No experimental evidence for violation of CPT theorem has been found, even a few of above predictions have been tested.

The most sensitive test so far arises from an indirect measurement of the mass difference of the K^0 and $\bar{K}^0$ [2].

$$\frac{m(K^0) - m(\bar{K}^0)}{m(K^0)} \leq 4 \times 10^{-18}. \tag{1}$$

The difference between gyromagnetic ratios of the electron and positron is [3]

$$\frac{g(e^+) - g(e^-)}{\text{average}} = (-0.5 \pm 2.1) \times 10^{-12}. \tag{2}$$

The difference between cyclotron frequencies of the proton and anti-proton in the same magnetic field is [4]

$$\frac{\omega_c(p) - \omega_c(\bar{p})}{\omega_c(p)} \leq 4 \times 10^{-8}, \tag{3}$$

which implies the mass ratio of anti-proton to proton of

$$\frac{M(\bar{p})}{M(p)} = 0.999999977(42). \tag{4}$$

It was obtained with an anti-proton Penning trap using an electron cooling method.

More precise measurement can be done with the anti-hydrogen atom. Feasibility studies of $\bar{H}$ formation have been proposed by several researchers including a method of radiative capture of positrons by anti-protons stimulated with laser beam.

The weak equivalence principle (WEP) requires that the gravitational acceleration of a falling body is independent of its composition. Although this principle has been tested to a precision of about 10^{-12} for normal matter, it has not yet been tested for anti-particles. In fact, gravitational force on $\bar{\mathrm{p}}$ is unknown and there is no reason to believe the equivalence of matter and anti-matter in a gravitational field. Therefore, it is necessary to check it experimentally. In order to measure the gravitational force, anti-protons must be decelerated to ultra-low energy and stored in a Penning trap. Then they are launched vertically up in a drift tube. It can be compared with H^- ions. The loss of the kinetic energy of the particles rising 1 m is 10^{-7} eV, which is much smaller than the thermal energy.

Recently, it was pointed out that the precise comparison of cyclotron frequencies of trapped protons and anti-protons in the same magnetic field is an excellent test of WEP [5].

Higher precision tests of WEP for protons and anti-protons may become available using anti-hydrogen. The natural line width of the $1s$-$2s$ two-photon transitions in hydrogen and anti-hydrogen can be possibly compared with a precision up to 10^{-15}.

Future developments in trapping of anti-particles and the spectroscopy of trapped anti-matter allow us the experimental tests of WEP.

3 Tests of Parity Non-conservation

In 1956 Lee and Yang [6] pointed out that the weak interations are not invariant under spatial invariance on the basis of the fact the K^+ could decay into 2π and 3π. In the next year Wu et al. [7] found an evidence of parity non-conservation in the β-decay of $^{60}\mathrm{Co}(J{=}5^+)\rightarrow{}^{60}\mathrm{Ni}^*\ (J{=}4^+){+}e^-{+}\bar{\nu}$, in which ^{60}Co nuclei were polarized. In fact, they observed electron emissions asymmetric to the direction of the nuclear spin. It implies parity violation, since the $\vec{\sigma}\cdot\vec{p}$ operator is pseudoscalar and changes the sign under parity reversal transformation. ($\vec{\sigma}$ and $\vec{p}$ are the spin and momentum operators).

It is known that the source of parity non-conservation (PNC) is in the weak sector of the electro-weak theory. The PNC occurs due to the presence of both vector and axial vector currents. The strengthes are determined for both charged and neutral currents in terms of the Fermi coupling G_{F} and the Weinberg angle θ_{W}. The current can be written as j(leptonic)+J(hadronic). Therefore, there are leptonic ($j^\mu j_\mu^+$), semi-leptonic ($j^\mu J_\mu^+ + j_\mu^+ J^\mu$) and non-leptonic ($J^\mu J_\mu^+$) interactions. Experimental data on leptonic and semi-leptonic interactions can be explained in the standard model, but the non-leptonic interaction is not well known.

Parity non-conservation in a electron-hadron interaction was found by means of the measurement of helicity asymmetries in scattering of polarized electrons by deuterons [8]. The experiment was carried out with electron beams of 16~22 GeV/c at SLAC. The parity-nonconserving symmetry, measured by the difference of cross sections for positive and negative helicity electrons were

$$A_L = \frac{\alpha_+ - \alpha_-}{\alpha_+ + \alpha_-} \simeq 10^{-4}q^2 \ \ (q^2 \ \text{in}(\mathrm{GeV}/c)^2). \tag{5}$$

The asymmetry is due to the interference between Z^0 and γ-exchange amplitudes. The total scattering amplitude consists of electromagnetic (A_{EM}) and weak (A_{WK}) amplitudes. Therefore, the cross section for e-N scattering is proportional to

$$\sigma_{eN} \propto |A_{EM} + A_{WK}|^2 = |A_{EM}|^2 \cdot (1 + \frac{A_{WK}A_{EM}^* + A_{WK}^* A_{EM}}{|A_{EM}|^2} + \frac{|A_{WK}|^2}{|A_{EM}|^2}). \quad (6)$$

The last term of the right-hand of the equation is negligible, because $|A_{EM}| \gg |A_{WK}|$. Therefore, the asymmetry can be shown as

$$A_L = \frac{A_{WK}A_{EM}^* + A_{EM}A_{WK}^*}{|A_{EM}|^2}. \quad (7)$$

Since $A_{EM} \simeq 4\pi\alpha/q^2$ and $A_{WK} \simeq G_{\mathrm{F}}$, we have

$$A_L \simeq \frac{G_{\mathrm{F}} q^2}{4\pi\alpha} \simeq \frac{q^2}{m_{\mathrm{Z}}^2}. \quad (8)$$

From the experiment we can obtain the Z^0 mass which is $m_{\mathrm{Z}} = 91.173 \pm 0.020$ GeV/c^2. This was the first measurement of the Weinberg angle (θ_{W}) which is

$$\sin^2 \theta_{\mathrm{W}} = 0.2325 \pm 0.0008. \quad (9)$$

In this interaction only a neutral weak current and electromagnetic current contribute to the scattering amplitude.

One of the recent interesting topics is parity violation in non-leptonic interactions. The simplest contribution to parity violation in the coupling between nucleons is single W or Z^0 exchange as shown in Fig. 1.

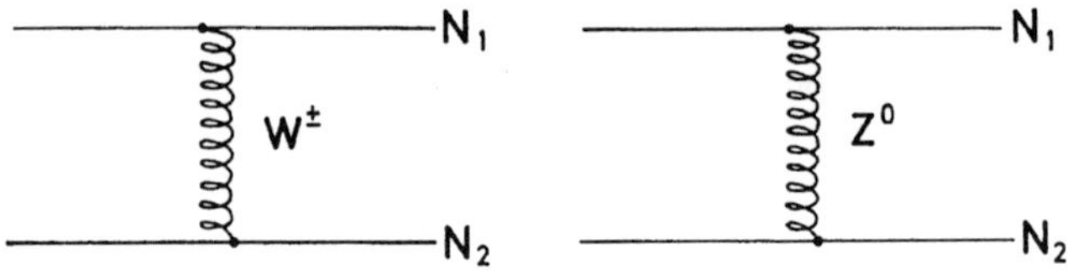

Fig. 1. The simplest contribution to parity violating interactions between nucleons by $W^{\pm}$ exchange and Z^0 exchange.

The above diagrams give rise to an effective pseudoscalar potentials between two nucleons. However, the effect of the potentials is greatly supressed, because the strong N-N interaction is repulsive at very short distance. Therefore, the exchange of light pseudoscalar and vector mesons with vertex of weak interactions as in Fig. 2 is the main contribution to parity-nonconserving reactions. The ratio of parity-violating (V_{PNC}) to parity conserving (V_0) potentials is estimated as

$$\frac{V_{PNC}}{V_0} \simeq G_{\mathrm{F}} m_\pi^2 \simeq 2 \times 10^{-7}. \quad (10)$$

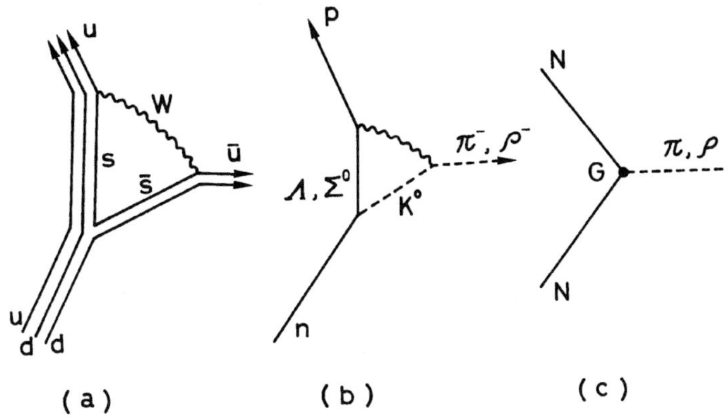

Fig. 2. One of the main contribution to the parity-nonconserving nucleon-nucleon reactions. (a) a gauge boson interaction, (b) weak interaction vertex on the hadron level, (c) weak vertex of NNπ and NNρ.

Therefore, the order of magnitude of the parity non-conserving helicity asymmetry in N-N scattering is about 10^{-7} for the energies below 1 GeV. The A_L^{tot} (helicity asymmetry in the total cross section) in p-p scattering arises from the interference between opposite parity amplitudes. The ${}^1S_0$-${}^3P_0$ amplitude is determined from measurements of A_L^{tot} at rather low energies [9, 10].

$$A_L^{tot} = -(0.17 \pm 0.08) \times 10^{-6} \quad \text{at 15 MeV} \tag{11}$$

$$= -(0.15 \pm 0.022) \times 10^{-6} \quad \text{at 45 MeV.} \tag{12}$$

On the other hand, the ${}^3P_2$-${}^1D_2$ amplitude contributes significantly to A_L^{tot} above 300 MeV. The experimental result

$$A_L^{tot} = (0.24 \pm 0.11) \times 10^{-6} \quad \text{at 800 MeV} \tag{13}$$

measured at Los Alamos can be reproduced well by this model [11]. In contrast,

$$A_L^{tot} = (2.65 \pm 0.6) \times 10^{-6} \quad \text{at 5.3 GeV,} \tag{14}$$

which was measured at Argonne, cannot be explained by simple models [12].

The experimental effort over the last decade has been devoted to the measurement of large PNC asymmetries in nucleon-nucleus interactions where PNC effects are enhanced, in particular, due to the interference between very close energy levels of the same spin but opposite parity. Very large effect was found in low energy neutron radiative capture reactions at p-wave resonances. A large helicity asymmetry of neutron in the neutron capture reactions was observed at the resonance of E_n = 0.734 eV for the n + ^{139}La reaction at Dubna [13], KEK [14] and LAMPF [15].

We present the experiments at KEK below [14]. Slow neutrons transmitted through a polarized proton filter become polarized, because the cross section for neutron-proton scattering in the singlet state is substantially larger than that in the triplet state. Protons in organic material containing Cr^V complexes are polarized by means of the dynamic method in which material is irradiated with microwave of 70 GHz in a 2.5 T magnetic field at a temperature lower than 0.5 K. Typical value of neutron polarization is about 75% at 1 eV. Neutrons are polarized transversely by a polarized proton filter and is rotated to the longitudinal direction.

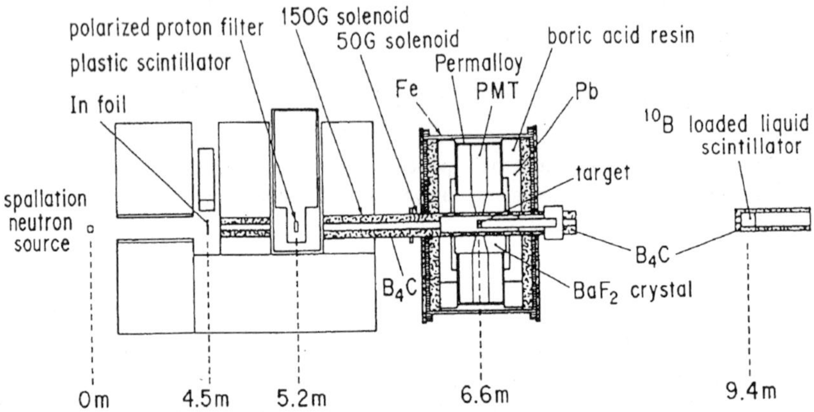

Fig. 3. The experimental setup for the measurement of helicity asymmetries of capture γ-rays using BaF_2 counters at KEK.

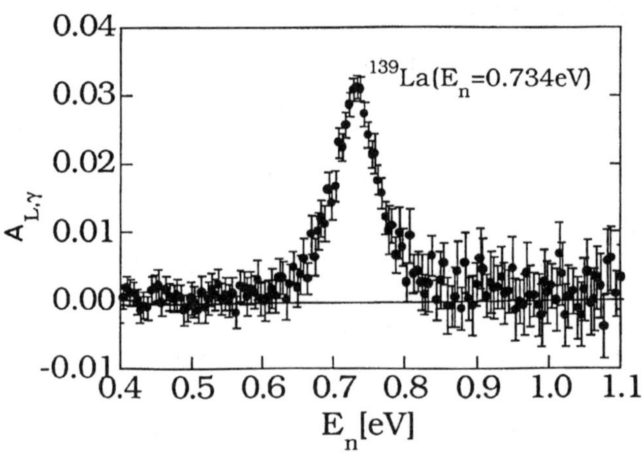

Fig. 4. The helicity asymmetry of capture γ-rays versus neutron energy near the p-wave resonance of ^{139}La at 90°.

Table 1. Experimental results of the helicity asymmetries of slow neutron for several target materials on p-wave resonances, as well as the energies of p-wave resonances and nearest s-wave resonances.

target nucleus	E_p [eV]	E_s [eV]	A_L	$\lvert xW \rvert$ [meV]
^{81}Br	0.88±0.01	101.10±0.14	0.021±0.001	2.6±0.1
^{93}Nb	35.9±0.1	119.2±0.2	0.003±0.005	
^{93}Nb	42.3±0.1	−105.39	−0.000±0.006	
^{108}Pd	2.96±0.01	33.10±0.17	0.002±0.002	
^{111}Cd	4.53±0.03	−4	$-(0.013^{+0.007}_{-0.004})$	$1.6^{+0.9}_{-0.5}$
^{124}Sn	62.0±0.1	−20	0.002±0.004	
^{139}La	0.734±0.005	−48.63	0.098±0.003	1.7±0.1

Longitudinal asymmetries in the cross section for the neutron-induced reaction on p-wave resonances of verious nuclei were examined by simultaneous measurements of the capture γ-rays and the neutron transmission. Several materials including ^{139}La, ^{81}Br, ^{111}Cd, ^{93}Nb, ^{108}Pd and ^{124}Sn were used as the targets. Capture γ-rays emitted from the target were detected with annular BaF_2 scintillation counters as shown in Fig. 3. A typical helicity asymmetry of neutron capture γ-rays on ^{139}La is shown in Fig. 4. Neutrons transmitted through the target were detected with a ^{10}B-loaded liquid scintillator. The results from the γ-ray detection experiment and the neutron transmission experiment are consistent each other. Experimental results of the helicity asymmetry are shown in Table 1. The helicity asymmetry of the γ-ray in the case of ^{139}La target was found to be independent of γ-ray energy and also independent of emitted angle. Several theoretical explanations for the enhancement of the parity violating effect have been proposed. One of them is that the PNC arises from large interference between s- and p-wave resonances [15]. The interference term becomes remarkably large due to the long lifetime of the compound state, a small difference in the resonance energies for s- and p-wave neutrons and a large centrifugal potential barriers for p-wave neutrons. The longitudinal asymmetry (A_L) is given as

$$A_L = \frac{2xW}{E_p - E_s}\sqrt{\frac{\Gamma^n_s}{\Gamma^n_p}}, \tag{15}$$

where E_p and E_s are the resonance energies, Γ^n_p and Γ^n_s are the neutron widths of p- and s-wave resonances, respectively. The W is the weak matrix element and x is given by $x^2 = \Gamma^n_{p1/2}/\Gamma^n_p$, where $\Gamma^n_{p1/2}$ is the partial neutron width for p-wave neutrons whose total angular momentum is 1/2. The parity mixing is enhanced if s-wave and p-wave resonances are close each other. It is "a dynamical enhancement factor". For instance, in the case of ^{139}La a p-wave resonance is located at 0.734 eV and a s-wave resonance is at −49 eV. In addition, there is "a structural enhancement factor" $\sqrt{\Gamma_s/\Gamma_p}$ due to the centrifugal barrier. Overall enhancement of 10^6 gives rise to asymmetry of 10^{-1} instead of typical nucleon-nucleon parity violating asymmetry of 10^{-7}.

Another theoretical approach is proposed in which the PNC effect is caused by the interference between p- and d-wave resonances [16]. Verification of the model is not so easy becuase the cross sections of d-wave resonances are very small.

These large enhancements have attracted the attention of nuclear and particle physics, because such enhancements are also expected in T-violation.

4 CP-Violation in K° and B° decays

After the discovery of parity violation in weak decay processes, it had been believed that the weak interactions were at least invariant under the CP-operation. In 1964, however, an experiment by Christenson et al. first demonstrated that the long-lived state of kaon could decay to $\pi^+\pi^-$ with a branching ratio of order of 10^{-3} [17]. The fact implied that CP-invariance is violated in K-decay. The experimental arrangement is shown in Fig. 5. The degree of CP-violation is usually shown as the amplitude ratio

$$|\eta_{+-}| = \frac{Ampl(\mathrm{K}_L \to \pi^+\pi^-)}{Ampl(\mathrm{K}_S \to \pi^+\pi^-)} = (2.268 \pm 0.023) \times 10^{-3}. \tag{16}$$

The state K_S consists principally of a CP=+1 amplitude, but with a little CP=-1 amplitude, and K_L consists vice versa. CP-violating effects are also found in the decay $\mathrm{K}_L \to 2\pi^0$, in which the degree of CP-violation is shown as η_{00}. The value of η_{+-} and η_{00} depend upon the amplitudes A_0 and A_2 for finding the final-state pions in a state of isospin I=0 or 2, and we can show

$$\eta_{+-} = \varepsilon + \varepsilon', \tag{17}$$

$$\eta_{00} = \varepsilon - 2\varepsilon', \tag{18}$$

where

$$\varepsilon' = \frac{i}{\sqrt{2}} \frac{\mathrm{Im}A_2}{A_0} e^{i(\delta_2 - \delta_0)}, \tag{19}$$

and δ_0 and δ_2 are the phase shifts in the case of I=0 and 2, respectively.

The origin of CP-violation is still unknown even after 29 years of its discovery. One of the possible interpretations was proposed by Wolfenstein [18]. This is a CP-violating process of ΔS=2(superweak interaction) which transforms K_L to K_S. The superweak coupling must be about 10^{-10} of the weak coupling. In this case the possibility of CP-violation in any other interaction is essentially zero. This model predicts that ε'=0 and thus $|\eta_{+-}| = |\eta_{00}|$.

There is a model involving 6 quark flavors. In this case mixing includes a possible finite CP-violating phase angle δ. It is the standard model proposed originally by Kobayashi and Maskawa. [19] A finite value for $\varepsilon'/\varepsilon \simeq 10^{-2} \sim 10^{-4}$ is expected in this model. The model predicts CP-violation in other systems, for example in the B-meson decay.

Recently, measurements of the ratio ε'/ε were performed at CERN and Fermilab. The results were

$$\mathrm{Re}(\varepsilon'/\varepsilon) = (3.1 \pm 1.1) \times 10^{-3} \quad \text{at CERN, 1988 [20]}, \tag{20}$$

$$\mathrm{Re}(\varepsilon'/\varepsilon) = (7.4 \pm 5.2[\mathrm{stat.}] \pm 2.9[\mathrm{syst.}]) \times 10^{-4} \quad \text{at Fermilab, 1993 [21]}. \tag{21}$$

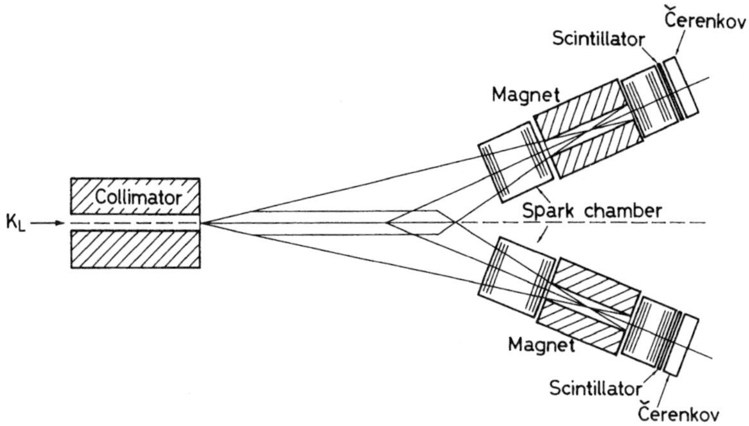

Fig. 5. Schematic view of the experimental arrangement for the measurement of $K_L^0 \to \pi^+\pi^-$ decay by Christenson et al. [17].

The $K_L^0 \to \pi^0 e^+e^-$ decay is a CP-violating process, if the e^+e^- pair is produced through Z^0 or one photon. The standard model predicts that the contribution to this decay mode from ε' is comparable to or larger than that from ε term. The branching ratio due to ε' is expected to be $\sim 10^{-10}$ from the values of the Kobayashi-Maskawa angles and the masses of t and c quarks. The $K_L^0 \to \pi^0 e^+e^-$ decay occurs also through a CP-conserving two photon exchange process whose branching ratio is calculated to be $10^{-11} \sim 10^{-14}$. Recently, the branching ratio of $K_L^0 \to \pi^0 e^+e^-$ was measured at Fermilab and the upper limit was determined to be 4.3×10^{-9} [22]. The measurements of the branching ratio to the level of 10^{-10} were proposed at Fermilab and at Kyoto-KEK [23]. The experimental apparatus at KEK includes an analyzing magnet, a gas Čherenkov counter and an electro-magnetic calorimeter of pure CsI scintillator. The mass resolution of π^0 and K^0 of the detector system is expected to be $5 \sim 6$ MeV. Data taking of the experiment will start in 1994.

The CP-violation effect is expected to be much larger in B-meson system than in K-decay in the standard model. Information on CP-violation in B-meson system can be provided by the time evolution of the neutral B-meson.

The cleanest way to produce B-mesons is e^+e^- collision on the peak of the resonance state (Υ) which decays into $b\bar{b}$ quark pair. There are projects of constructing asymmetric e^+e^- colliders at KEK and at SLAC. B-mesons produced by the collider are boosted along the beam direction causing them to move before decay as shown in Fig. 6. Therefore, we can measure two decay vertex points separately and observe the time evolution of neutral B-mesons. CP-violation can be observed due to the interference between decay amplitudes of B^0 and $\bar{B}^0$ in B^0-$\bar{B}^0$ mixing. We can find the CP-violation effect in

$$\frac{R(B^0(\Delta t) \to \psi K_S) - R(\bar{B}^0(\Delta t) \to \psi K_S)}{R(B^0(\Delta t) \to \psi K_S) + R(\bar{B}^0(\Delta t) \to \psi K_S)} = \sin(\Delta m \cdot \Delta t) \cdot \sin(2\delta), \tag{22}$$

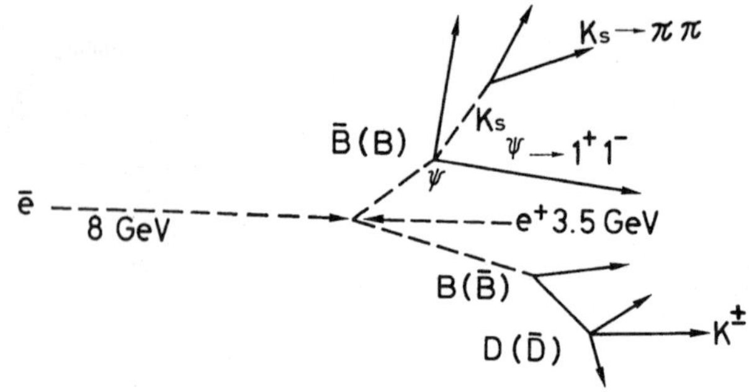

Fig. 6. Decay of B-mesons produced by an asymmetric collider, which are boosted along the beam direction.

where Δm is the mass difference between two neutral B-meson mass eigenstates and δ is the CP-violating phase. The construction of the collider at KEK [24] will start in 1994.

5 Tests of Time Reversal Invariance in Nuclear β-decays

The test of time reversal invariance has become very important in recent years, because the fundamental origin of CP-violation in K^0-decay has remained unclear. The reason is that the CP- and T-transformations are connected by the CPT-theorem, which is one of the most important principles of quantum field theory, even though the CPT theorem is not so experimentally solid. Thus, CP non-invariance implies time reversal invariance violation.

The transformations of several operators under parity and time reversal are shown in Table 2.

Table 2. The transformations of several quantities under spatial inversion P and time reversal T.

	T	P	
$\vec{\sigma} \cdot \vec{B}$	+	+	Magnetic Moment
$\vec{\sigma} \cdot \vec{E}$	−	−	Electric Dipole Moment
$\vec{\sigma} \cdot \vec{p}$	+	−	Longitudinal Polorization
$\vec{\sigma} \cdot (\vec{p_1} \times \vec{p_2})$	−	+	Transverse Polorization
$\vec{\sigma_1} \cdot (\vec{p} \times \vec{\sigma_2})$	−	−	Correlation of Two-spins and a Momentum
$\vec{\sigma_1} \cdot (\vec{p} \times \vec{\sigma_2}) \cdot (\vec{p} \cdot \vec{\sigma_2})$	−	+	Five Vector Correlation

$\vec{\sigma}$: spin
$\vec{p}$: momentum

In this section, recent topics on the parameters for $\vec{\sigma} \cdot (\vec{p_1} \times \vec{p_2})$ and $\vec{\sigma_1} \cdot (\vec{p} \times \vec{\sigma_2})$ in β-decays are presented.

In nuclear β-decays the test of T-invariance has been performed through measurements of the triple T-odd angular correlations $D{\cdot}\vec{I}{\cdot}(\vec{p_e}\times\vec{p_\nu})/E_eE_\nu$ and $R{\cdot}\vec{I}{\cdot}(\vec{p_e}\times\vec{\sigma_e})/E_e$, where $\vec{I}$, $\vec{\sigma_e}$, $\vec{P_e}$ and $\vec{P_\nu}$ are the nuclear spin vector, the electron spin vector and momenta of electron and neutrino, respectively. The D-parameter is odd under time reversal and even under parity, while the R-parameter is odd under both time reversal and parity.

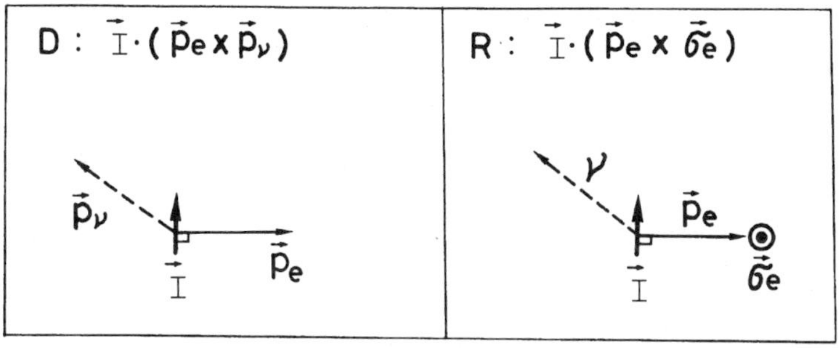

Fig. 7. Schematics of D- and R-parameters in nuclear β-decays.

5.1 Measurements of $D \cdot \vec{I} \cdot (\vec{p_e}\times\vec{p_\nu})/E_eE_\nu$ in nuclear β-decays

The D-parameter is an interference term between imaginary components of vector and axial-vector couplings [25], if tensor and scalar couplings are small.

$$D \propto \mathrm{Im}(C_S C_T^* - C_V C_A^* + C_S' C_T'^* - C_V' C_A'^*). \tag{23}$$

A Princeton group [26] measured the D-parameter in the nuclear β-decay ^{19}Ne $\rightarrow$ ^{19}F + e$^+$ + ν, which is a mixture of Fermi ($\mathrm{M_F}$) and Gamov-Teller ($\mathrm{M_{GT}}$) transitions. Polarized ^{19}Ne atomic beams were captured in a holding cell, where decay positrons and recoil ions were observed. Positrons were measured with scintillators. Recoil F$^-$ ions were accelerated by electric field and struck the inside surface of the cell emitting secondary electrons which were accelerated into an electron multiplier (Fig. 8). The value for the asymmetry parameter is $D = 0.0004 \pm 0.0008$. It is consistent with zero, and, therefore, consistent with T-invariance, if the final state interaction is negligible. The D-parameter from the effect of final state interaction (D_{FSI}) is estimeted to be as small as 2×10^{-4}, since the effect vanishes in leading order owing to a cancellation [27]. Improvement of the sensitivity up to $5{\times}10^{-5}$ will be done by the Princeton group in near future by means of higher polarization and higher intensity of ^{19}Ne.

Measurements of the D-parameter in β-decay of the polarized free neutron were performed at the Institute Laue-Langevin (ILL), Grenoble [28] and at Leningrad Nuclear Physics Institute, Gatchina [29]. In these cases, the triple correlation among the neutron polarization vector, momentum vectors of electron and anti-neutrino were observed(Fig. 9). Cold neutrons were used for the experiments. Electrons were detected

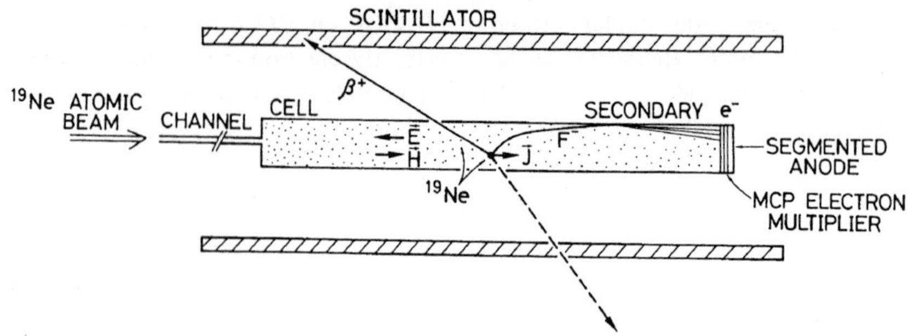

Fig. 8. The experimental arrangement of the measurement of the D-parameter in nuclear β-decay of ^{19}Ne. (see the text)

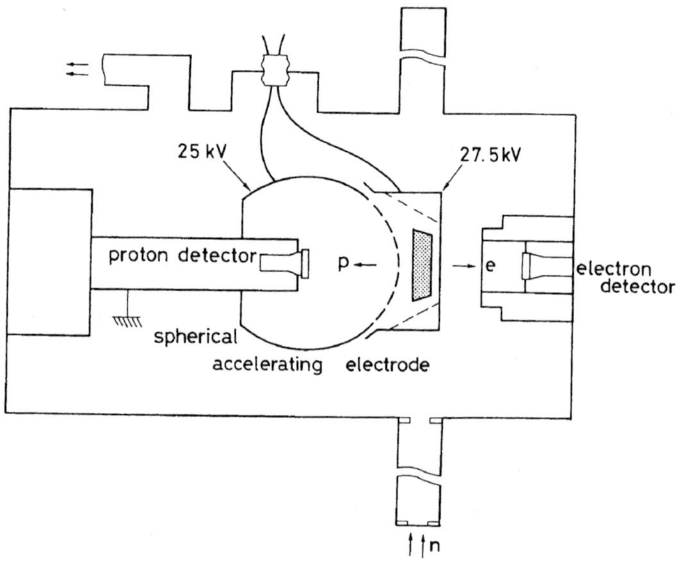

Fig. 9. The experimental arrangement of the measurement of the D-parameter in the β-decay of the polarized free neutron at Leningrad.

by scintillators. Recoil protons after drifting through field free region were accelerated and counted by scintillation detectors. The experimental results are

$$D = (-1.1 \pm 1.7) \times 10^{-3} \text{ at ILL, 1976,} \tag{24}$$

$$D = (2.2 \pm 3.0) \times 10^{-3} \text{ at Leningrad, 1978 .} \tag{25}$$

The values are consistent with zero. The value of D_{FSI} is about 2×10^{-5} in the case of neutron β-decay. It is much smaller than other nuclear β-decays. Better sensitivity will be obtained in ILL and St. Petersburg using high intensity ultra cold neutrons.

T-violation arises only in second order for nuclear β-decay in the standard model. Thus, the value for the D-parameter is order of 10^{-10} which is much lower than sensitivities of the current experiments. On the other hand, left-right symmetric models, in which a right handed vector boson is introduced, predict that T-violation occurs in the first order and the D-parameter is $\lesssim 10^{-4}$. [27] The value will be compared with experimental results improved in near future.

5.2 Measurements of $R \cdot \vec{I} \cdot (\vec{p_e} \times \vec{\sigma_e})/E_e$ in nuclear β-decay

The $R{\cdot}\vec{I}{\cdot}(\vec{p_e}\times\vec{\sigma_e})/E_e$ involves the nuclear spin $\vec{I}$, the spin of electron $\vec{\sigma_e}$ and the electron momentum $\vec{p_e}$ in β-decay. It arises from an interference between imaginary part of the scalar and axial-vector couplings, in the case of admixture of Fermi and Gamow-Teller interactions.

$$R \propto \mathrm{Im}(C_S C_A'^* + C_S' C_A^*). \tag{26}$$

In the pure V-A interaction it should be zero. Therefore, nonvanishing R-parameter shows violation of the time reversal symmetry in the scalar coupling.

The Princeton group [30] measured the parameter in the β-decay of $^{19}\mathrm{Ne} \rightarrow {}^{19}\mathrm{F} + e^+ + \nu$ with a scattering polarimeter of positrons outside the cell, utilizing the same atomic beam apparatus as used for the measurement of the D-parameter. The value obtained is

$$R = (7.9 \pm 5.3) \times 10^{-2}. \tag{27}$$

Accuracy of 10^{-3} could be realized with future improvement of the polarimeter.

In the standard Higgs model, C_S should be $\frac{m_e m_{u,d}}{m_H^2} \sim 10^{-8}$. However, if there is exta-Higgs doublets, imaginary scalar coupling is in the first order and $R = 0.1$ may be possible [27].

Recently, the R-parameter for the decay of $^8\mathrm{Li} \rightarrow {}^8\mathrm{Be}$ (2.9 MeV)+$e^- + \bar{\nu}_e$ was measured at ETH in Zurich [31]. It is the transverse polarization of electrons emitted in the decay of polarized ^{8}Li in the plane perpendicular to the axis of the nuclear spin(Fig. 10). Since there is an isospin change (ΔT=1) in this transition, Fermi matrix element vanishes, and essentially pure Gamow-Teller transition can be measured. Therefore, the largest contribution to R comes from an imaginary axial vector-tensor interference term as

$$R \propto \mathrm{Im}(C_T C_A'^* - C_T' C_A^*) \sim \mathrm{Im}(C_T + C_T'). \tag{28}$$

The value should be zero in standard model. However, the exchange of spin-2 bosons of coupling to spin-0 leptoquarks could generate a charged weak tensor interaction.

Polarized ^{8}Li nuclei were produced by the polarization transfer in the reaction ^{7}Li(d,p)^{8}Li initiated by 10 MeV deuterons. Transverse components of the electron polarization were analyzed by large angle Mott scattering in the electrostatic field of high Z nuclei (Fig. 11). The end point energy of electrons is $E_0 \approx$14 MeV. The obtained result $R = 0.000 \pm 0.004$ is consistent with the time-reversal invariance and provides the first direct determination of the imaginary charged weak tensor interaction

$$\mathrm{Im}\frac{C_T + C_T'}{C_A} = 0.000 \pm 0.012. \tag{29}$$

This is the most precise determination of the transverse spin of leptons emitted from polarized particle/nuclei. Fortunately, the effect of final state interaction is rather small ($\sim 7 \times 10^{-4}$) because of high electron energy. Further improvement of the value of R to 1×10^{-3} is expected in near future.

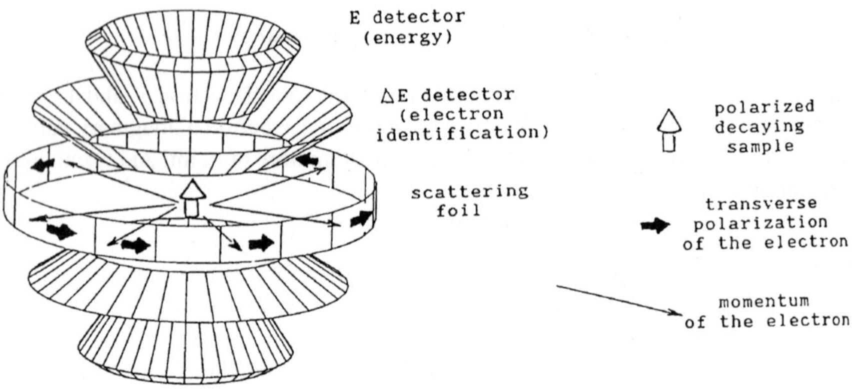

Fig. 10. Schematic view of the polarimeter for the measurement of the R-parameter in the decay of polarized ^{8}Li at ETH [31].

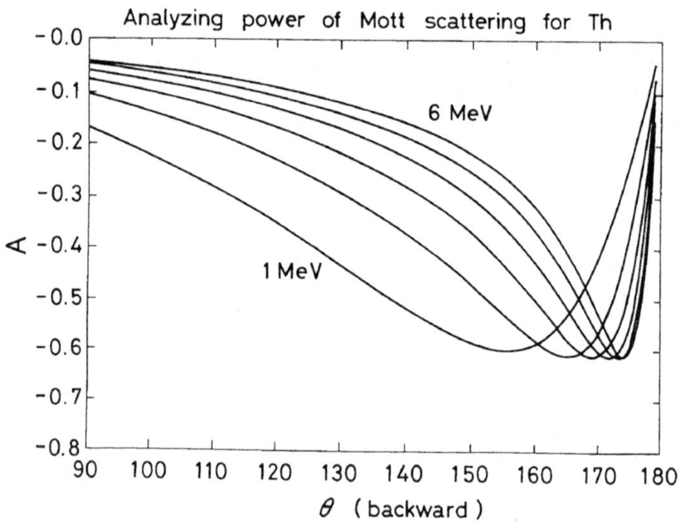

Fig. 11. Analyzing power of the Mott scattering of the electron for Thorium (point approximation).

6 Muon Polarization in $K^+ \longrightarrow \pi^0\mu^+\nu$

The muon polarization normal to the decay plane (P_T) in $K^+ \to \pi^0\mu^+\nu(K_{\mu3})$ decay is a T-violating parameter, that is

$$P_T = \frac{\vec{\sigma_\mu} \cdot (\vec{p_{\pi^0}} \times \vec{p_\mu})}{|\vec{p_{\pi^0}} \times \vec{p_\mu}|}, \tag{30}$$

where $\vec{\sigma_\mu}$ and $\vec{p_{\mu(\pi^0)}}$ are the muon spin vector and the muon (pion) momentum vector, respectively [32]. It is shown that P_T is proportional to an imaginary part of the parameter ξ which is defined as a ratio of two effective form factors, F_+ and F_- in the $K_{\mu3}$ matrix element ($\xi(q^2) \equiv F_-(q^2)/F_+(q^2)$). The P_T in $K_{\mu3}$ decay does not have contribution from standard model and higher order effects are negligibly small. The superweak model and the left-right symmetry model also predict zero P_T, while scalar interaction models like Weinberg's multi-Higgs doublet model and leptoquark models predict non-zero P_T. Recent calculations [33] based on the Weinberg's three Higgs doublet model [34] predict P_T of as large as 10^{-3}, with the present constraints from neutron electric dipole moment d_n and ε in kaon decay. The final state interaction mainly due to electromagnetic interaction between the pion and muon is $\sim 10^{-6}$ in $K^+ \to \pi^0\mu^+\nu$ decay, while $\sim 10^{-3}$ in $K^0 \to \pi^-\mu^+\nu$.

An experiment to measure P_T was carried out at BNL in 1983. They used in-flight decay of K^+s with an unseparated beam and obtained $P_T = (-3.1\pm5.3)\times10^{-3}$.

A new experiment to search for P_T of muon in $K^+_{\mu3}$ decay at rest is in progress at KEK [32, 35]. The muons from $K^+_{\mu3}$ decay of stopped positive kaons are momentum analyzed by a toroidal magnet with 12 gaps and stopped in polarimeters made of alumimum, in which polarization is fully held (Fig. 12). Muon polarization is determined by detecting positrons in the $\mu^+ \to e^+\nu\bar{\nu}$ decay which are emitted preferentially along the muon spin. Neutral pions are identified by a photon detector surrounding the target with large solid angle. In this experiment a sensitivity of 5×10^{-4} to P_T will be obtained.

7 Electric Dipole Moment of Neutron

The existence of an electric dipole moment (EDM) of the neutron implies violation of both T- and P-invariances. Since the electric dipole moment can be measured with good precision, it provides a very sensitive test of T-invariance. The EDM of the neutron has been searched for after Purcell and Ramsey pointed out the importance of the measurement in 1950 [36]. In the early stage of this kind of experiments, Ramsey et al. constructed a neutron beam resonance apparatus with a strong electric field applied in the region between two separated oscillating fields [37]. In their experiments a reactor was used as a source of thermal neutrons. Then, in order to make the experiment more sensitive, cold neutron of 180 m/sec were selected and polarized by means of magnetized iron mirrors[38]. In this method transversely polarized neutron beams traverse a region in which there is a homogeneous magnetic field $\vec{B_0}$ with a parallel and reversible static electric field $\vec{E}$ (Fig. 13). In the absence of the electric field, neutrons precess at the Larmor frequency corresponding to the strength of the magnetic field. The frequency can be measured by oscillating magnetic fields at the end of the homogeneous field region. The

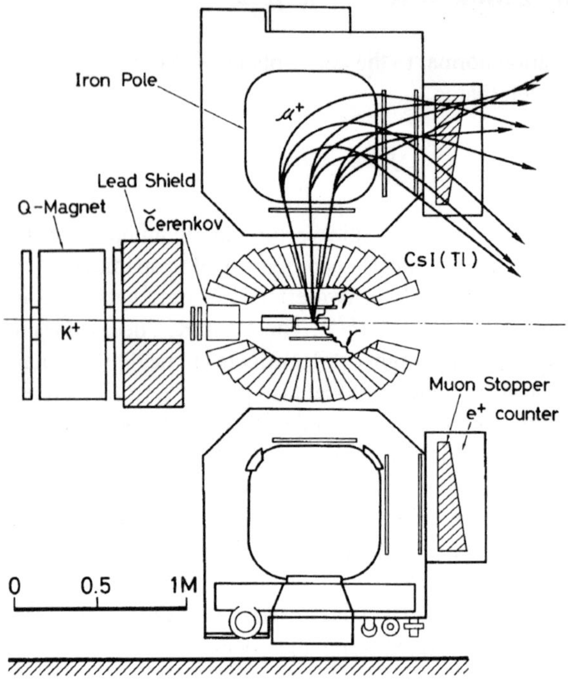

Fig. 12. Schematic view of the experimental setup for the measurement of T-violating muon polarization in $K^+ \to \pi^0\mu^+\nu$ decay using stopped kaon at KEK [32].

oscillating magnetic fields, which are perpendicular to $\vec{B}_0$, introduce spin-flip transition of neutrons. If the neutron has an EDM associated with its spin, the presence of the electric field would cause a shift of the frequency of the Larmor precession according to the expression for the Hamiltonian:

$$H = -\vec{\mu} \cdot \vec{B} - \vec{p} \cdot \vec{E}, \tag{31}$$

where $\vec{\mu}$ is the magnetic dipole moment, $\vec{B}$ the magnetic field, $\vec{p}$ the EDM, and $\vec{E}$ the applied electric field. The frequency of the precession is the transition energy between the upper and lower spin states given by

$$h\nu = -2\mu B - 2pE. \tag{32}$$

The EDM can be obtained by the measurement of the precession frequency with the electric field parallel and antiparallel to the magnetic field. The experimental result of neutron EDM measurement using this method at ILL was $(4\pm15) \times 10^{-25} e\cdot\text{cm}$ [39].

Recent experiments on the neutron EDM involve the ultra cold neutron (UCN), whose velocity is less than 7 m/sec, trapped in a box by neutron reflection on the surface of the box. Two groups are engaged in the EDM search with UCN using reactors. One of them is at the Leningrad Institute of Nuclear Physics, Gatchina and the other one is at ILL, Grenoble.

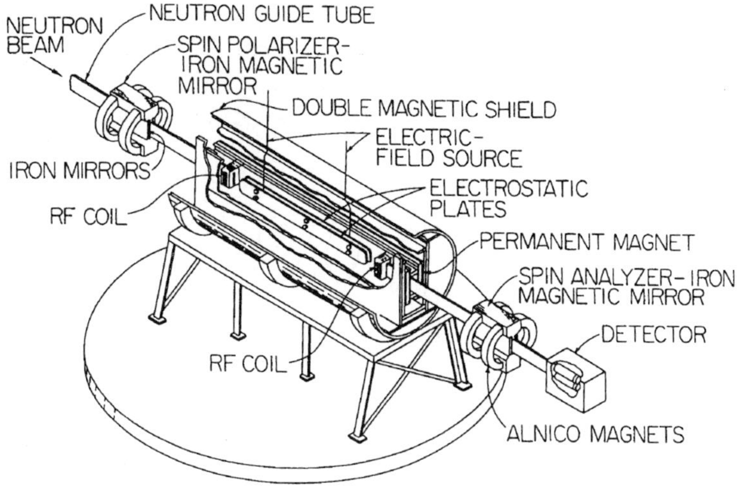

Fig. 13. Schematic view of the apparatus used by Dress et al. to measure the EDM of the neutron using polarized neutron beam [39].

Cold neutrons are produced in a liquid deuterium moderator in the experiment at ILL. The UCNs are obtained by using a turbine developed by Steyerl et al. which slows down the neutron velocity [40]. Polarized UCNs are guided and stored in a cylinder of 20 cm in diameter and 10 cm high with the end plates made of metallic beryllium. After the neutron bottle is filled with UCNs, a shutter is closed, and UCNs are stored for about 100 sec. The oscillating field is applied at the beginning and at the end of storage. The precession frequency is measured during their return passage through a polarization analyzer after the shutter is opened (Fig. 14).

The stored UCNs offer two particularly important advantages. The neutron magnetic resonance curve with UCNs is almost 10^4 times narrower than in the beam experiment. In addition, it is not difficult to eliminate an apparent magnetic field ($=-\vec{v}\cdot\vec{E}/c^2$), since the average value of v is very small. Recent experimental results are:

$$(-0.3 \pm 0.5) \times 10^{-25} e\cdot\text{cm} \qquad \text{at ILL, 1990 [41],} \tag{33}$$

$$(0.26 \pm 0.42[\text{stat.}] \pm 0.16[\text{syst.}]) \times 10^{-25} e\cdot\text{cm} \text{ at Leningrad, 1992 [42].} \tag{34}$$

Further improvement can be done by precise monitoring of a magnetic field change correlating with electric field. In addition, higher density UCN will help better statistics.

Golub and Pendlebury proposed a method to cool the neutrons by super-fluid helium [43]. On the intersection of neutron dispersion relation with that of superfluid ^{4}He quasi-particles, phonons in super helium take away the heat of incoming cold neutrons (Fig. 15). The advantage of this method is that the density of the UCN increases with time, because the UCN are trapped and stored in a bottle. The method allows helium to be near the cold source. Therefore, a spallation neutron source can be used for production of UCN, since the source of spallation neutrons is small and large solid angle can be covered by a helium converter. Such a UCN source was proposed at KEK [44]. In

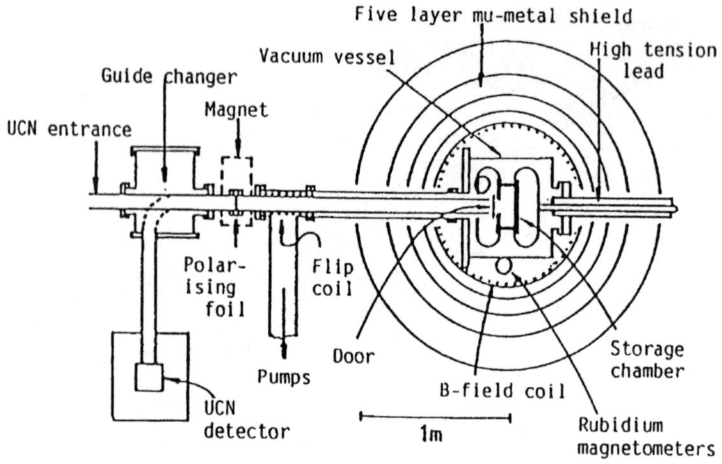

Fig. 14. Schematic view of the apparatus to search for the neutron EDM using UCN at ILL [41].

order to test the feasibility of the method, a bottle of helium II cooled by a ^{3}He cooling system was constructed and tested at JAERI. It was confirmed that the UCNs were produced on the intersection of the dispersion curves of the neutron and the superfluid ^{4}He quasi-particle at 8.78±0.06 Å as predicted (Fig. 15) [44]. It was the first success of the observation of UCN production with monochromatic cold neutrons. The method is expected to improve the EDM measurement.

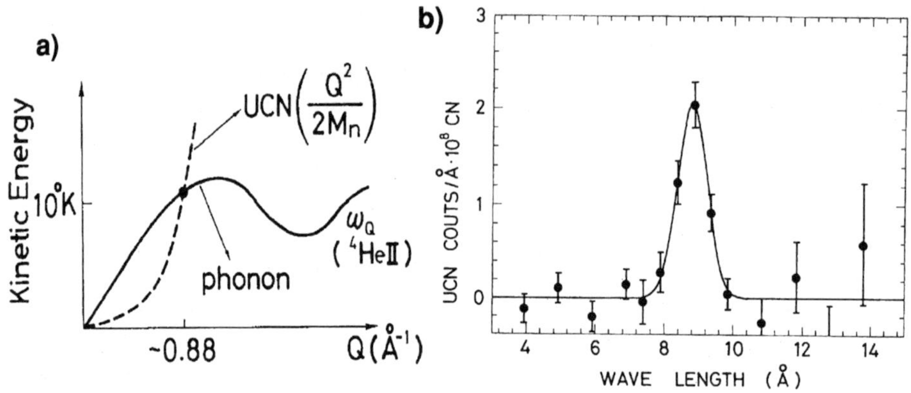

Fig. 15. (a) The neutron dispersion relation and that of a superfluid ^{4}He quasiparticle ω_Q. There is an intersection between both curves at 8.78±0.06 Å. (b) Counting rate of UCN produced in superfluid ^{4}He versus the wave length of incident neutrons measured at JAERI [44].

The theoretical predictions for the value of the neutron EDM are $10^{-19} \sim 10^{-33}$

$e$cm. Several predictions have already been abandoned due to the disagreement with experimental results. For example, a milliweak force predicted values larger than $10^{-23}e$cm. Some of the recent predictions by the popular models are:

$$\begin{aligned} \text{Kobayashi-Maskawa}: &\ 10^{-31} \sim 10^{-32}\ e\cdot\text{cm}, \\ \text{Super-symmetry}: &\ \sim 10^{-26}\quad e\cdot\text{cm}, \\ \text{Left-right Symmetry}: &\ \sim 10^{-26}\quad e\cdot\text{cm}, \\ \text{Multi-Higgs boson}: &\ \sim 10^{-26}\quad e\cdot\text{cm}. \end{aligned}$$

8 Electric Dipole Moments of Atoms

Electric dipole moments (EDM) of atoms are due to EDMs of electrons and nucleons. The EDM of an atom (d_A) can be written as

$$d_A = R_e d_e + R_n d_n, \tag{35}$$

where R_e and R_n are the enhancement/suppression factors of the EDMs for electrons and nucleons, respectively, which depend on the atoms. The d_e and d_n are the EDM of an electron and a nucleon. Schiff showed that the EDM of a non-relativistic atom vanishes even in the case that the atomic constituents have EDMs (Schiff's theorem). It is based on two assumptions [45]:
1. the atom consists of nonrelativistic particles, which interact only electrostatically,
2. the distribution of the electric dipole moment of each constituent is identical to its charge distribution.

The theorem works well for the ground state of the hydrogen atom, whereas, it fails badly for some atoms. For instance, there are enhancements of the EDM of a constituent by 2 orders of magnitude in heavy atoms.

8.1 EDM of electron (d_e)

States with opposite parities can mix with each other through P-violating interactions and cause enhancement of EDM [45]. Since the first excited state of hydrogen, $2S_{1/2}$ and $2P_{1/2}$ are splitted only by the Lamb shift, the enhancement factor is very large, that is, $\Delta E_{\rm LS}/\Delta E_{\rm Lamb} \sim 1/\alpha$=137. Enchancement occurs also in heavy atoms with an unpaired electron. In such an atom a valence electron feels an unshielded strong Coulomb field, when it comes close to the nucleus. In that case the electron velocity is comparable to the velocity of light in the inner core region of a heavy atom. Thus, contribution of d_e to d_A is enhanced strongly, because the electric dipole interaction makes the mixing between opposite-parity states. In the case of the Cs atom, the enhancement factor is about 100, which is a theoretically reliable value because of a tightly bound electronic core.

For atoms with paired electons, total electron EDMs are zero naively, but a hyperfine interaction prevents complete cancellation.

The origin of the EDM of the electron has been discussed in various models [45]. If neutrinos are massless, no T-violating coupling occur among leptons in standard model.

On the other hand, the supersymmetric models predict $d_e \lesssim 10^{-2} d_n$ if masses of all the SUSY particles are almost the same. There are many predictions on magnitude of electron EDM using multi-Higgs model. Among them, Barr and Zee show a possibility of a substantial EDM at the level of the present experimental sensitivity [45].

Contrary to other models, the leptoquark models predict d_e larger than d_n, in which leptoquarks are bosons of spin=0 or 1 that turn leptons into quarks and vice versa.

8.2 Nuclear EDM

Since T-violation takes place in the quark sector in gauge models, the EDM of atom arises mainly from the nuclear EDM in usual cases, and that of electrons is expected to be at least three orders of magnitude less.

Because of screening by electron shell, atomic EDM generated by the nuclear EDM vanishes. For heavier atoms, due to relativistic corrections, usually it becomes $10^{-2} \sim 10^{-4}$ of nucleus. However, there are some enhancement factors which can compensate the above factor. Sometimes enhancement factors become 20$\sim$100 caused of the interaction of an unpaired nucleon with a nuclear core. It is a dynamic origin of the enhancement.

Nuclear structure effects are other causes of enhancement of atomic EDM. If two nuclear states with opposite parity and the same spin are so close energetically, significant enhancement can be expected due to mixing of nuclear levels. Possible enhancements of 10$\sim$1000 have been suggested by Haxon and Henley for specific nuclei [46]. In cases of some bi-atomic molecules, very high internal electric field can become a cause of enhancement of observed EDM. The amplification factor for the molecule of ^{161}DyI is expected to be almost 10^6 [47].

8.3 Experimental search for parmanent EDM of atoms

Experiments to search for the EDMs of ^{199}Hg and ^{129}Xe are being performed by a Seattle group [48, 49]. The ground state of ^{199}Hg has an electric configuration of $^1S_0$ and nuclear spin of $I = 1/2$. Thus, it is specified by nuclear-spin projection quantum number $M_I = \pm 1/2$. The ^{199}Hg nucleus becomes polarized through optical pumping using 253.7 nm resonance light from ^{204}Hg. The polarization is driven by an oscillating magnetic field (B_x) to precess at frequency ω=18.0 Hz which is close to the Larmor frequency (Fig. 16). Phase shifts in the NMR frequency produced by applied external electric fields (5$\sim$11 kV/cm) were measured. The difference of the phase shift with the electric field of opposite directions is sensitive to the EDM. A recent experiment shows EDM<$5\times10^{-27}e$cm. Even though the electric field experienced by the nucleus is screened by electrons, the limit on nuclear EDM is lower than that on neutron EDM. The EDM of ^{199}Hg is also sensitive to an intrinsic electron EDM with the suppression factor $R_e = -1.4 \times 10^{-2}$ [45].

They also measured the atomic EDM of ^{129}Xe and got [45]

$$d_A(^{129}\mathrm{Xe}) = (-0.3 \pm 1.1) \times 10^{-26} e\cdot\mathrm{cm}. \tag{36}$$

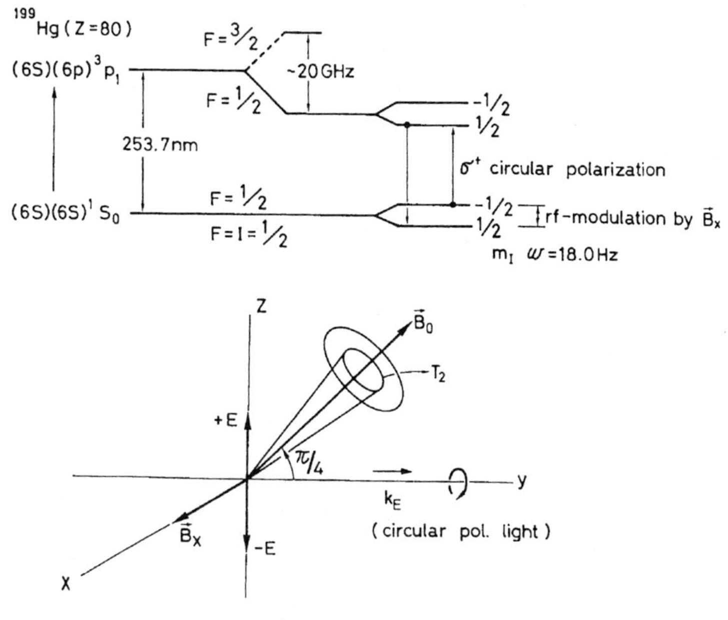

Fig. 16. Scheme of the measurement of ^{199}Hg EDM at Seattle. (see the text).

The EDM of the ground state of ^{133}Cs was measured using a two-laser method that does not require the presence of external magnetic field [50]. Light from a circularly polarized single-mode laser excites the $6S_{1/2}(F = 3) \to 6P_{1/2}$ transition in Cs in the presence of electric field. Optical pumping, spin exchange and polarization transfer through the excited state create an orientation of the $6S_{1/2}(F = 4)$ level. The measured value d_A(Cs)=$(-1.8 \pm 6.7 \pm 1.8) \times 10^{-24} e$cm implies that the electron EDM is

$$d_e = (-1.5 \pm 5.5 \pm 1.5) \times 10^{-26} e\cdot\text{cm}. \tag{37}$$

The EDM of the tallium atom was measured by means of the observation of change in the NMR frequency with the reversal of the direction of electric field in a tallium floride molecular beam. For tallium, where the $6^2P_{1/2}$ state mixes with $7^2S_{1/2}$, enhancement (R_e) of 500$\sim$700 has been predicted. The current value of d_e from an experiment on the EDM of tallium is

$$d_e = (0.1 \pm 3.2) \times 10^{-26} e\cdot\text{cm} \quad [51]. \tag{38}$$

9 T-violation Tests with Resonance Neutron Absorption

Possibility of the test of time reversal symmetry violation has been searched for at KEK, using large enhancement which happens for parity violation effects in neutron radiative resonance capture reactions [52]. As mentioned in Section 3, in the case of the neutron radiative capture reaction on ^{139}La, the helicity asymmetry is about 10^{-1} on a p-wave resonance of 0.734 eV. It can be explained by 2 kinds of enhancements, that is, "dynamical enhancement" and "structural enhancement". If there is the same order of resonance enhancement for T-violation as for P-violation, it is hopeful to observe the T-violation effect in neutron radiative capture reactions. T-violation can be observed through measurements of T-odd spin correlations involving neutron absorption reactions. The measurement of the asymmetry of the transmitted neutron in the reaction of the transversely polarized neutron ($\vec{\sigma}_n$) on the transversely polarized ^{139}La nucleus ($\vec{I}$), whose spin direction is perpendicular to the neutron spin, is a simple example of the test of T-violation. The parameter can be written as $D \cdot \vec{\sigma}_n \cdot (\vec{I} \times \vec{k}_n)$. It is a P-odd, T-odd term. It looks an excellent experiment, because a remarkable enhancement factor ($\sim 10^6$) for the time reversal violating interaction is expected. In addition, there is no final state interaction because of 0° scattering. However, the following problem is pointed out. If there is a small component of $\vec{\sigma}_n$ parallel to $\vec{I}$, the real part of $\vec{\sigma}_n \cdot \vec{I}$ interaction precesses around $\vec{I}$ in the first half of the target, so that in the second half the $\vec{\sigma}_n \cdot \vec{k}_n$ interaction produces a transmission asymmetry. Although both interactions are time reversal invariant, overall effect looks as same as the time reversal violating interaction.

Yamaguchi [16, 53], Stodolsky [54], Kabir [55], Bunakov et al [56] and Calaprice [57] proposed several possible experiments of the T-invariance test. One of them is the measurement of the difference between the probability of the helicity flip (+ → −) and that of (− → +) during the transmission in the transversely polarized target (Fig. 17). It is proportional to the imaginary part of the interference between the magnetic interaction term ($B \cdot \vec{\sigma}_n \cdot \vec{I}$) and the time reversal violating term ($D \cdot \vec{\sigma}_n \cdot (\vec{I} \times \vec{k}_n)$).

$$\text{Prob.}(+ \to -) - \text{Prob.}(- \to +) = 4 \;\; \text{Im}\; BD^*. \tag{39}$$

It is a "spin detailed balance". Another example is the measurement of the difference between the probability of the spin non-flip (+ → +) of vertically polarized neutron during the transmission in the horizontally polarized (perpendicular to the beam direction) target and that of spin non-flip (− → −) in the same target. It is proportional to the real part of the interference between the strong interaction term (A) and the time reversal violating term ($D \cdot \vec{\sigma}_n \cdot (\vec{I} \times \vec{k}_n)$).

$$\text{Prob.}(+ \to +) - \text{Prob.}(- \to -) = 4 \;\; \text{Re}\; AD^*. \tag{40}$$

It looks like "backward run of the film."

For these experiments polarized ^{139}La targets are indispensable. Sizable polarization of ^{139}La nucleus has been observed in the crystal of $La_2Mg_3(NO_3)_{12}\cdot 24H_2O$ with a little Nd^{3+} by the dynamic method. However, the crystal is not useful for the La target, because the n-p scattering cross section is much larger than that of the n-La capture reaction at the p-wave resonance of E_n=0.734 eV. Recently, a new polarized La target of Nd^{3+}:$LaAlO_3$ was proposed by Takahashi et al [58]. The observed ESR line width is narrow enough to expect a resolved solid effect. An attempt to polarize the La nucleus in $LaAlO_3$ crystal using the dynamical method has been made successfully at Kyoto University.

There is another problem in the time reversal violation experiment using slow neutrons. Neutrons precess around the axis of the polarizaiton direction of the target nucleus, even in the case of zero external magnetic field [59]. Pseudomagnetic field (H^*) can be shown in the equation

$$H^* = -4\pi N_0 \mu^* P, \tag{41}$$

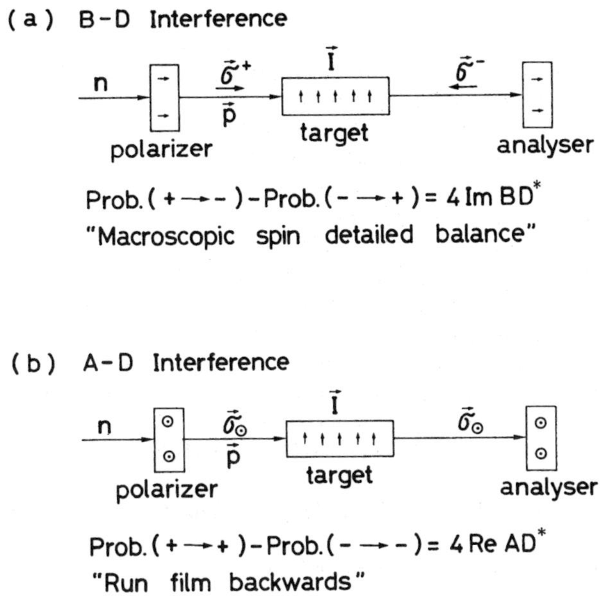

Fig. 17. Possible measurements of the T-invariance test using slow neutrons. The A, B, C, D are the coefficients of strong, magnetic, parity violating and time reversal violating terms, respectively (see the text).

where N_0 is the number density of nucleus and P is the polarization of nucleus. The μ^* is an effective (or hypothetical) magnetic moment that would magnetically scatter the neutron with the same probability as the nuclear interaction. From the analogy of magnetic scattering we can expect that neutrons "see" a pseudomagnetic field (H^*). In the case of $LaAlO_3$ the H^* is 2,120 KG for thermal neutrons if the La nuclei are polarized completely. It means that the neutron spin rotates 5 times in a target of 1 cm in thickness at 0.734 eV. Therefore, we must compensate H^* by external magnetic field. The experiment will start in 1994.

A test of time reversal invariance in the forward elastic scattering amplitude for 2 MeV polarized neutrons on aligned holmium was carried out at TUNL [60]. The experiment is sensitive to the five-hold correlation term $\vec{\sigma}_n \cdot (\vec{k}_n \times \vec{I}) \cdot (\vec{k}_n \cdot \vec{I})$ which is parity conserving but odd under time reversal, where $\vec{\sigma}_n$ and $\vec{k}_n$ are the spin and momentum of the neutron and $\vec{I}$ is the spin of ^{165}Ho. Since the five-hold correlation term is quadratic in $\vec{I}$, the target must be aligned rather than polarized — a preferential spin axis must exist rather than a spin direction. The analyzing power was found to be $(1.0{\pm}6.3)\times10^{-4}$, which is consistent with time reversal invariance.

The Kyoto-KEK-TIT group has proposed to carry out a measurement of the same parameter with polarized epithermal neutrons. In this case we can expect an enhancement of the T-violating effect on some resonances.

References

1. R.J. Hughes, *Proceedings of the Antihydrogen Workshop* (Munich, Germany 1992) p3.
2. R. Carosi et al., *Phys. Lett.* **B237** (1990) 303.
3. R.S. Van Dyck Jr. et al., *Phys. Rev. Lett.* **59** (1987) 26.
4. G. Gabrielse et al., *Phys. Rev. Lett.* **65** (1990) 1317.
5. R.J. Hughes et al., *Phys. Rev. Lett.* **66** (1991) 854.
6. T.D. Lee and C.N. Yang, *Phys. Rev.* **104** (1956) 254.
7. C.S. Wu et al., *Phys. Rev.* **105** (1957) 1413.
8. C.Y. Prescott et al., *Phys. Lett.* **77B** (1978) 347; **84B** (1979) 524.
9. E.G. Adelberger and W. Haxton, *Ann. Rev. Nucl. Parti. Sci.* **35** (1985) 501.
10. S. Kistryn et al., *Phys. Rev. Lett.* **58** (1987) 1616.
11. V.W. Yuan et al., *Phys. Rev. Lett.* **57** (1986) 1680.
12. N. Lockyer et al., *Phys. Rev. Lett.* **45** (1980) 1821; *Phys. Rev.* **D30** (1984) 860.
13. V.P. Alfimenkov et al., *Nucl. Phys.* **A398** (1983) 93.
14. Y. Masuda et al., *Nucl. Phys.* **A478** (1988) 737C; **A504** (1989) 269.
    H.M. Shimizu et al., *Nucl. Phys.* **A552** (1993) 293.
15. O.P. Sushkov and V.V. Flambaum, *Usp. Fiz. Nauk.* **136**(1982) 2 [*Sov. Phys. Usp.* **25** (1982) 1].
16. Y. Yamaguchi, *Prog. Theor. Phys.* **85** (1991) 101.
17. J.M. Christenson et al., *Phys. Rev. Lett.* **13** (1964) 138.
18. L. Wolfenstein, *Phys. Rev. Lett.* **13** (1964) 562.
19. M. Kobayashi and T. Maskawa, *Prog. Theor. Phys.* **49** (1973) 652.
20. H. Burkhardt et al., *Phys. Lett.* **B206** (1988) 129.
21. L.K. Gibbons et al., *Phys. Rev. Lett.* **70** (1993) 1203.
22. T. Yamanaka, private communication.
23. N. Sasao, private communication.
24. *KEK Report* **90-7, 90-23, 90-24** (1990), **92-3** (1992).
25. J.D. Jackson et al. *Phys. Rev.* **106** (1957) 517; *Nucl. Phys.* **4** (1957) 206.
26. A.L. Hallin et al., *Phys. Rev. Lett.* **52** (1984) 337.
27. F.P. Calaprice, World Scientific Publishing Co. Pte. Ltd. and Yamada Science Foundation (1989) pl44.
28. R.I. Steinberg et al., *Phys. Rev.* **D13** (1976) 2469.
29. B. Erozolimski et al., *Sov. J. Nucl. Phys.* **28** (1978) 48.
30. M.B. Schneider et al., *Phys. Rev. Lett.* **51** (1984) 1239.
31. M. Allet et al., *Phys. Rev. Lett.* **68** (1992) 572; *Proceedings of 10th International Symposium on High Energy Spin Physics (Nagoya)* (1992) p753.
32. Y. Kuno, *KEK Preprint* **92-207** (1993).
33. R. Garisto and G. Kane, *Phys. Rev.,* **D44** (1991) 2038.
    G. Bélanger and C.Q. Geng, *Phys. Rev.* **D44** (1991) 2789.
34. S. Weinberg, *Phys. Rev. Lett.* **37** (1976) 657.
35. J. Imazato et al., *KEK-PS Research Proposal* (1991).
36. E.M. Purcell and N.F. Ramsey, *Phys. Rev.* **78** (1950) 807.
37. J.H. Smith et al., *Phys. Rev.* **108** (1957) 120.
38. I.S. Altarev et al., *Phys. Lett.* **102B** (1981) 13.
39. W.B. Dress et al., *Phys. Rev.* **D15** (1977) 9.
40. A. Steyerl et al., *Phys. Lett.* **A116** (1986) 347.
41. K.F. Smith et al., *Phys. Lett.* **234B** (1990) 191.
42. I.S. Altarev et al., *Phys. Lett.* **276B** (1992) 242.
43. T. Golub and J.M. Pendlebury, *Phys. Lett.* **53A** (1975) 133.

44. H. Yoshiki et al., *Phys. Rev. Lett.* **68** (1992) 1323; H. Yoshiki, private communication.
45. W. Bernreuther and M. Suzuki, *Rev. Mod. Phys.* **63** (1991)313;
L.R. Hunter, *Science* **252** (1991) 73.
46. H.C. Haxton and E.M. Henley, *Phys. Rev. Lett.* **51** (1983) 1937.
47. O. Sushkov et al., *JETP* **87** (1984) 1521.
V.M. Lobashev, *Nucl. Phys.* **A478** (1988) 245c.
48. S.K. Lamoreaux et al., *Phys. Rev. Lett.* **59** (1987) 2275;
J.P. Jacobs, *Proc. Intersection between Particle and Nuclear Physics* (Tucson 1991) p981.
49. T.G. Vold et al., *Phys. Rev. Lett.* **52** (1984) 2229.
50. S.A. Murthy et al., *Phys. Rev. Lett.* **63** (1989) 965.
51. K. Abdullah et al., *Phys. Rev. Lett.* **65** (1990) 2347.
52. Y. Masuda et al., *KEK-preprint* **92-124** (1992)1, ibid. **92-123** (1992) 1.
A. Masaike, *Hyperfine Int.* **74** (1992) 139.
H.M. Shimizu, *PhD-thesis, Kyoto Univ.* (1992).
53. Y. Yamaguchi, *J. Phys. Soc. Jpn* **57** (1988) 1518 (L); ibid 1522 (L); ibid 1525 (L); ibid 2331; ibid 3339; ibid 3344.
54. L. Stodolsky, *Nucl. Phys.* **B197** (1982) 213; **B172** (1986) 5; *Proc. of Workshop on Tests of Time Reversal Invariance in Neutron Physics*, (Chapel Hill, 1987) p.12.
55. P.K. Kabir, *Phys. Rev.* **D25** (1982) 2013; *Phys. Rev. Lett.* **60** (1988) 686.
56. V.E. Bunakov and V.P. Gudkov, *Nucl. Phys.* **A401** (1983) 93.
57. F.P. Calaprice, private communication.
58. Y. Takahashi et al., private communication.
59. H. Glättli, *J. de Phys.* **40** (1979) 629.
60. J.E. Koster et al., *Phys. Rev. Lett.* **B267** (1991) 23.

Lecture given at the JSPS-INS International Spring School, March 1993 at Shimoda

Exotic Aspects of Hadronic Atoms - Anomalous Quasi-Stabilities

Toshimitsu Yamazaki

Institute for Nuclear Study
University of Tokyo
3-2-1, Midori-cho, Tanashi, Tokyo 188, Japan

Abstract
Recently revealed, but hitherto unknown, new aspects of hadronic atoms, namely, anomalous quasi-stabilities of negative hadrons, are discussed. One is on long-lived antiprotonic helium atoms, characterized as "atomic exotic halo" and the other is on deeply bound pionic atoms, characterized as "nuclear exotic halo".

1 Introduction

The hadronic atom is a special atom composed of a negatively-charged hadronic particle X^- and a nucleus of charge $+Ze$. They are combined by the Coulomb attraction, forming Bohr orbitals with principal quantum number n and orbital angular momentum l. The binding energy is expressed by a well known Bohr formula

$$E(n,l) = -Z^2 \frac{M}{m_e} \frac{1}{n^2} Ry \tag{1}$$

where

$$Ry = \frac{m_e e^4}{2\hbar^2} = 13.6 \text{ eV}. \tag{2}$$

The energy levels are scaled by the reduced mass M of X^-, and proportional to Z^2. They are degenerate for a given n and different $l = 0, 1, 2, ..., n-1$. This $n-$fold degeneracy is characteristic of a two-body Coulomb atom with a point nucleus, and is removed when any change of this situation takes place, such as the effect of atomic electrons and of finite nuclear size.

The size of the Bohr orbital is given by

$$\langle r_X(n) \rangle = n^2 \frac{m_e}{M} \frac{1}{Z} a_B, \tag{3}$$

where

$$a_B = \frac{\hbar^2}{m_e e^2} = 0.53 \times 10^{-8} \text{cm}. \tag{4}$$

It is scaled by m_e/M and by $1/Z$. This means that a typical hadronic Bohr orbital lies somewhere between the electronic orbital and the nuclear surface.

The exotic atom is formed by a capture reaction as

$$X^- + [Z]^0 \rightarrow (X^- - [Z]^+)_{nl} + e^-, \tag{5}$$

namely, an incoming X^- ionizes one of the atomic electrons of a neutral target atom $[Z]^0$ to an ionized state $[Z]^+$ by which X^- is bound to form a state (n, l). The most likely state initially formed is often said to be given by the condition that the orbital size of X^- is equal to that of the ionized electron $\langle r_X(n) \rangle \sim \langle r_e \rangle$. If a K-shell electron is ionized, the equality

$$n^2 \frac{m}{M} = 1 \tag{6}$$

yields the famous estimate for the most probable n:

$$n = n_0 \equiv (M/m)^{1/2}. \tag{7}$$

This condition is equivalent to another condition that the binding energy of the formed X^- orbital be equal to that of the ionized electron.

The above estimate has, however, not been tested so far, because those initial states are usually *invisible*. They proceed to lower bound states via further Auger processes and eventually via radiative transitions, whichever faster. What we can observe is a series of x-ray cascade after X^- reaches a lower bound state.

The precision spectroscopy of muonic x-rays has provided detailed knowledge of the nuclear charge radius and its distribution because μ^- can cascade down to the ground state, which is located in the nuclear surface region. In the case of hadronic atoms, however, the situation is quite different; as soon as X^- feels the strong interaction from the nucleus, it is quickly absorbed by the nucleus and the x-ray cascade terminates there. The "last orbital", where the x-ray cascade terminates, is characterized by the condition that the absorption width Γ_{abs} be equal to the x-ray cascade width Γ_{rad}. Since the strong interaction has an imaginary part, whose strength is about 10 MeV, a hadronic orbital which is located as far as 10 times the nuclear surface has an absorption width as much as 10 keV, which eats up the radiative width. So, $\langle r_X \rangle = 10 R_N$ is a rough criteria for the last orbital. X-ray spectroscopy for last orbitals provides experimental information on the strong-interaction shift ΔE_{nl} and width Γ_{nl}. These are the most important sources of information on the strong interaction potential.

It is obvious that no x-ray spectroscopy is possible for inner bound states. For instance, pionic x-ray spectroscopy has found that the $4f$ (or possibly $3d$) state is the last orbital in ^{208}Pb. Then, how and where do the $2p$ and $1s$ states exist? This very natural question had not been raised presumably due to the prevailing misbelief that no such inner states do exist. Toki and Yamazaki [1,2] pointed out in 1988 that such inner states are expected to exist with narrow widths for pions and can be produced by suitable nuclear reactions.

The present situation of hadronic atoms is illustrated in Fig. 1. So far, the intermediate region has well been studied experimentally, but neither the initial region $\langle r_X \rangle \sim \langle r_e \rangle$ nor the inner region $\langle r_X \rangle < 10 R_N$ has been touched upon. Recently, the Tokyo group discovered new phenomena that the initially formed hadronic states in helium media are metastable [3–5]. These discoveries have provided the first opportunity to study the initial region. In this lecture the author describes these new developments.

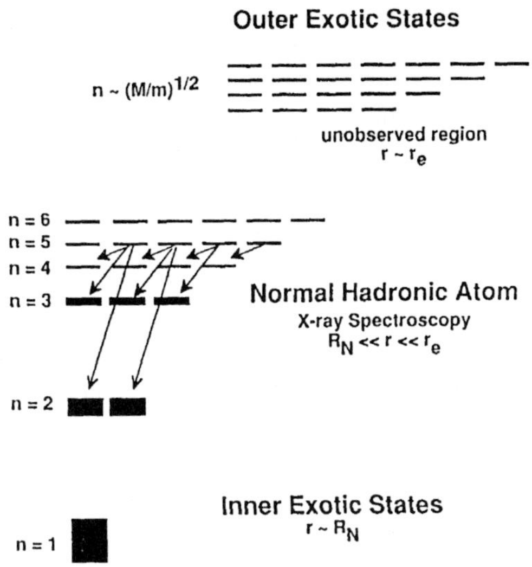

Fig. 1. Overview of hadronic atoms. The entry region in the case of helium constitutes metastable "exotic atomic halo", while the deeply bound region in heavy nuclei may probably constitute Coulomb-assisted "exotic nuclear halo" of anomalously narrow widths, if the negative hadron feels repulsion from the nucleus.

2 Discovery of Metastability of Hadronic Helium Atoms

While the SKY group of the University of Tokyo at KEK was searching for Σ hypernuclei to be produced by (stopped K^-, π) reactions in liquid helium [6], they encountered a suprising fact that striking peaks appeared at 205 MeV/c and 235 MeV/c in a momentum spectrum of negatively-charged particles [3]. They were immediately identified as characteristic two-body decay products of K^-; $K^- \to \pi^- + \pi^0$ and $K^- \to \mu^- + \bar{\nu}_\mu$, respectively. Any negative hadron is subject to common fate to immediate death after strong nuclear capture, which usually takes place in a picosecond after hadronic-atom formation. Therefore, there is only a very little time for any negative hadron stopped in matter to undergo free decay of lifetime τ_{free}. Contrary to this belief, they found the existence of free-decay products of about 2 % branching [3].

Later they learned that old experiments in helium bubble chambers had found anomalous free decays of π^- and K^- of 2 % fraction. See Hayano *et al.* [7] for historical accounts of this problem. At that time these fractions f were used to deduce "average cascade times" in such a way that $T_{av} = \tau_{free} \times f$. The T_{av} thus deduced from this "homogeneous-process" interpretation turned out to be 10^{-10} sec, which is by two orders of magitude slower than the theoretical expectation. On the other hand, Condo [8] proposed a totally different interpretation (can be called "inhomogeneous-process" model) that negative hadrons of this fraction are trapped by long-lived atomic orbitals.

Condo pointed out this possibility with reasoning that large-l states of the initially formed neutral hadronic atoms, $e^-X^-He^{++}$ (to be often described as X^-He^+), cannot decay by fast Auger transitions, because the $\Delta l = 1$ transition energies are smaller than the electron ionization energy, $I_0 = 25$ eV, of helium atom. Furthermore, the neutrality of this atom including one electron makes the Stark mixing with other helium atoms less likely. Russell [9] made theoretical calculations for the structure and transition probabilities based on the variational method. He found that the near circular orbitals of X^-He^+ have long lifetimes (in the case of $\bar{p}$ the lifetime is as long as 1 μsec), but of course this calculation should apply to ideally isolated atoms, not to those in condensed matter like liquid helium. Since $\bar{p}p$ atoms are known to quench in hydrogen gas [10] as well as the metastable $2s$ μ^-He atoms [11] in helium gas, one hardly believed such metastability to persist in liquid helium.

So, this interesting possibility had remained untouched experimentally until the accidental encounter in 1989 by the SKY experiment [3]. They found the two monoenergetic components in a delayed spectrum (Fig. 2), revealing that the lifetime is around 10 nsec compared with the free lifetime of $K^{\pm}$ of 12 nsec. The overall lifetime of the trapping states was thus deduced to be 50 nsec or so. Outa *et al.* [12] found a shorter lived component from K^- absorption products.

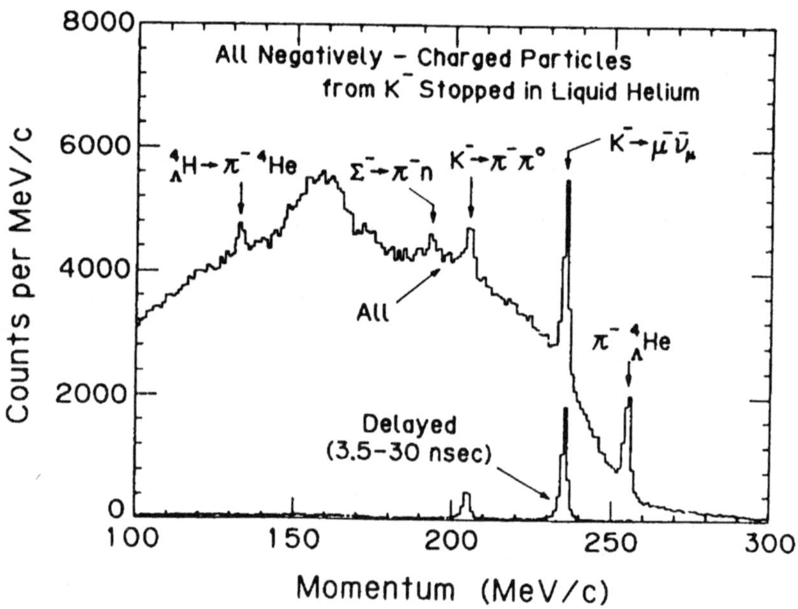

Fig. 2. Prompt and delayed momentum spectra of all negatively charged particles emitted from K^- stopped in liquid helium. The delayed spectrum revealed the existence of metastable states of overall trapping lifetime of 50 nsec [3].

Nakamura *et al.* [4] searched for a similar phenomena for π^- at TRIUMF. They found that delayed protons are emitted after π^- absorption in liquid helium. The overall trapping lifetime was determined to be 10 nsec.

What would happen for antiprotons? Spectacular phenomena might be revealed, because the mass of $\bar{p}$ is large and furthermore, $\bar{p}$ is stable in free space, and thus, has an unlimited time window. Iwasaki *et al.* [5] searched for delayed annihilation products (charged pions) after $\bar{p}$ stopping in liquid helium using an antiproton beam from KEK. They discovered that about 3 % of stopped antiprorons are subject to delayed annihilation, showing a multi-component exponential decay with an average lifetime of about 3 μsec. Antiprotons do survive as long as 10 μsec.

The observed metastability is certainly connected with the structure of X^-He^+. Yamazaki and Ohtuki [13] made further theoretical accounts of the phenomena. The metastable $\bar{p}He^+$ atom and its level structure are illustrated in Fig. 3. The initial quantum number n_0 is around 38 in the case of $\bar{p}$ in ^4He. The electron in its $1s$ state is quantum-mechanically distributed, while the $\bar{p}$ of $(n, l) = (38, 37)$ is localized and is circulating around the nucleus with a frequency much smaller than the electron frequency. The electron cloud tends to follow the motion of $\bar{p}$, and is slightly polarized in the direction opposite to $\bar{p}$. This $\bar{p}He^+$ atom looks like i) an exotic helium atom with replacement of one e^- by $\bar{p}$ at high quantum states keeping the same energy, ii) an exotic hydrogen with a "giant pseudo proton" of $\bar{p}He^{++}$, and iii) an exotic diatomic molecule with two "nuclei", one is He^{++}, the other, $\bar{p}$, has a charge of -1. We called this exotic atom "atomcule". Its presence in condensed matter and its reaction with other atoms/molecules are very intriguing subjects to study. The KEK experiment raised a number of questions.

1. Why are the metastable states stable against collisions in liquid helium? What is the effect of dense medium on the lifetimes and trapping fractions?
2. Why are the metastable states populated? What gives the trapping fraction f?
3. What is the precise shape of the time spectrum? How is it explained theoretically?
4. Is there any isotope effect between ^4He and ^3He?
5. Any such metastability in other atoms?

In order to obtain further information on this exotic metastability we have to study the phenomena in various phases of helium. For this purpose a series of new experiments have been undertaken and carried out by using the Low Energy Antiproton Ring (LEAR) at CERN.

3 CERN Experiment PS205

A new experimental group PS205 was formed, which started a series of experiment at CERN in 1991. LEAR provides an ideal antiproton beam for this purpose. It is monoenergetic and low energy. This means that we can stop antiprotons in a localized space even in a gas target. With a beam of 100 MeV/c momentum the total range is about 50 g/cm^2 and its distribution is about 0.5 mg/cm^2. It is only a few cm in 3-bar helium gas! The beam spot was as small as a few mm. This fantastic beam permitted us to carry out the measurement of time spectra of annihilation pions in various gas medium.

One beam spill contains $10^8 \sim 10^9$ $\bar{p}$'s, which are slowly extracted over 1000 sec, yielding $10^5 \sim 10^6$ $\bar{p}$'s/sec. We used a small fraction of the beam as a parasitic user. Since our time window spans 50 μsec, we did not need more than 10^4 $\bar{p}$/sec. The experimental set up used in 1992 is shown in Fig. 4. The incoming $\bar{p}$ is detected by a very

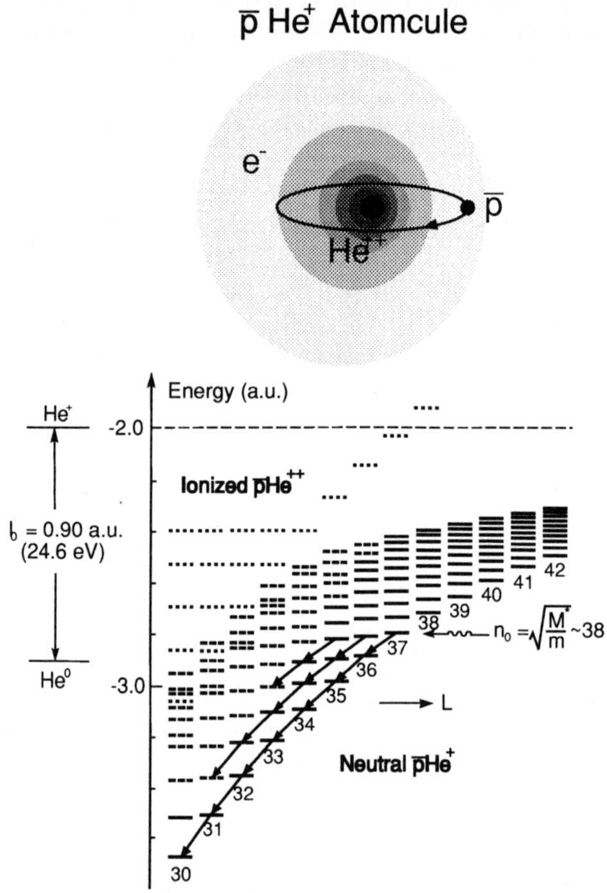

Fig. 3. Energy diagram of metastable antiprotonic helium atomcule. The levels expressed by bold lines are metastable, while those by broken lines are fast decaying states. Calculated by K. Ohtsuki. Taken from [14].

thin (100 μm thick) plastic scintillator, which is located just before a target gas chamber. Pions emitted from the target were detected by arrays of scintillator telescopes. The time between the arrival of p̄ and the emission of pions was measured by using multi-stop Time to Digital Converters (Lecroy TDC4208).

This seemed to be straightforward in principle, but in practice we had to solve the following problems. The most serious is how to remove fake delayed events. Prompt annihilation events which occur within the time gate of 50 μsec produce fake "delayed pions" ($n_{\text{fake}}/N_{\text{beam}} = N_{\text{beam}} \cdot T_{\text{gate}}$ = order of 1, while the true delayed fraction is of order of 0.03!). To cut this type of fake events we used the information of "second p̄", namely, when a second p̄ arrives within the time gate, such an event was rejected. However, since the efficiency of the counter B is not 100 %, some escaped from this rejection.

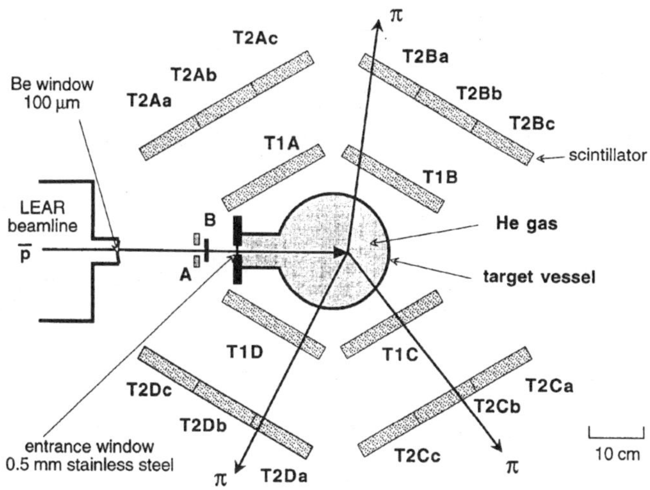

Fig. 4. Setup of the PS205 experiment at CERN-LEAR in 1992.

Another type of fake events arose from secondary e$^+$ emitted from prompt π^+ stopped somewhere in the vicinity of the target and the counters via the $\mu^+ \to$e$^+$ process with a mean life time of 2.2 μsec. To remove this we required coincidence of 2 or 3 simultaneous delayed particles.

The first CERN experiment thus carried out was very successful. We obtained time spectra of delayed annihilation of p̄ stopped in gaseous helium. A brief report of the results was published [14]. It took only one hour to obtain a good enough time spectrum (at KEK it took one week of beam time even for liquid helium!) and the fake deleyed events were completely suppressed. Some typical time spectra are shown in Fig. 5 and 6.

The most surprising observation is that the time spectrum for pure ^{4}He gas looks very much like that taken at KEK for liquid helium. The average lifetime was 3 μsec, but the good statistics allowed us to recognize that the time spectrum is not a sum of expenential functions, but shows a growth-decay type bending that is characteristic of chain decay through 2 or more metastable states. Phenomenologically, we define the average lifetime for a given time interval $t_0 < t < t_{max}$ as follows.

$$T_{av}(t_0, t_{max}) = \frac{\int_{t_0}^{t_{max}} t \cdot N(t)dt}{\int_{t_0}^{t_{max}} N(t)dt} - t_0 \tag{8}$$

The lifetime does not depend much on the pressure.

We observed a significant isotope effect between ^{4}He gas and ^{3}He gas, as shown in Fig. 5. The lifetime for ^{3}He is 13 % shorter than for ^{4}He. This observation is in good agreement with that expected from the difference in the reduced mass of p̄ (see section 7).

Another interesting finding is that the delayed component is quenched by adding a very small amount of hydrogen and other impurity gasses. The effect is shown in Fig. 6.

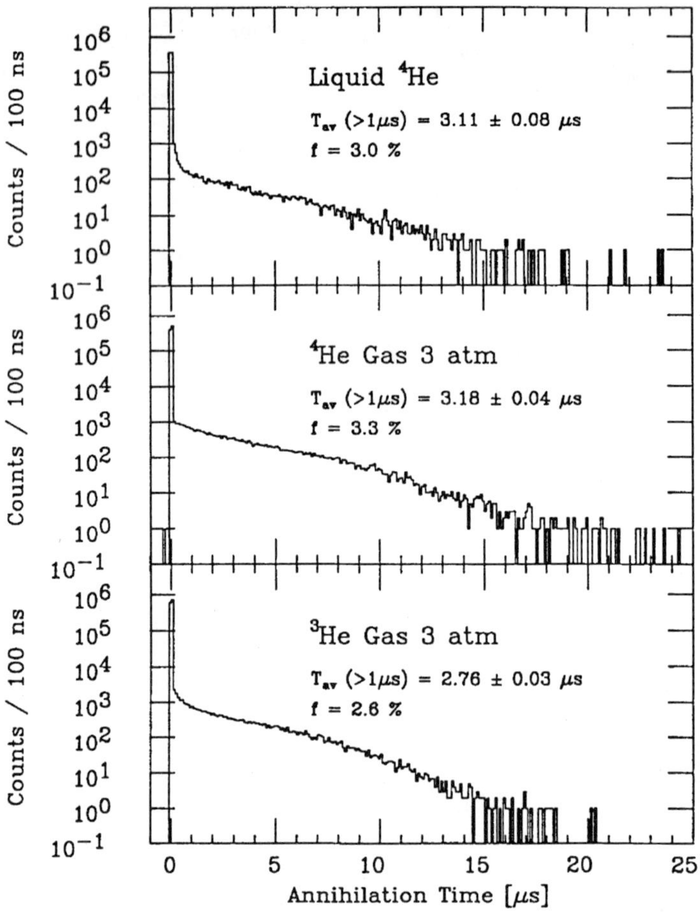

Fig. 5. Time spectra of antiproton annihilation in gas ^{4}He, gas ^{3}He, liquid ^{4}He.

The delayed fraction is unchanged, but the lifetime is shortened. This can be interpreted in terms of collision of the metastable atoms with impuriry atoms/molecules. Assuming that $\bar{p}He^+$ atoms are in thermal equilibrium, we obtain the quenching cross section to be 10^{-15} cm^2 [14].

In 1992 run the PS205 group carried out more systematic and comprehensive studies of this phenomenon [15]. They measured delayed annihilation time spectra in solid He, high-pressure liquid He, low-temperature gas He, gas He with mixture of Ne, Ar, Kr, Xe, H_2, N_2, O_2, *etc.* The longest lifetime was observed in low-temperature He gas.

In 1993 we are planning to begin laser spectroscopy for the metastable states. A high-power pulsed excimer laser combined with dye laser will be used to induce resonance transition between a metastable state and a short-lived state. The proposed methodology is described by Morita *et al.* [16].

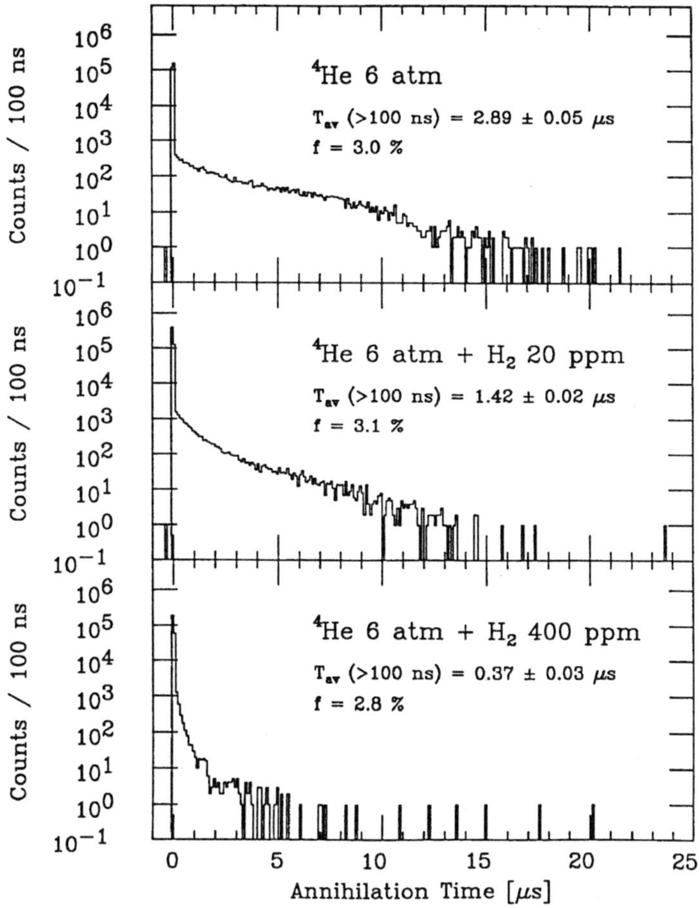

Fig. 6. Time spectra of antiproton annihilation in gas ^{4}He with mixture of H_2 gas.

4 Atomic Yrast Bands

In the following sections let us discuss the interesting features of the metastable $\bar{p}He^+$ atomcule. The most distinct aspect of this particular exotic atom that is not seen in usual atoms (even in hadronic atoms) is that it possesses large angular momenta. The metastable states can be called atomic yrast bands corresponding to the popular word for high-spin states in nuclear physics.

The usual two-body exotic atom consisting of a nucleus and an exotic particle X^- of a reduced mass M without electron contribution is described by the one-dimensional Schrödinger equation with a potential

$$U(r) = -\frac{Ze^2}{r} + \frac{l(l+1)}{2Mr^2}. \tag{9}$$

The $U(r)$ consists of an attractive long-range Coulomb potential and an repulsive short-range centrifugal potential. The potential $U(r)$ resembles a Morse potential, when the exotic particle has a large angular momentum (hence large n), and thus the system looks like a molecule, where rotational (J) and vibrational (v) quantum numbers could be uniquely assiged as $J = l, v = n - l - 1$, in the same way as Shimamura [17] showed in the case of the three-body system X^-He^+. Namely, a circular orbit ($l = n-1$) is a rotating state with a nodeless radial wavefunction, which corresponds to a vibrational quantum number $v = 0$, a next-circular state ($l = n - 2$) with a single node, corresponding to $v = 1$, and so on.

These atomic yrast states have interesting selection rules when they decay via E1 transitions; the transitions without changing the radial node are favoured, namely, $\Delta n = \Delta l = 1$. Thus, the radiative decay sequence proceeds as shown by arrows in Fig. 3, though the transitions of $\Delta n = 2$ or more have higher energies than these. This is the result of well localized radial wavefunctions in these yrast states. In the terminology of rotation-vibration we can say that a selection rule for the vibrational quantum number, $\Delta v = 0$, holds.

The Bohr orbitals do not, however, possess the usual properties of rotation and vibration; this vibrational motion is far from harmonicity and the rotational energy is not proportional to $J(J+1)$ at all (the moment of inertia is J-dependent). Here, the rotation-vibration coupling is maximal ($\hbar\omega \sim J\hbar/\mathcal{I}_{rot}$) and the moment of inertia keeps increasing with J (large centrifugal stretching). This peculiar "rotation-vibration mode" reflects the fact that the system is not like a molecule, but a Coulomb atom. In spite of this, the assignment of rotational-vibrational quantum numbers to a two-body exotic atom is a quite sufficient and useful description, although, of course, it is not a necessary one.

Now, when an electron is coupled to this large-l circular (or near-circular) orbitals, the electron motion can be treated adiabatically and the molecular approach (thus the assignment of rotation-vibration quantum numbers) does play an important role, as fully discussed by Shimamura [17]. In the adiabatic approximation the potential is now given by

$$V(R) = U(R) + \epsilon_\gamma(R) \tag{10}$$

where $\epsilon_\gamma(R)$ is the adiabatic potential arising from the electron involved. Shimamura [17] and previously Ahlrichs *et al.* [18] and recently Greenland and Thürwächter [19] solved this equation. Certainly, this molecular approach is more elegant and powerful than any other procedure starting from the atomic configurations which ought inevitably to take into account various configuration mixing effects among them. The potential shape as calculated by Shimamura is shown in Fig. 7.

Here, I would like to give some remarks on the molecular approach. Although the potential $V(R)$ looks like an interatomic Morse potential in shape, it is essentially different from the molecular case. In the present exotic helium case, the long-range attractive part comes from the Coulomb interaction and the short-range repulsion is from the centrifugal barrier, while in real molecules the long-range part comes from the molecular interatomic binding force mediated by the electron(s) and the short-range component is from the Pauli and internuclear repulsion. In the former, both the centrifugal and the Coulomb interactions together form a potential minimum where bound states are accommodated, but the electron energy $\epsilon_\gamma(R)$ is a slowly varying function which plays no essential role in binding the two objects, and only modifies the binding energies. The effect of the electron is most elegantly and accurately taken care of by the molecular approach, but physically, the binding is essentially of "atomic origin", and as shown below, most of the results of the molecular approach as derived by Shimamura [17] such as the propensity rule ($\Delta v = 0$ dominance) do exist already in the two-body system without electron.

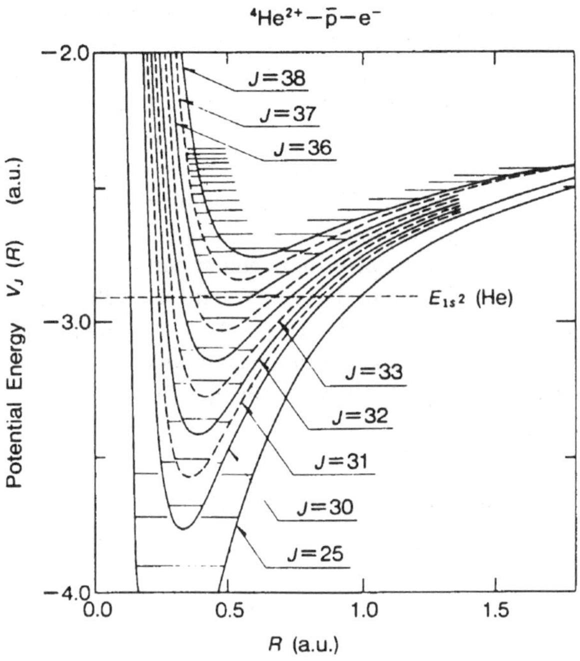

Fig. 7. Potential shape $V_J(R)$ and energy levels, as calculated by Shimamura [17].

The interesting theoretical fact that the large-l circular orbitals behave like bound states in a Morse-type potential has not manifested itself experimentally except in the case of the present metastable exotic helium atom. It is thus very interesting to consider the correspondence between the atomic approach and the molecular approach. The former is physically transparent, but mathematically troublesome and we need "perturbational" thinking. The latter is mathematically more valid, though the structure is essentially of atomic origin. Yamazaki and Ohtsuki [13] have attempted some possible discussions in a qualitative and intuitive way.

5 Atomic Core Polarization

Yamazaki and Ohtsuki [13] pointed out that there exists an effect similar to nuclear core polarization, which is a generic concept in nuclear physics, symbolizing various phenomena in which low-energy transitions (or static moments) of multipolarity ($M\lambda$ or $E\lambda$) are enhanced or suppressed due to the coherent coupling with core excitation of the same multipolarity even if the core excitation energy is high. The most famous effect is seen in nuclear magnetic moments, as first clarified by Arima and Horie [20]. Leon and Seki [21] discussed such effects in the case of exotic atoms. Here, the core in its physical essence is not a "spatially inner" region but an "energetically deeper" region (in a nuclear system all the nucleons are confined within the same nuclear boundary, where the core implies deeper bound states, not spatially inner).

In this sense, the simplest "atomic core polarization" is seen in the present antiprotonic helium atom, where the E1 transition energies ($\approx$ 2 eV) are by an order of magnitude smaller compared with the electronic excitation energy ($\approx$ 18 eV). Here, the $1s$ electron constitutes the simplest core with respect to the $\bar{p}$. The important and essential point is that the residual interaction causes a special type of mixture of configurations that gives a coherent contribution to the transition rates to the first order of the mixing amplitude. This effect was not taken into account in the calculation of Russell [9], but is naturally accommodated in the molecular treatment [17–19].

The dominant configuration of $\bar{p}He^+$ of quantum numbers (N, L) is, of course, the antiprotonic $|n, l\rangle$ with the electronic $1s; n = N$ and $l = L$. The most important mixed configurations are the antiprotonic $|n, l\rangle = |N-1, L-1\rangle$ and $|N+1, L+1\rangle$ with the electronic np, which contribute to the transition amplitude $L \to L \pm 1$ to the first order. Yamazaki and Ohtsuki [13] calculated the mixing amplitudes defined as

$$\Psi(L+1, L) \cong \phi_{1s}^{(e)}|L+1, L\rangle_{\bar{p}} + c_L \phi_{2p}^{(e)}|L, L-1\rangle_{\bar{p}} + d_L \phi_{2p}^{(e)}|L+2, L+1\rangle_{\bar{p}} + \ldots \quad (11)$$

Specifically, the mixing coefficients are expressed by

$$c_L = -\sqrt{\frac{L}{3(2L+1)}} \left\langle \frac{r_<}{r_>^2} \right\rangle \frac{1}{\Delta E_{2p-1s}},$$

$$d_{L-1} = \sqrt{\frac{L}{3(2L-1)}} \left\langle \frac{r_<}{r_>^2} \right\rangle \frac{1}{\Delta E_{2p-1s}}. \quad (12)$$

The typical numerical values are $c_{37} = -0.127$ and $d_{36} = 0.128$. The magnitudes of these mixings are of the order of 1%. Nevertheless, the mixing causes a dramatical change in the transition rate.

The transition matrix element is in general

$$< f\|\mathbf{r}^{(1)}\|i > = < (l_e^f l_{\bar{p}}^f)L'\|\mathbf{r}_e^{(1)} + \mathbf{r}_{\bar{p}}^{(1)}\|(l_e^i l_{\bar{p}}^i)L >$$

$$= (-)^{l_e^f + l_{\bar{p}}^i + L + 1} \begin{Bmatrix} l_e^f & 1 & l_e^i \\ L & l_{\bar{p}}^i & L' \end{Bmatrix} \sqrt{(2L+1)(2L'+1)} (l_e^f \|\mathbf{r}_e^{(1)}\| l_e^i) < l_{\bar{p}}^f \mid l_{\bar{p}}^i > \quad (13)$$

$$+(-)^{l_e^f + l_{\bar{p}}^i + L' + 1} \begin{Bmatrix} l_{\bar{p}}^f & 1 & l_{\bar{p}}^i \\ L & l_e^f & L' \end{Bmatrix} \sqrt{(2L+1)(2L'+1)} (l_{\bar{p}}^f \|\mathbf{r}_{\bar{p}}^{(1)}\| l_{\bar{p}}^i) < l_e^f \mid l_e^i >,$$

where

$$(l'\|\mathbf{r}^{(1)}\|l) = (-)^{l'} \sqrt{(2l'+1)(2l+1)} \begin{pmatrix} l' & 1 & l \\ 0 & 0 & 0 \end{pmatrix} < f'(r)|r|f(r) > \delta(l' \pm 1, l). \quad (14)$$

The unperturbed amplitude for a typical transition $(n, l) \to (n-1, l-1)$ is

$$< (s\ L-1)L-1\|\mathbf{r}_{\bar{p}}^{(1)}\|(sL)L >= -\sqrt{L} < f_{L-1}(r)|r|f_L(r) >_{\bar{p}} \tag{15}$$

and the first-order contribution is

$$c_L < (s\ L-1)L-1\|\mathbf{r}_e^{(1)}\|(p\ L-1)L > +d_{L-1} < (pL)L-1\|\mathbf{r}_e^{(1)}\|(sL)L >$$

$$= c_L\sqrt{\frac{2L+1}{3}} < s\|\mathbf{r}_e^{(1)}\|p > +d_{L-1}\sqrt{\frac{2L-1}{3}} < p\|\mathbf{r}_e^{(1)}\|s >, \tag{16}$$

$$= \left[-c_L\sqrt{\frac{2L+1}{3}} + d_{L-1}\sqrt{\frac{2L-1}{3}}\right] < f_{2p}(r)|r|f_{1s}(r) >_e$$

which has a destructive contribution to the unperturbed amplitude.

The typical numerical value for the unperturbed amplitude is

$$< 37, 36\ \|\mathrm{E1}\|\ 38, 37 >_{\bar{p}} = -3.07, \tag{17}$$

while the 1st-order amplitude is

$$\left(-c_{37}\sqrt{\frac{2L+1}{3}} + d_{36}\sqrt{\frac{2L-1}{3}}\right) < f_{1s}(r)|r|f_{2p}(r) >_e = +1.20 \tag{18}$$

The radial matrix element was evaluated by using the calculated basis functions. The second term gives rise to substantial reduction of the total transition amplitude, yielding a suppression factor, $1 - \alpha \approx (1.87/3.07)^2 \approx 0.37$.

6 Distribution of Angular Momentum

How much is the angular momentum to be brought in exotic-atom formation? Let us make a naive and intuitive estimate. The angular momentum is given essentially by a classical estimate, namely, the momentum times the impact parameter, because the wavelength of the incoming X^- is much smaller than the collision size. The maximum angular momentum L at an incident c.m. energy E is

$$L(E) = a\sqrt{2ME}, \tag{19}$$

with a being the maximum impact parameter. Since the bound X^- is formed most likely at a radius, which is around the electron radius, we can put $a = a_e$.

In the energy loss process of X^- the final energy of X^- at the exotic atom formation is expected to be distributed uniformly as

$$\begin{aligned} w(E) &= 1 \text{ for } E < I_0 \\ &= 0 \text{ for } E > I_0. \end{aligned} \tag{20}$$

because the X^- of energy $E > I_0$ still undergoes ionization loss, whereas the X^- of energy $E < I_0$ forms a bound exotic atom. For a crude estimate of the initial population of l, $P(l)$, let us assume a uniform distribution of $w(E)$ below $E = I_0$.

In the simplest case of hydrogen, the maximum angular momentum for the highest energy $E \sim I_0$ becomes

$$l_{max} \cong (M/m_e)^{1/2} = n_0 \tag{21}$$

because $I_0 = me^4/2\hbar^2$ and $a_e = \hbar^2/me^2$. This is an interesting relation. Namely, it is just at $E = I_0$ that the circular orbital $l = n_0 - 1$ can be marginally populated.

For the sake of simplicity we assume that the atomic-capture cross section is given by

$$\sigma(l, E) = (2l+1)\pi\left(\frac{\lambda}{2\pi}\right)^2 = (2l+1)\frac{\pi}{2ME} \tag{22}$$

for

$$l < L(E) = a(2ME)^{1/2}, \tag{23}$$

where λ is the wavelength of X^-, namely,

$$\lambda = 2\pi(2ME)^{-1/2}. \tag{24}$$

This expression gives the total capture cross section to be a constant $\approx$ geometrical cross section

$$\sum_{l=0}^{l_{max}} \sigma(l, E) \approx \pi a^2. \tag{25}$$

Cohen and Padial [22] and Dolinov *et al.* [23] actually show that the formation cross section of hydrogenic and helium exotic atoms is nearly constant in the energy range $E < I_0$.

Then, the distribution $P(l)$ for a uniformly distributed energy spectrum $w(E)$ in the range $0 < E < I_0$ is

$$P(l) = \int_{E_l}^{I_0} \sigma(l, E)\frac{w(E)}{I_0} dE \tag{26}$$

with

$$E_l = \left(\frac{l}{a}\right)^2 \frac{1}{2M}$$

$$I_0 = \left(\frac{l_{max}}{a}\right)^2 \frac{1}{2M} \tag{27}$$

which leads to a simple formula

$$P(l) \propto (2l+1)\ln(l_{max}/l). \tag{28}$$

This gives a bell shape distribution with a smooth cut off toward l_{max}. Baker [24] calculated the initial distribution $P(n, l)$ in hydrogenic exotic-atom formation and obtained a similar bell-shape form, though no clear description on the energy distribution at capture was given.

Since the energy distributes from 0 to only I_0, there is a certain limit of l; the statistical distribution of l for unlimited l is inconceivable. No precise experimental information on the initial population of l is available from any earlier experiments on exotic atoms because they are not directly related to the initial distribution of l.

Sometimes, a non-statistical distribution with a cut-off parameter, $P(l)$ = $(2l+1)$ exp($-\alpha l$), is introduced without real justification. The above bell-shape distribution formula results from definite and reasonable physical assumptions, but the value of l_{max} is being introduced as a somewhat adjustable parameter to allow the actual l distribution to differ from the crude one given in eq.(11) as the effect of the Coulomb interaction. Theoretically better-founded formula may certainly be possible and would indeed be welcome.

The l distribution has a scaling factor $\sqrt{M}$ as n_0 and l_{max} do, and thus, the shape of $P(l)$ is a universal function of l/l_{max}. The critical angular momentum L_0 which divides the metastable and the "prompt" states (namely, a state with $l > L_0$ is metastable) can be crudely estimated as follows. i) The binding energy of the formed atomic levels of $\bar{\text{p}}\text{He}^+$ at $n = n_0$ (with respect to He^{++}) is roughly given by

$$E_n \approx 39.5 \frac{M}{m} \frac{1}{n^2} + 39.5 \text{ eV} \tag{29}$$

where the number 39.5 is half the total binding energy ($I_0 + 4Ry$ = 79 eV) of helium atom which is assumed to be shared by X^- and e^-. The critical angular momentum is given by setting the binding energy of circular orbitals of ionized $\bar{\text{p}}\text{He}^{++}$,

$$4 \frac{M}{m} \frac{1}{(l+1)^2} Ry, \tag{30}$$

to 79 eV, which yields

$$\frac{L_0}{n_0} \approx 0.83 \tag{31}$$

The actual critical angular momentum may be somewhat larger than this crude estimate, but what is important here is that L_0 is scaled by n_0, and thus, by $M^{1/2}$. This means that the delayed fraction is independent of the mass of X^-. This is consistent with the observation that the delayed fractions for π^-, K^- and $\bar{\text{p}}$ are all around 3 % [3–5].

7 Isotope Effect

Yamazaki and Ohtsuki [13] considered the isotope effect on the overall lifetime of He+$\bar{\text{p}}$ atom on the following grounds. Let us define the ratio of the reduced masses of $\bar{\text{p}}$ as R = $M(\bar{\text{p}}^4\text{He})/M(\bar{\text{p}}^3\text{He})$ = $16/15$. In the atomic picture the energy spacing is approximately given by differentiating eq.(29) as

$$\Delta E \approx \frac{79}{n_0} \tag{32}$$

for the most probable $n = n_0 = (M/m)^{1/2}$. The simplest estimate using eq. (29) gives

$$\frac{\Delta E_{n_0}(^3\text{He})}{\Delta E_{n_0}(^4\text{He})} = R^{1/2} = 1.03 \tag{33}$$

The calculation by the molecular approach gives this ratio to be 1.025, showing that the crude estimate is not bad. The E1 matrix element depends on the reduced mass, but as stressed in Yamazaki and Ohtsuki [13], the transition matrix element for $n = n_0$ is independent of the reduced mass. The reason is clear: the spatial distribution of $\bar{\mathrm{p}}$ is the same at the most probable n, whatever the reduced mass is. It was shown that the unperturbed transition rates at the most likely n are the same for ^{3}He and ^{4}He. In the first-order correction the transition amplitudes remain equal to each other because the reduction factor α is independent of M. So, we consider only the M dependence of $(\Delta E)^{-3}$, giving

$$\tau_{n_0}(^4\mathrm{He})/\tau_{n_0}(^3\mathrm{He}) = R^{3/2} = 1.09. \tag{34}$$

The molecular approach gives a value, 1.09, equal to the above simple estimate.

The overall lifetime should be the typical level lifetime multiplied by an average cascade number of metastable states. Since the metastable states are expected to distribute over a given energy interval, the number of metastable cascades is proportional to the level density, namely, to $(\Delta E)^{-1}$. Multiplying this factor Yamazaki and Ohtsuki [13] obtained

$$\tau \propto (\Delta E)^{-4} \propto M^2. \tag{35}$$

This crude estimate gives $\tau(^4\mathrm{He})/\tau(^3\mathrm{He}) = R^2 = 1.13$ which turns out to agree with the recently observed value [14].

The present crude estimate may not be good, because effects other than the radiative transitions are neglected. Better theories would take into accounts Auger transitions, Stark effects, *etc.* This is an open question.

8 Deeply Bound Pionic Atoms Predicted

In 1987 Yamazaki *et al.* [25] pointed out an interesting situation for a strongly-absorptive negative hadron; it feels both the attractive long-range Coulomb potential and a short-range strong interaction. The bound states formed are not entirely "nuclear", but lie somewhere between nuclear and atomic regions. They are called "Coulomb assisted hybrid bound states". Since the negative hadron resides beyond the nuclear surface with density distribution $\rho_{nl}(r)$, the absorption width of a state nl,

$$\Gamma_{nl} = -2\int \rho_{nl}(r)W(r)d\tau \sim -2W_0 \int \rho_{nl}(r), \tag{36}$$

is reduced compared with $\Gamma_0 = -2W_0$, which would be the width when the hadron were entirely confined inside the nucleus. Yamazaki *et al.* [25] suggested the existence of reasonably narrow bound states of Σ^- for some assumed potential shape. When it is repulsive, the hadron is pushed outward, and the level width is expected to be very much reduced. Such a situation may take place in the case of π^-, as Toki and Yamazaki [1] showed.

The potential parameters in the pion-nucleus interaction [26, 27] have been empirically determined by using pionic atom data (shallow atomic states). The most important fact is that the central potential is repulsive; this was indicated by small positive energy shifts of pionic atoms, observed even in pionic hydrogen atom. The absorption potential W_0 is known to have a magnitude of 10 MeV, which would mean that the expected level width is enormously large; $\Gamma_0 \sim 20$ MeV! This would be true only when the π^- were confined within the nuclear boundary.

Toki and Yamazaki [1,2] solved the Klein-Gordon equation for a negative pion bound by heavy nucleus, showing that the deeply bound states have narrow absorption widths compared with the level spacings. The calculated level scheme for π^- in ^{208}Pb is shown in Fig. 8. The reason why the level widths are so small is illustrated in Fig. 9, where the pion optical potential and the state densities $\rho_{nl}(r)$ of the $1s$, $2p$ and $2s$ states are shown. Obviously, the potential pocket around the nuclear surface accommodates those bound states. They are all *halo* states. This mechanism had been suggested by Friedman and Soff [28] who considered π^- bound states in very heavy nuclei, but it had been completely discarded in physics community. One of the reasons is the prevailing misbelief on the absorption width. Another reason is probably that they did not conceive any experimental means to detect those bound states.

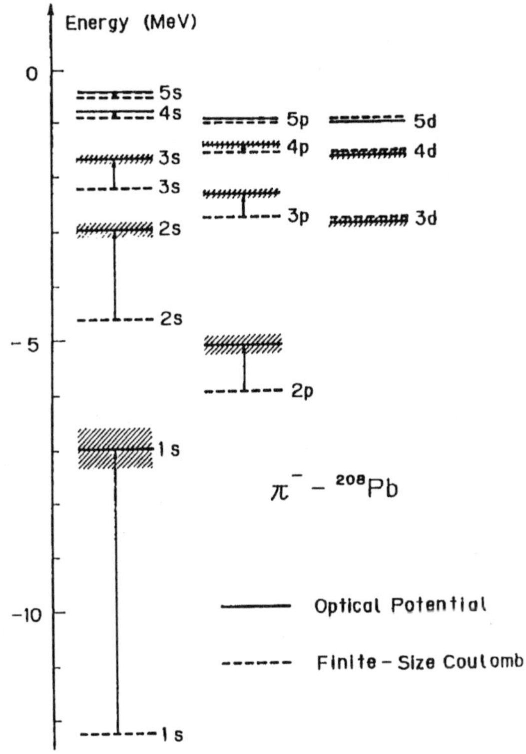

Fig. 8. Energy diagram of deeply bound pionic atom in ^{208}Pb, taken from Toki and Yamazaki [1].

9 Pion Transfer Reactions

Toki *et al.* [1,2] proposed some ways to populate those deeply bound pionic atoms, called *pion-transfer reactions*. The most straightforward idea was to make use of the

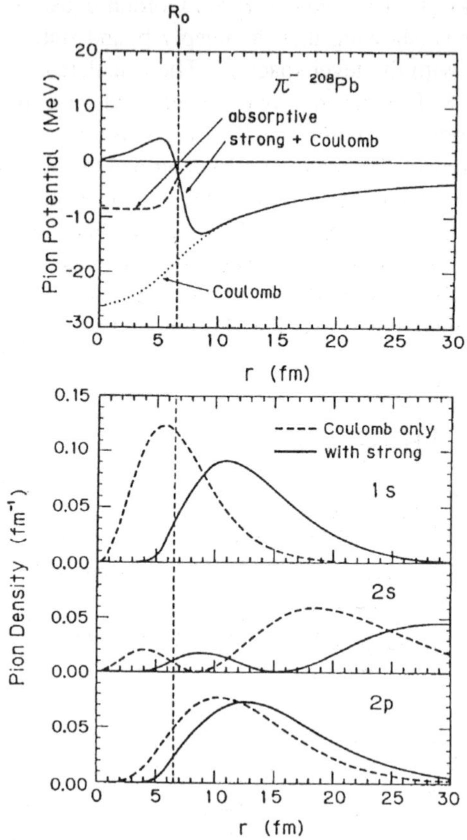

Fig. 9. The π^--nucleus optical potential for ^{208}Pb and the pion densities of the lowest three states. Taken from Toki *et al.* [2].

Yukawa vertex itself; namely, $n \rightarrow p + \pi^-$. This process does not take place in free space, because this does not satisfy the energy-momentum conservation. However, when a free neutron hits a heavy nucleus, the produced π^- can be bound by the nucleus, leaving a free proton as a spectator. Thus, in principle, this (n,p) type charge-exchange reaction produces deeply bound pionic states.

The produced states are resonance nuclear states of quantum numbers J^π satisfying $\Delta S = \Delta I = 1$, which correspond to pionic-atom quantum numbers l; $\pi = (-)^{l+1}$. The deeply bound pionic states are regarded as a new family of Gamow-Teller resonances. The (n,p) type reaction for the production of deeply bound pionic states has a large momentum transfer, q = 200 MeV/c. This makes the production cross sections very much reduced by the nuclear distortion effect [29, 30]. Search experiments have been carried out for ^{208}Pb(n,p) [29] and ^{208}Pb(d,^{2}He) [31]. Only a smoothly varying continuous spectrum was observed.

Toki *et al.* [32] have studied theoretically proton-pick-up reactions such as (n,d) and

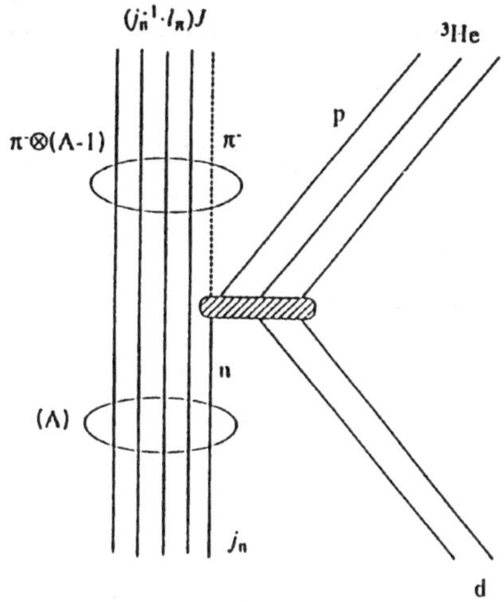

Fig. 10. Diagram for the pion-transfer (d, ^{3}He) reaction.

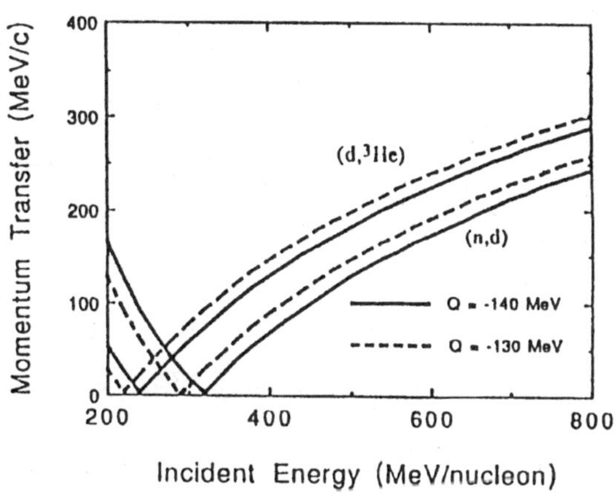

Fig. 11. Momentum transfers of (n,d) and (d, ^{3}He) reactions as functions of the incident energy.

(d,^{3}He). The diagram and the momentum transfer are presented in Fig. 10 and Fig. 11, respectively. These reactions are found to be suitable for the population of the $1s$ and $2p$ states with a neutron hole of j_n with configurations of $[(1s)_\pi \cdot j_n^{-1}]J$ and $[(2p)_\pi \cdot j_n^{-1}]J$ when the momentum transfer q matches the required angular momentum transfer, namely, $J \sim qR_o$. When the matching condition is fulfilled, the distortion effect is not too large. In the (n,d) and (d, ^{3}He) reactions the momentum transfer can be tuned by selecting the incident energy. The cross sections were calculated [32] for various pionic states on the low-lying neutron hole states in ^{208}Pb, as shown in Fig. 12. The optimum energy for (d,^{3}He) leading to the $(1s)_\pi(i_{13/2})_n^{-1}$ states is 600 MeV/u. On the other hand, at 300 MeV/u, where the momentum transfer is small, special pionic states of "substitutional" configurations, such as $(2p)_\pi(3p)_n^{-1}$, are preferentially populated.

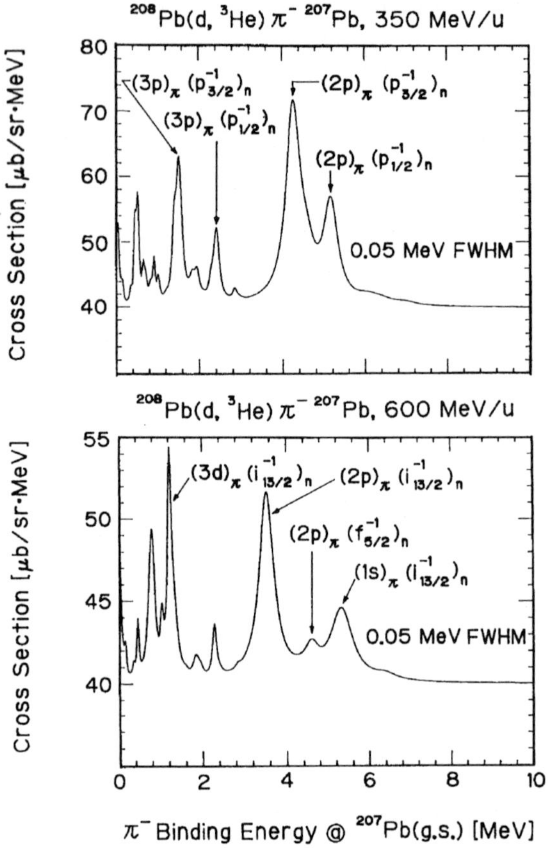

Fig. 12. Calculated ^{208}Pb(d,^{3}He) cross sections for deeply bound π^- states.

Very recently, an experiment to measure the ^{208}Pb(n,d) spectrum at excitation energies in the pion-mass region at T_n = 400 MeV has been carried out at TRIUMF [33], which showed up a large continuum characterized by the quasi-free production of π^- on a smooth background of 300 μb/sr·MeV, as shown in Fig. 13. In addition, a significant bump below the π^- emission threshold is observed. The integrated cross section over the bound π^- region is about 800 μb/sr, which is comparable with the predicted cross section. The expected spectrum in the bound π^- region is also shown in Fig. 13. Thus, this experiment gives the first experimental indication of the formation of deeply bound π^- states, but the statistics and resolution obtainable with the secondary neutron beam were not sufficient to resolve discrete peaks of the deeply-bound π^- states. To obtain further information with sufficient statistics the use of an alternative reaction (d, ^{3}He) is most desirable. The expected spectra including the quasi-free continuum have been calculated by Hirenzaki and Toki [34].

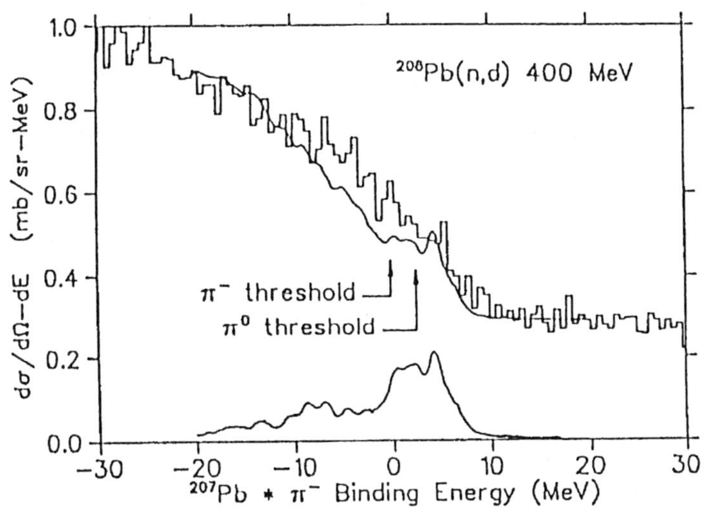

Fig. 13. Inclusive deuteron spectrum from ^{208}Pb(n,d) at 0° and 400 MeV, measured in the TRIUMF Experiment E628 by Trudel *et al.* [33]. The smooth curve is a calculation by Hirenzaki and Toki [34] for ^{208}Pb(n,d) ^{207}Pb$\otimes\pi^-$ for the pionic bound states and the free pion continuum convoluted with 1.5 MeV FWHM resolution for comparison with the data. The lower curve shows just the bound state calculation.

The Q value of the (d, ^{3}He) reaction is

$$\begin{aligned} -Q &= \omega + S_n(j_n) - [M_n + M_d - M(^3\text{He})] \\ &= m_\pi - B.E.(\pi^-) + S_n(j_n) - 6.787\ \text{MeV}, \end{aligned} \tag{37}$$

where ω is the excitation energy of the pionic state with respect to the neutron-hole j_n and $S_n(j_n)$ is the neutron separation energy. The four neutron orbitals, $p_{1/2}(S_n = 7.367$ MeV), $f_{5/2}(S_n = 7.937$ MeV), $p_{3/2}(S_n = 8.246$ MeV) and $i_{13/2}(S_n = 9.000$ MeV) are taken into account. Hereafter, the π^- binding energy will be defined with respect to the ground state of ^{207}Pb, and thus it corresponds to the (d, ^{3}He) reaction Q value as :

$$-Q = 140.147\ \text{MeV} - B.E.(\pi^-) \tag{38}$$

The expected line profile and their angular dependence of the ^{208}Pb(d,^{3}He) reaction at T_d = 600 MeV with an instrumental resolution of 50 keV FWHM are shown in Fig. 14. It has a steep angular dependence, peaked at 0 degree. Above the π^- emission threshold there appears a quasi-free continuum, while discrete states are expected below the threshold. The constant background of ^{208}Pb (n,d) which does not contribute to the pion production has recently been measured at TRIUMF to be 300 μb/sr/MeV [30]. The constant background for (d,^{3}He) is assumed to be 40 μb/sr/MeV without angular dependence. There is no experimental information yet on this background in the case of (d, ^{3}He).

The (d, ^{3}He) reaction has a very good intrinsic calibration coming from the two-body final state p(d, ^{3}He)π^0 when a target including hydrogen is used; the (d, ^{3}He) spectrum on a $(CH)_n$ target exhibits a distinct peak in the same momentum region as in the ^{208}Pb(d, ^{3}He) reaction. The position of this peak moves with the angle of ^{3}He.

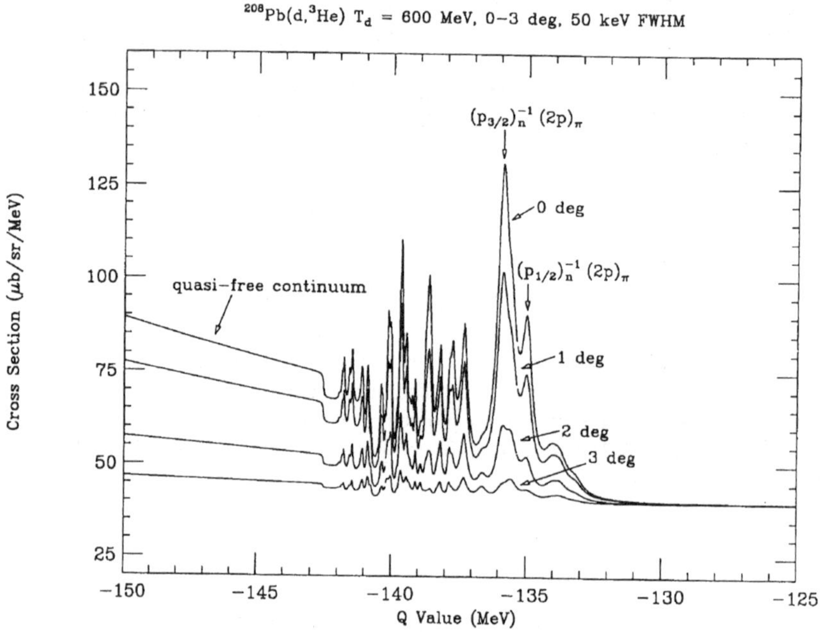

Fig. 14. Calculated spectrum and its angular dependence for ^{208}Pb(d,^{3}He) at T_d = 600 MeV with FWHM resolution of 50 keV. From the DWIA calculation of Hirenzaki and Toki [34].

Recently, a test experiment of the ^{208}Pb(d, ^{3}He) reaction was carried out at Saturne [35], where the SPES1 spectrometer system was used to detect ^{3}He particles. Although the ^{3}He particles were well identified and discriminated from other particles such as p, d and ^{4}He, the detection of (d, ^{3}He) spectrum at small angles (<3.5 deg) was prevented by a huge (d, ^{3}He) background caused by beam particles hitting the spectrometer wall. This background at small angles was difficult to suppress because the SPES1 spectrometer has a single-bend structure, while the formation cross section of deeply bound pionic states is forward peaked. It is indispensable to use a spectrometer with two (or more) bending magnets. Thus, we are proposing to measure the (d,^{3}He) reaction spectra on ^{208}Pb using the Fragment Separator [36] in combination with a deuteron beam from the SIS18 heavy-ion synchrotron of GSI.

10 Proposed (d, ^{3}He) Experiment

We propose to measure the (d, ^{3}He) reaction spectra at T_d = 0.6 and 1.2 GeV. The relevant kinematical quantities are shown in Table 1. We will use the Fragment Separator (FRS) system [36] with which we will measure the momentum of ^{3}He in the range of about 2 GeV/c. Since FRS is composed of 4 bending magnets, most of the background events caused by the incident deuteron beam inside the first bending magnet will be eliminated already at the intermediate focal plane, where particle tracking MWPC's and scintillation counters are placed to provide informations on p/Z, Δt and $\Delta E/\Delta x$. The particles with p/Z in a momentum bite Δp pass through these counters and are refocussed again in the final focal plane after the 3rd and 4th bending magnets. The p/Z as well as the trajectory of each particle are precisely determined by using the information from the MWPC's at the middle focal plane. At the final focal plane additional scintillation counters are placed. The combined information of the time of flight Δt and $\Delta E/\Delta x$ and p/Z is sufficient to identify individual particles.

Since the field value p/Z for ^{3}He is very close to $p_d/2$, the momentum of protons from deuteron break-up, we have to reject protons in the trigger stage to keep the trigger rate to be reasonably small. This is possible because the energy losses of the p and the ^{3}He are quite different (by a factor of 6).

A special mode for the operation of FRS [37] (large acceptance $\sim$ 3 msr and large mid-plane dispersion $\sim$ 4.2cm/%) is preferable. The overall momentum resolution of FRS including the beam momentum spread may be achieved to be 2×10^{-4}, but here, it is safely assumed to be

$$\frac{\delta p}{p} = 5 \times 10^{-4}. \tag{39}$$

This corresponds to the energy resolution in the final state after (d,^{3}He) as follows,

$$\begin{aligned} \delta\omega &= (p_{\rm He})^2/E_{\rm He} \times (\Delta p/p)_{\rm He} \\ &= 0.43 \text{ MeV for } T_d = 0.6 \text{ GeV} \\ &= 0.90 \text{ MeV for } T_d = 1.2 \text{ GeV} \end{aligned} \tag{40}$$

The resolution is deteriorated by the energy loss of the ^{3}He particle in the target. The FWHM width due to this effect is estimated to be

$$\Delta E = 6.4 \text{ keV/mg/cm}^2 \tag{41}$$

in the case of T_d = 600 MeV. For instance, ΔE = 0.32 MeV for a 50 mg/cm^2 target. So, using this thickness, we can obtain an overall energy resolution of 0.5 MeV FWHM for T_d = 600 MeV. For a better resolution we have to use a thinner target at the expense of a lower event rate.

We simulated expected spectra for a typical instrumental resolution of 0.5 MeV FWHM, as shown in Fig. 15. The yield estimate assumes:

$$\begin{aligned} I_d &= 5 \times 10^{10} \text{ /sec} \\ N_{\rm target} &= 1.5 \times 10^{20} \text{ atoms/cm}^2 \text{ (50 mg/cm}^2\ ^{208}\text{Pb)} \\ \Omega &= 3 \times 10^{-3} \text{ sr } (\Delta\theta_x \sim \pm 40 \text{ mr}, \Delta\theta_y \sim \pm 20 \text{ mr}) \end{aligned} \tag{42}$$

$$\text{momentum bite } \frac{\Delta p}{p} = 0.032$$

Since the differential cross section decreases rapidly with angle, the cross section was calculated by integrating the differential cross section over the spectrometer acceptance $A(\theta_x, \theta_y)$ as

$$\sigma = \int \frac{d\sigma}{d\Omega} A(\theta_x, \theta_y) d\Omega. \tag{43}$$

They are shown in Fig. 15 (upper). Typically, $\sigma = 0.1 - 0.2$ μb/MeV.

The event rate was calculated as

$$R = I_d \times N_{target} \times \sigma. \tag{44}$$

The typical value in the above condition is

$$R = 0.75 \sim 1.5 \text{ events/MeV/sec.} \tag{45}$$

The simulation presented in Fig. 15 assumes a 24-hour beam time. When the beam intensity, the spectrometer efficiency, or the calculated cross section is less than the assumed one, we have less statistics accordingly.

The energy band corresponding to the momentum bite $\Delta p/p = 0.032$ is

$$\begin{aligned} \Delta\omega &= 26 \text{ MeV for } T_d = 600 \text{ MeV} \\ \Delta\omega &= 54 \text{ MeV for } T_d = 1200 \text{ MeV}. \end{aligned} \tag{46}$$

This is very suitable in covering the energy region of present interest. We obtain 20–40 events/sec over the 26 MeV bite, which can be acquired by the present data taking system, when only the ^{3}He events were selected to trigger.

The background from break-up protons is most serious. Its yield was estimated by using the double differential cross sections given by Jafar *et al.* [38]. It is also forward peaked, and the same procedure as above was taken. The total cross section integrated over the momentum byte of the spectrometer is 50 mb. This background gives a huge counting rate in a middle-plane scintillation counter of as much as 4×10^5/sec. Taking into account the macroscopic duty factor of SIS ($\sim$0.5) we estimate the instantaneous intensity is nearly $\sim$1 MHz. The signal to background ratio is thus 1×10^{-5}. However, the energy deposit of these break-up protons is a factor of 6 smaller than that for ^{3}He.

So, we can discriminate these protons and the ^{3}He particles in the hard logic stage. We can use special drift chambers which were designed and developed for a high-rate measurement such as (π^+, K^+) reactions. Another way is to divide the mid-plane scintillation counter into several pieces.

The simulation shows that the presence of the main peaks, $(p_{1/2})_n^{-1}(2p)_\pi$ and $(p_{3/2})_n^{-1}$ $(2p)_\pi$, will be firmly demonstrated, if they exist.

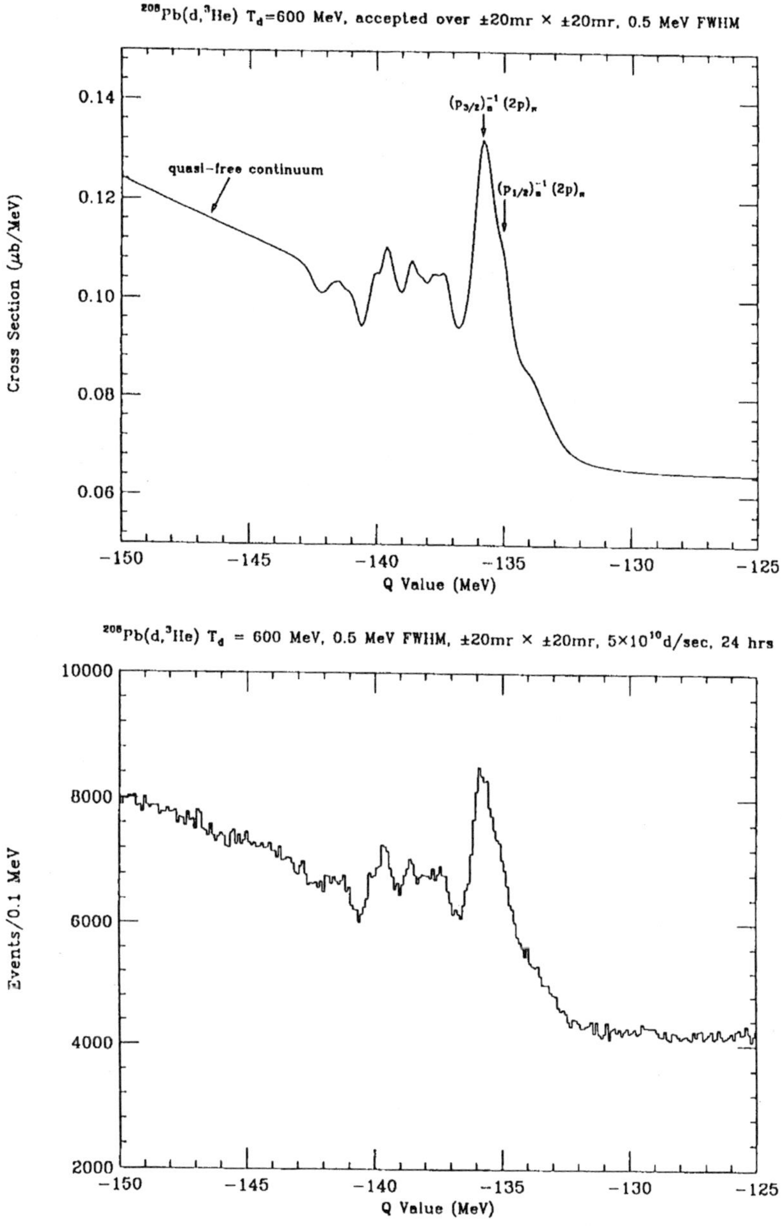

Fig. 15. Simulated ^{208}Pb(d,^{3}He) spectrum at T_d=600 MeV. A FWHM resolution of 0.5 MeV is assumed.

11 Concluding Remarks

The discovery and subsequent studies of metastable antiprotonic helium atomcules have provided a new domain of physics and chemistry. The following problems are raised, but mostly unsolved; open questions to be studied in the future.

i) Formation and structure of $\bar{p}He^+$ in very dilute medium.
ii) Structure and lifetimes of $\bar{p}He^+$ in condensed matter.
iii) Laser spectroscopy of energy levels of $\bar{p}He^+$.
iv) Photon emission spectroscopy of $\bar{p}He^+$.
v) Chemical reactions of $\bar{p}He^+$ with other atoms/molecules.
vi) Gate way to antihydrogen formation.

This interesting domain of physics and chemistry involving one baryonic antiparticle will develop in the years to come both experimentally and theoretically.

The problem of deeply-bound pionic atoms has only recently raised and it is still in a premature stage. Since it is a cluster of sharp resonance nuclear states just below the pion mass, its experimental identification will be by itself important, even surprising; it is the first nuclear state in which hadrons other than the nucleons are bound. Once this is discovered, its spectroscopy will develop, and it will be more fascinating. A high-resolution spectroscopy of deeply bound pionic atoms will provide, first of all, sets of pion bound states, which will be used to solve the question as to the potential that π^- feels from inside of the nucleus, the effective mass of π^- in nuclear medium, *etc.* It is also connected with the nuclear shape, as discussed by Toki *et al.* [39] in connection with neutron skin.

A few years ago we proposed experiments of high-resolution spectroscopy using the nuclear reactions of inverse kinematics, namely, d(^{208}Pb,π^{-208}Pb)^{2}He, d(^{208}Pb, π^{-207}Pb) ^{3}He, with a cooled beam and an internal target in the ESR of GSI [40, 41]. The proposed experiment needs more detailed consideration of beam conditions and experimental arrangements to achieve the required luminosity [42], and is thus to be realized in the future. This method will permit formation of deeply-bound pionic atoms on unstable nuclei which are circulating in a storage ring.

The author would like to thank Drs. R.S. Hayano, H. Toki, S. Hirenzaki, K. Ohtsuki, E. Widmann, J. Eades and P. Kienle for many collaborative works. The present work is supported by the Grant-in-Aid for Specially Promoted Research of Japanese Ministry of Education, Science and Culture.

References

1. H. Toki and T. Yamazaki, Phys. Lett. **B213** (1988) 129.
2. H. Toki, S. Hirenzaki, T. Yamazaki and R.S. Hayano, Nucl. Phys. **A501** (1989) 653.
3. T. Yamazaki *et al.*, Phys. Rev. Lett. **63** (1989) 1590.
4. S.N. Nakamura *et al.*, Phys. Rev. **A45** (1992) 6202.
5. M. Iwasaki *et al.*, Phys. Rev. Lett. **67** (1991) 1246.
6. R.S. Hayano *et al.*, Phys. Lett. **B231** (1989) 355.

7. R.S. Hayano, M. Iwasaki and T. Yamazaki, in *"Perspectives of Meson Science"* (ed. T. Yamazaki, K. Nakai and K. Nagamine, North-Holland, Amsterdam, (1992) Ch.13.
8. G.T. Condo, Phys. Rev. **9** (1964) 65.
9. J.E. Russell, Phys. Rev. Lett. **23** (1969) 63.
10. G. Reifenröther and E. Klemt, Nucl. Phys. **A503** (1989) 885.
11. G. Reifenröther, E. Klemt and R. Landua, Phys. Lett. **B191** (1987) 15.
12. H. Outa *et al.*, to be published.
13. T. Yamazaki and K. Ohtsuki, Phys. Rev. **A45** (1992) 7782.
14. T. Yamazaki *et al.*, Nature **361** (1993) 238.
15. E. Widmann *et al.*, Nucl. Phys. **A558** (1993) 679c.
16. N. Morita, K. Ohtsuki and T. Yamazaki, Nucl. Instr. Meth. **A330** (1993) 439.
17. I. Shimamura, Phys. Rev. **A46** (1992) 3776.
18. R. Ahlrichs, O. Dumbrajs, H. Pilkuhn and H.G. Schlaile, Z. Phys. **A306** (1982) 297.
19. P.T. Greenland and R. Thürwächter, Hyperfine Interactions, **76** (1993) 355, P.T. Greenland, private communication (1993).
20. A. Arima and H. Horie, Prog. Theor. Phys. **12** (1954) 623.
21. M. Leon and R. Seki, Nucl. Phys. **A298** (1978) 333.
22. J.S. Cohen and N.T. Padial, Phys. Rev. **A41** (1990) 3460.
23. V.K. Dolinov, G. Ya Korenman, I.V. Moskalenko and V.P. Popov, Muon Cat. Fusion 4, (1989) 169.
24. G.A. Baker, Jr., Phys. Rev. **117** (1960) 1130.
25. T. Yamazaki, R.S. Hayano, O. Morimatsu and K. Yazaki, Phys. Lett **B207** (1988) 393.
26. M. Ericson and T.E.O. Ericson, Ann. Phys. **36** (1966) 496.
27. R. Seki and K. Masutani, Phys. Rev. **C27** (1983) 2799.
28. E. Friedman and G. Soff, J. Phys. **G11** (1985) L37.
29. M. Iwasaki *et al.*, Phys. Rev. **C43** (1991) 1099.
30. J. Nieves and E. Oset, Nucl. Phys. A518 (1990) 617.
31. R.S. Hayano, in *"Pions in Nuclei"* (ed. E. Oset *et al.*, World Scientific, Singapore, 1992) p.330; R.S. Hayano *et al.*, to be published.
32. H. Toki, S. Hirenzaki and T. Yamazaki, Nucl. Phys. **A530** (1991) 679; S. Hirenzaki, H. Toki and T. Yamazaki, Phys. Rev. **C44** (1991) 2472.
33. A. Trudel *et al.*, TRIUMF Progress Report 1991.
34. S. Hirenzaki and H. Toki, to be published.
35. T. Yamazaki *et al.*, Saturne Proposal no.254 (1992).
36. H. Geissel *et al.*, Nucl. Instr. Meth. **B70** (1992) 286.
37. A. Schröter, PhD Thesis, Technische Universität München, July 1993 (unpublished).
38. J.D. Jafar, H.B. Van der Raay, D.G. Ryan, J.A. Stiegelmair and R.K. Tandon, Nucl. Phys. **A161** (1971) 105.
39. H. Toki, S. Hirenzaki and T. Yamazaki, Phys. Lett. **B249** (1990) 391.
40. T. Yamazaki, R.S. Hayano, H. Toki and P. Kienle, Nucl. Instr. Meth. **A292** (1990) 619.
41. T. Yamazaki, R.S. Hayano and H. Toki, Nucl. Instr. Meth. **A305** (1991) 406.
42. T. Yamazaki, *Proc. 19th INS International Symposium on Cooler Rings and their Applications* (ed. T. Katayama and A. Noda, World Scientific, Singapore, 1991) p.138.

Lecture given at the JSPS-INS International Spring School, March 1993 at Shimoda

Cosmic Rays and Nuclear Physics

Jiro Arafune

Institute for Cosmic Ray Research, University of Tokyo
Tanashi, Tokyo 188, Japan

Abstract Introduction to cosmic ray physics is given with several typical examples of cosmic ray researches to show how it is related nuclear/particle physics and astrophysics.

1 Introduction

1.1 Cosmic Rays

Cosmic rays were discovered about eighty years ago with balloon-borne detectors [1, 2]. They are composed of various kinds of particles from space such as protons, nuclei, electrons, gamma rays, anti-particles, neutrinos, or any other extraterrestrial signals including gravitational waves, or axions.

1.2 Flux and Spectrum

The charged particle cosmic rays are mainly protons with flux about $\sim 1/\text{cm}^2\text{sec str}$ in our Galaxy. The flux at the top of the atmosphere is reduced from it by an order of magnitud due to the interaction with the solar or terrestrial magnetic field. The flux decreases with energy rapidly as $dN/dE \sim -E^{-\beta}$ with $\beta \simeq 2.6$ for $E \leq 10^{16}$ eV and $\beta \simeq 3.0$ for 10^{16} eV $\leq E \leq 10^{19}$ eV and above it the power index β may decrease. The break of power index at around 10^{16} eV is called 'knee', and the break at around 10^{19} eV is called 'ankle' sometimes [3].

An energetic cosmic ray, after colliding with air nuclei many times, produces numerous secondary particles such as pions or electron-positron pairs in the air. Such a bundle of particles is called an air shower. The observation of air showers is a powerful method to detect high energy cosmic rays with using the natural atmosphere as calorimeter material. Cosmic rays observed at the ground level are composed mainly of secondary muons produced in the atmosphere, and are considered as one of the main sources of the natural radioactivities against mankind.

2 Properties of Cosmic Rays

2.1 Origin of Cosmic Rays

Origin of cosmic rays is not well known in spite of the long history of the cosmic ray research. It is widely accepted, however, that their energy is from supernova explosion, since one percent of it is enough to compensate the energy loss of cosmic rays due to the leakage from our Galaxy. It is not clear though whether the acceleration should be made by the shock wave or by the spin rotaion of the pulsar after the explosion. The shock wave acceleration is attractive, since it can beautifully explain the power spectrum of cosmic ray flux, while it is considered not enough to explain cosmic rays above the knee region [4]. Active galactic nuclei are considered as a candidate of extragalactic cosmic ray sources. Much efforts are being paid to try to confirm these ideas by identifying point sources of high energy gamma rays or neutrinos.

2.2 Chemical or Isotopic Composition of Cosmic Rays

Chemical or isotopic composition is deeply related with the origin and the propagation of cosmic rays, and it is one of the most important subjects of the present cosmic ray studies. It is generally similar to that of solar elements, but it is remarkably different for sub-carbon elements such as Li, Be, and B as well as for sub-iron elements such as Sc, Ti, V, Cr, and Mn. Those elements are much more abundant in cosmic rays than in solar abundance, which is considered as due to the spallation products of the primary heavier cosmic rays in collision with interstellar gas. The elemental ratio B/C is found to decrease with energy as $\sim E^{-\delta}$ with $\delta \simeq 0.6$ [5], suggesting the path length of cosmic rays in our Galaxy, $\sim 6\,\mathrm{gram/cm^2}$ below 1 GeV/n, decreases with this power.

The abundance of the unstable isotopes ^{10}Be and ^{26}Al are used to study the propagation of cosmic rays, giving the path length as well as the average interstellar gas density along the cosmic ray trajectories, which is found to be 1/3 of the average density of the whole interstellar gas. Isotopic abundance deviation from the solar composition is further used to study the cosmic ray acceleration origins [6].

2.3 Anti-particles

The observation of the very small composition of anti-proton is very attractive, since it will represent the effect of the proton propagation in the Galaxy, and moreover there are many exotic candidates for the origins such as remote anti-galaxies, Hawking radiation from mini black holes, annihilation of monopoles or that of dark matter particles like photinos, and cosmic strings, all of which are extraordinarily interesting [7]. Several results have been obtained so far, and the BESS experiments at the low energy region will be quite promising in the near future. Positrons and anti-helium nuclei are also related important subjects to study further.

2.4 Gamma Rays and Point Sources

Gamma ray observation should give direct evidence of sources for high energy cosmic ray acceleration , since the gamma rays propagate without losing the directional information.

Crab nebula is the first such celestial object that was observed to emit high energy gamma rays above 10^{12} eV. The discovery [8] was made by Whipple Observatory in US with the mirror telescope to detect cerenkov light produced by the air shower of gamma rays. The observation of the active galactic nuclei MK421 is also claimed by the same mirror. The second pulsar, PSR1706-44, was observed [9] to emit TeV gamma rays with CANGAROO mirror and photo-tubes brought to Australia from Japan. These suggest that the astronomy with TeV gamma rays is going to begin.

2.5 Highest Energy Cosmic Rays

It is predicted that there should be an upper bound at around 10^{20} eV on the cosmic ray energy, since such high energy proton or nuclei will make inelastic collision with cosmic microwave background photons to lose energy by producing pions or electron-positron pairs. This upper bound is called 'Greisen, Kuzmin & Zatsepin cut off', and the cutt off energy depends on the distance between the source and our Galaxy.

The observation of such cut off should give us information on the source distance and source spectrum of highest energy cosmic rays. At present only two events above 10^{20} eV have been observed in the world, one in US [10] and the other in USSR [11].

Cosmic rays in the highest energy region are very important, since the sources of such high energy particles are very restricted and there are considered very few candidates for the sources, such as active galactic nuclei, cosmic strings, annihilation of monopoles or evaporation of mini black holes, all of which are attractive. AGASA in Japan, Fly's Eye in US, and Yakutsk Array in Russia, each with an effective area between 18 and 100 km^2 are now in operation. Future projects with an effective area of 10^4 km^2 are planned in international collaboration.

It should be noted here that the high energy air shower observation can derive the information on proton-proton collision cross section. The AGASA experimental results predict the p-p cross section to be rising with energy up to $\simeq$ 130 mb at the SSC energy region.

3 Neutrinos

The electron neutrinos from a reactor and mu neutrinos with an accelerator were observed in 1953 and in 1962, respectively [12, 13], and the atmospheric neutrinos were observed with deep underground detectors in 1965 [14, 15].

The small reaction cross section of neutrino has made it difficult to observe in astrophysics study or to examine their mass properties in laboratory physics research. The supernova neutrino burst from SN 1987A in Large Magellanic Cloud were observed in 1987 [16, 17], and it has clearly shown that the neutrino is a very powerful tool for astrophysics. The observation directly gave the information on the surface of the neutrino sphere made at the center of the supernova, the temperature of about 4 MeV, the radius of about 30 km, and the cooling time of about 10 sec, and also showed that about 99% of the explosion energy went away in the form of neutrinos. It further gave us the bounds on the masses of not only the electron neutrino [18, 19] but also the other neutrinos, $m_{\nu_\tau}, m_{\nu_\mu} < 30$ keV [20], as well as on the mass of the axion , $m_{axion} < 10^{-3}$ eV, or $m_{axion} > 1$ eV [21].

The solar neutrinos have been observed by the pioneering work of Davis *et al.* with ^{37}Cl targets [22]. They have observed only about a third of the theoretical prediction. This has been a puzzle for many years. The detector is sensitive to pp chain neutrinos from ^{8}B and ^{7}Be and CNO cycle neutrinos.

Kamiokande detector has added to it the new data [23] of the solar neutrinos with electrons in water as targets, and it is remarkable that it has measured the direction, the spectrum, and the real time of neutrino for the first time. It is sensitive only to ^{8}B neutrinos, and has observed only a half of the theoretical prediction. The new data by

GALLEX and SAGE with ^{71}Ga targets have observed neutrinos less than the prediction [24, 25] but it may be too early to conclude. Caliburation experiments with radioactive source are expected in the future, and will be very helpful. If we take the deficit results of all the three types of experiments seriously, we may have only the MSW neutrino oscillation model [26] left as a good explanation. We expect that the future plans of SNO [27] with 1,000 tons of heavy water target and SuperKamiokande [28] with 50,000 tons of water target will give very important data for this problem.

The atmospheric mu neutrinos were observed many years ago in India [14] and in South Africa [15], but the recent measurements of the atmospheric neutrino ratio ν_μ/ν_e have given a serious problem. The ratio found by Kamiokande [29] is about 60% of the theoretical prediction, and it is consistent with IMB experiments [30], suggesting a possibility of neutrino oscillation. The both groups are now doing calibration experiments using 12 GeV proton synchrotron at KEK. The data on energy dependence of the ratio will be helpful to understand the phenomena, and be obtained with SuperKamiokande [28] in the near future.

There are several projects under construction to detect high energy celestial neutrinos of around 1 TeV, such as DUMAND at deep under water near Hawai [31], AMANDA at deep ice of the Antarctica [32], Lake Baikal Project at deep under water of the lake [33], and NESTOR deep under the Mediterranean Sea [34]. They aim to detect neutrino point sources, such as active galactic nuclei, or to observe possible oscillation of neutrinos from the atmosphere or from accelerators, and will start full operation in the near future.

4 Proton Decay and the Origin of the Universe

The universe consists of matter and not of anti-matter, which needs the three famous conditions by Sakharov:
(1)baryon number conservation should be violated,
(2)the symmetries C, and CP should be violated,
(3)the process should not be in thermal equilibrium.

The second condition is well known to be satisfied in laboratory experiments, and the third condition is satisfied in the rapidly expanding universe. Though the first condition is not yet observed in laboratory experiments, there are theoretically known two such possibilities . One is the standard weak interaction theory, where there are chiral gauge anomalies which can break the baryon number conservation, though it is not clear if the symmetry breaking force is strong enough or not. The other is GUT, though there are umbiguities about what should be the ultimate symmetry at the GUT energy scale. The latter in general predicts the proton decay life time accessible by underground detectors if they have enough volume of material. The simplest GUT model of SU(5) predicting the life time shorter than 10^{32} years has already been rejected by IMB and Kamiokande [35, 36]. The simple modifications of the model SO(10) or SUSY SU(5) can be tested by the coming SuperKamiokande experiments. If the decay is observed, it will make a tremendous progress of the particle theory, and will give us a deep understanding of the universe as early as the possible inflation period.

5 Unstable Isotopes Produced by Cosmic Rays

Many kinds of unstable isotopes are produced by cosmic rays, and used for various fields of science. Here let me give examples of ^{14}C and ^{10}Be.

The isotope ^{14}C is well known for the dating of ancient organic remains. It is produced in the atmosphere with the reaction $^{14}N(n,p)^{14}C$ with the neutron produced in the air by cosmic ray reactions. It is not only useful in dating, but also giving us important information of ecology such as circulation of CO_2 on the earth, the solar activities in the past, and the cosmic ray intensities in the past.

The unstable isotope ^{10}Be, which is also produced by cosmic rays in the air, has a longer life time, $t_{1/2} \sim 1.6$ million years, and is used for various purposes. The isotope is known to be found in rather young lava. It is considered as first being accumulated deep at the bottom of the sea, and being carried to deep below the land along the plate tectonic movement, and finally climbing within lava up to volcanos. The systematic correlation of ^{10}Be with abundance with the distance from the volcanic front is observed , giving the velocity of the plate consistent with other observations [37, 38] .

Raisbeck *et al.* observed a strong peak of ^{10}Be abundance in the ice core of 35,000 years ago taken at Vostok [39], Antarctca. It may perhaps be the evidence of the nearby explosion of supernova in the past [40]. If so, it will be extremely interesting, and worthy of more study to confirm it.

6 Summary

In summary we see the cosmic rays are closely related to particle/nuclear physics as well as to astrophysics/cosmology and various other fields of basic science, and we can expect fruitful progress of these fields of science in the future. The author would like to thank Professor S. Homma for encouragement.

References

1. V. F. Hess, Phys. Zeits. **11**, (1911) 998; *ibid.* **13**, (1912) 1084.
2. W. Kolhörster, Phys. Zeits. **14**, (1913) 1153.
3. S. Hayakawa, see the textbook, *Cosmic Ray Physics*, Wiley-Interscience John Wiley & Sons, (1969); see also the recent review by M. Teshima, *Proceedings of 23rd International Cosmic Ray Conference, Calgary* (1993).
4. R. Blandford and D. Eichler, review in Phys. Rep. **154**, (1987) 1; and references therein.
5. M. Garcia-Munoz *et al.*, Ap. J. Supp. **64** (1987) 269.
6. S. Yanagita *et al.*, Proceedings of 21st International Cosmic Ray Conference, Adelaide (1990), **4** 44.
7. K. Anraku *et al., Proceedings of 23rd International Cosmic Ray Conference, Calgary* (1993), ed. D. A. Leathy, **2** 156; see also *Proceedings of 3rd Workshop on Balloon-borne Experiments with Superconducting Magnet Spectrometers, KEK* (1992), ed. A. Yamamoto.
8. T. C. Weekes *et al.*, Ap. J. **342**, (1989) 379.
9. T. Kifune *et al., Proceedings of 23rd International Cosmic Ray Conference, Calgary* (1993), ed. D. A. Leathy, **1** 392.
10. D. Bird *et al.* , Phys. Rev. Lett. , to be published.

11. N. N. Efimov *et al.*, *astrophysical Aspects of the Most Energetic Cosmic Rays*, ed. M. Nagano & F. Takahara, World Scientific, (1990) 20.
12. F. Reines and C. L. Cowan, Phys. Rev. **92**, (1953) 830.
13. G. Danby *et al.*, Phys. Rev. Lett. **9**, (1962) 36.
14. KGF Group, C. V. Achar *et al.*, Phys. Lett. **18**, (1965) 196; *ibid.* **19**, (1965) 78.
15. F. Reines *et al.*, Phys. Rev. Lett. **15**, (1965) 429.
16. Kamiokande Collab., K. Hirata *et al.*, Phys. Rev. Lett. **58**, (1987) 1490.
17. IMB Collab., R. M. Bionta *et al.*, Phys. Rev. Lett. **58**, (1987) 1494.
18. J. Arafune *et al.*, Phys. Rev. Lett. **59**, (1987) 367.
19. J. L. Rosner *et al.*, Phys. Rev. Lett. **58**, (1987) 1906.
20. A. Burrows *et al.*, Phys. Rev. Lett. **68**, (1992) 3634.
21. see the review by P. Sikivie in *dark Matter in the Universe, Springer* (1988), ed. H. Sato and H. Kodama, 94; and the references therein.
22. R. Davis, Jr. *et al.*, Phys. Rev. Lett. **20**, (1968) 1205; see also *Frontiers of Neutrino Astrophysics, University Academy Press, Tokyo* (1992) 47.
23. Kamiokande Collab., K. Hirata *et al.*, Phys. Rev. Lett. **65**, (1990) 1297.
24. Gallex Collab., P. Anselmann *et al.*, Phys. Lett. **B285**, (1992) 376, 390.
25. SAGE Collab., A. I. Abazov *et al.*, Phys. Rev. Lett. **67**, (1991) 3332.
26. L. Wolfenstein, Phys. Rev. **D17** (1978) 2369, *ibid.* **D20** (1979) 2634; S. P. Mikheyev and A. Yu. Smirnov, Nouvo Ciento **C9** (1986) 17; H. A. Bethe, Phys. Rev. Lett. **56** (1986) 1305.
27. G. T. Ewanrwick , *Frontiers of Neutrino Astrophysics, University Academy Press, Tokyo*, (1992) 147.
28. M. Takita , *Frontiers of Neutrino Astrophysics, University Academy Press, Tokyo* (1992) 135.
29. K. S. Hirata *et al.*, Phys. Lett. **B205**, (1988) 416; see also T. Kajita, *Frontiers of Neutrino Astrophysics, University Academy Press, Tokyo* (1992) 293.
30. IMB Collab., R. Becker-Szendy *et al.*, *Frontiers of Neutrino Astrophysics, University Academy Press, Tokyo* (1992) 303.
31. DUMAND Collab., Proceedings of 23rd International Cosmic Ray Conference, Calgary (1993), ed. D. A. Leathy, **4** pp. 515-549.
32. S. W. Barwick *et al*, *Frontiers of Neutrino Astrophysics, University Academy Press, Tokyo* (1992) 329.
33. I. A. Belolaptikov *et al.*, Proceedings of 23rd International Cosmic Ray Conference, Calgary (1993), ed. D. A. Leathy, **4** 573.
34. E. G. Barinov, *et al.*, Proceedings of 23rd International Cosmic Ray Conference, Calgary (1993), ed. D. A. Leathy, **4** 554.
35. IMB Collab., R. Becker-Szendy *et al.*, Phys. Rev. ,**D42** (1990) 2974.
36. Kamiokande Collab., K. Hirata *et al.*, Phys. Lett. **B220**, (1989) 308.
37. L. Brown *et al.*, Nature, **299** 718.
38. M. Imamura *et al.*, Nucl. Inst. & Meth., **233** (1984) 211.
39. G. M. Raisbeck *et al.*, Nature, **326** (1987) 326.
40. G. E. Kocharov *et al.*, Proceedings of 22nd International Cosmic Ray Conference, Dublin (1991), ed. D. O'Sullivan, **2** 388.

Lecture given at the JSPS-INS International Spring School, March 1993 at Shimoda

Physics with Radioactive Nuclear Beams

Isao Tanihata

RIKEN (The Institute of Physical and Chemical Research)
2-1 Hirosawa, Wako, Saitama 351-01, Japan

1 Introduction

A principal constraint for studying nuclear physics and other nuclear science was the confinement of both beam and target species to the line of stability, the well-balanced nuclei. This restriction is a fundamental one for nuclear physics, not only because it precludes access to a vast region of bound nuclei, but also because it limits the type and extent of phenomena that can be studied in the nuclei that can be reached. Because of this, it is only for the narrow band of stable nuclei that we have detailed experimental information on radii, magnetic moments, single-particle structures, giant resonance, the entire field of scattering processes, and reaction mechanisms and so on.

Recent development of radioactive nuclear beams provides us new opportunities to extend our study using wider range of nucleus. Studies using radioactive nuclear beam has been remarkably developed in these few years and new knowledge have been gained in nuclear physics and astrophysics. Here in this lecture, I review methods of production and some selected recent use of the radioactive nuclear beams for studies of structure of nuclei far from the stability line.

The principle of the production of secondary radioactive nuclear beams is shown in the following section. Then I present the recent studies on structure of neutron rich nuclei, in particular the nuclear radii and the density distributions. Discoveries of thick neutron skins in ^{6}He and ^{8}He nuclei and the neutron halo in ^{11}Li and ^{11}Be will be presented.

Separators, which are used for production of radioactive nuclear beams, are also extensively applied for other studies such as new isotope searches, mass measurements, measurements of magnetic moment of β radioactive nuclei, and others. But these studies are not covered in this lecture.

Many articles have been published on the subject and other applications of the radioactive nuclear beams can be found in recent proceedings of international conferences and in review articles [1–3]. Readers, who need a more detailed description of a specific subject, are recommended forward to these references and references therein.

2 Production of Radioactive Nuclear Beams

In principle any nuclear reactions, which produce radioactive nuclei, can be used for production of secondary radioactive nuclear beams. However it only became practical to use radioactive nuclear beams in reaction studies after the discovery of "projectile fragmentation" in high-energy heavy-ion collisions. In fact the first systematic study with radioactive beam was made using the secondary beams from the projectile fragmentation. Here in this section we briefly review the principle of the production of secondary beams of radioactive nuclei.

2.1 Projectile fragmentation of heavy ions

At a beam energy well higher than a binding energy of a nucleon, nuclear fragments are emitted copiously into the narrow forward cone. This phenomena is called as the projectile fragmentation and the emitted nuclei are called as projectile fragments. The projectile fragmentation has three important characters in producing radioactive beams;

1. Various nuclides including nuclei far from the stability line are produced with large cross sections. In fact many extremely neutron (and proton) rich nuclides were discovered by the projectile fragmentation.
2. All projectile fragments are emitted with almost same velocity as the projectile. Therefore the separation of nuclides is simple and a high collection efficiency can easily be obtained.
3. A fragmentation cross section is almost independent of the incident energy.

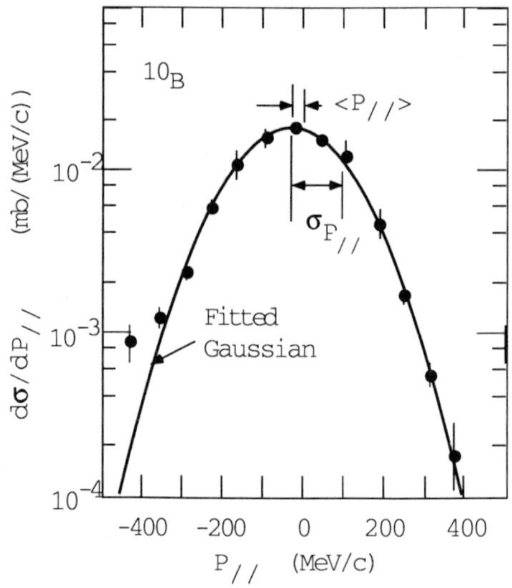

Fig. 1. Momentum distribution of ^{10}B fragment from ^{12}C reaction at 2.1*A* GeV. The distribution is plotted in the rest frame of the incident particle.

Now we see the momentum distribution of the projectile fragments in the following, since it is essential for the radioactive beam production. Figure 1 shows the momentum distribution, in the projectile rest frame, for ^{10}Be fragments from ^{12}C fragmentation at 2.1*A* GeV as an example [4]. As can be seen, a Gaussian shape provides a good fit to the spectra observed for all isotopes regardless of beam, energy, or target. It is also found that the momentum distribution are almost the same in both parallel (//) and perpendicular ($\perp$) directions. A spectrum is characterized by its central momentum $\langle P_{//} \rangle$ and standard deviation $\sigma(P_{//})$. The value of $\langle P_{//} \rangle$ was found to be in the range of -10 to -130 MeV/c for various fragments. (A negative sign for $\langle P_{//} \rangle$ indicates that the fragment speed is less than that of the projectile.) A fragment has almost the same velocity as the incident projectile nucleus because the total momentum of the projectile fragment is much larger than $\langle P_{//} \rangle$.

The width of the momentum spread $\sigma(P_{//})$, as well as $\langle P_{//} \rangle$, are found to be essentially independent of target mass and beam energy but do depend on the mass number of the projectile (A_P) and of the fragment (A_F). The dependence of $\sigma(P_{//})$ on A_P and A_F can be expressed as

$$\sigma(P_{//}) = \sigma_0 \sqrt{A_F(A_P - A_F)/(A_P - 1)} \tag{1}$$

where $\sigma_0 = 90$ MeV/c [5], $\sigma(P_{//})$ takes its maximum value when $A_F = A_P/2$. The width $\sigma(P_\perp)$ of the transverse momentum distribution of the fragment is equal to $\sigma(P_{//})$, it thus shows isotropic production of fragments in a frame moving at $\beta_{//} = -\langle P_{//} \rangle / E$ in the projectile frame. The value of $\sigma(P_{//})$ is known to remain constant down to energies as low as $20A$ MeV/nucleon [6]. In contrast it is found that $\sigma(P_\perp)$ behaves differently at low energies. Empirically, it was shown that, at projectile energies below 200 MeV/nucleon, the $\sigma(P_\perp)$ is fitted by [7],

$$\sigma^2(P_\perp) = \sigma_0^2 A_F(A_P - A_F)/(A_P - 1) + \sigma_1^2 A_F(A_P - A_F)/A_P(A_P - 1) \tag{2}$$

where $\sigma_1 = 200$ MeV/c.

In order to minimize broadening of the product momentum it is important to use a primary beam of mass number close to that of the desired nucleus but, especially at high energy, to have both the proton and the neutron numbers of the projectile nucleus to be larger than those of the desired product nucleus.

2.2 Separation technique of secondary beams

In principle any type of recoil mass separator with high collection efficiencies can be used to deliver a secondary beams. However, special attention is required for efficient separation. Those are;

1. Production target has to be selected so that a produced nuclide has a small momentum and an angular broadening.
2. The momentum and the angular acceptance of a separator should be as large as possible to obtain the highest intensity of a separated beam.
3. It is desirable, in most cases, that the separated beam is delivered achromatically (position of focus is independent of a particle momentum) to a reaction target.

The requirement 1 favors projectile fragmentations of high-energy heavy-ions and some cases of low-energy transfer reactions. In general the lightest and easy to handle element Be provides the best selection as a target for projectile fragmentation.

Now we see the principle of the separation.

2.2.1 Separation method

The necessary nuclear separation depends on the specific experiment. For example, if one wishes to measure the reaction cross-section at high energy, the admixture of other nuclei in the secondary beam can be effectively rejected by identifying the nuclei one by one. On the other hand, if one wishes to study low-energy reaction having small cross-section, a single-nuclide beam is essential because it is difficult to identify a particle under the required strong intensity.

Taking advantage of the properties of the projectile fragmentation, the first beam line was constructed and used in the reaction studies at Berkeley [8]. The magnetic rigidity R of a fully stripped ion is written as,

$$R = \frac{P}{Z} = \frac{\beta\gamma m A}{Z}, \tag{3}$$

where P is the total momentum of an ion, m is the nuclear mass unit and β, γ are Lorentz factors. Because velocities of fragments (β) are same, a selection by the magnetic rigidity gives an A/Z separation. For many types of experiment the A/Z separation is sufficient because Z can be identified easily particle by particle before incident on a reaction target. As seen in Fig. 2 this type of separator has two sections.

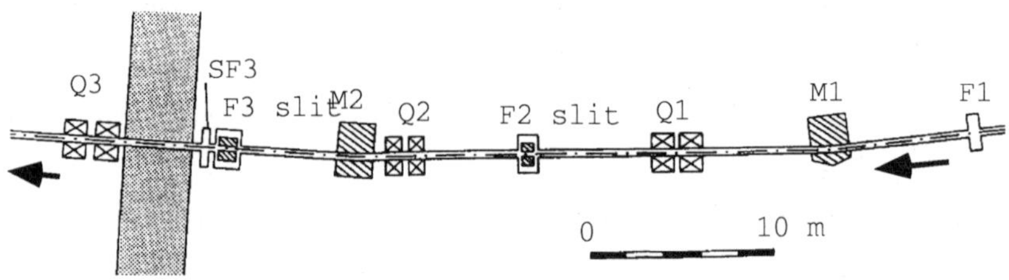

Fig. 2. The secondary beam line at LBL, the first beam line used for reaction studies with high-energy radioactive nuclear beams. Secondaries are produced at F1 and separated at F2.

First, the primary beam from an accelerator is focused on the first focus (F1) where a production target is placed. Fragments produced at the target are analyzed at F2 – a focus with large momentum dispersion. A separation slit is located at F2 and nuclei within certain range of rigidity are selected. Then the separated beam is focused achromatically at F3. The cleaning collimator is placed there at F3 to remove any beam halo. The separated beam is then delivered to an experimental area.

A certain type of experiment requires a beam without admixture of other nuclide than an objective one. An additional separation can be obtained by an energy degrader because a dE/dx is proportional to Z^2 and is different from A/Z. In particular "energy-loss achromat" method, an energy degrader technique developed at LISE of GANIL, provides a simple and efficient method of separation [9]. Other separators, the RIPS at RIKEN, the separator at MSU, and the FRS at GSI, also use this technique.

We now discuss the principles of this separation briefly. It is known that at intermediate energy the range R of the ^{A}Z nucleus having an energy per nucleon E can be empirically expressed as,

$$R = \kappa E\gamma A/Z^2 + CA, \tag{4}$$

where κ, γ, and C are constants depending only on the stopping material. After some manipulation of this equation, one calculate the rigidity R_2 of the nucleus after passage through an absorber as

$$R_2(A, Z, d, R_1) = R_1 \left(1 - \frac{dA^{2\gamma-1}}{\kappa Z^{2\gamma-2}} R_1^{-2\gamma}\right)^{1/2\gamma} \tag{5}$$

The ions not separated in terms of R_2 are those having the same ratio $A^{2\gamma-1}/Z^{2\gamma-2}$. Using $\gamma = 1.75$ as an example (the value for an aluminum absorber) this ratio may be written as $A^{2.5}/Z^{1.5}$. This system thus provides a selection on A and Z that is different from the first selection which is sensitive only to A/Z. This difference in selectivity is shown in Fig. 3. One can see that any single nucleus can, in principle, be separated using this approach. Such a separator, called RIPS, is being used in RIKEN for various experiment. Figure 4 shows a plan view of the RIPS.

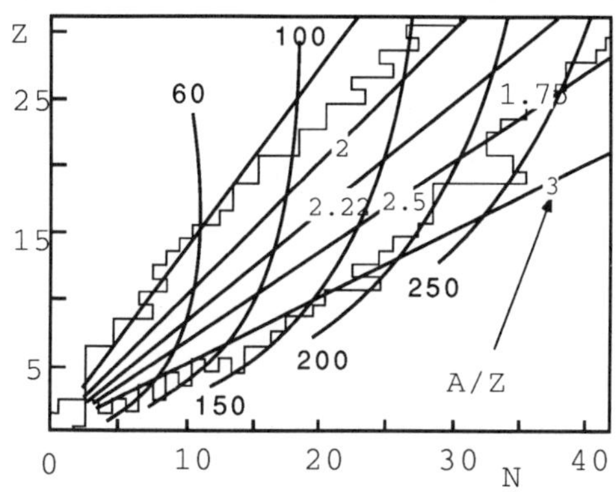

Fig. 3. Selectivity of the separation magnet and combined energy degrader. Straight lines show the selections made by the first part of the secondary beam line. Curved lines indicates the selection by a combination of achromatic degrader and the last half of the magnets. Application of both provide a separation of single isotope.

2.2.2 Facilities

High and intermediate energy facilities have been constructed in several laboratories such as LBL, GSI, MSU, RIKEN, GANIL, and are under construction in many laboratories [1].

Among those the RIPS at RIKEN has to be mentioned because of its unique setup (see Fig. 4), the beam swinger, in front of the production target [10]. Using the beam swinger, the scattering angle of projectile fragments can be selected by changing the incident angle of a primary beam up to 10°. It then provides a tool to select polarized nuclei. Polarized nuclei have been used for determination of their magnetic moments [11].

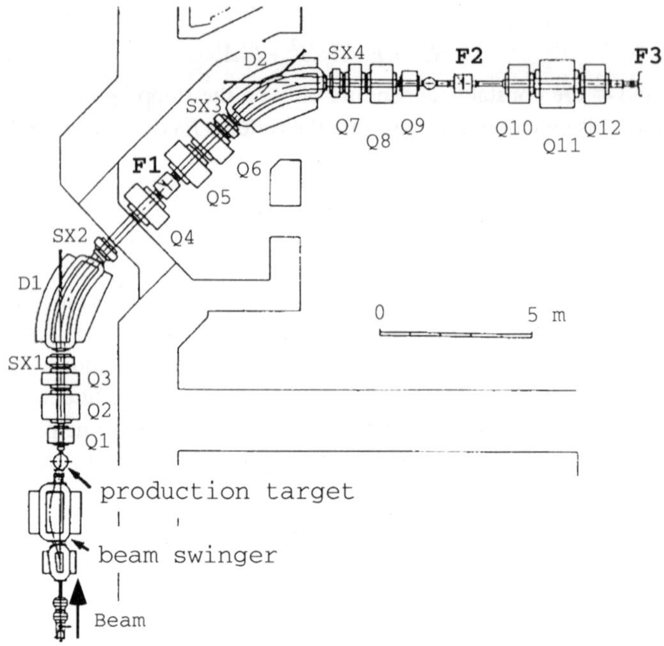

Fig. 4. The fragment separator RIPS at RIKEN Ring Cyclotron Facility. It uses the energy achromat technique using wedge degrader at F1. Its unique feature is the beam swinger by that it can deliver polarized radioactive nuclear beams.

3 Nuclear Radii

3.1 Measurements of interaction and fragmentation cross sections

To study properties of light exotic nuclei, measurements were made on interaction cross sections, momentum distributions of projectile fragments, and electromagnetic-dissociation (EMD), mostly at 0.8*A* GeV and some at lower energies 0.4*A* and 0.6*A* GeV. Typical experimental setups are shown in Fig. 5. Interaction cross sections, σ_I were measured by a transmission method. Here the σ_I is defined as the total cross section for the change of proton and/or neutron number of the projectile nucleus. The HISS spectrometer at the LBL was used to identify the non-interacting ones after the target, by its magnetic rigidity, *dE/dx*, and time-of-flight. When the same nucleus was detected as observed before the target, the event was counted as non-interacting. The cross section was then calculated by the formula,

$$\sigma_I = (1/N_T) \log(\gamma_o/\gamma) \tag{6}$$

where N_T is the number of target nuclei per cm^2, γ is the number ratio of non-interacting to incoming ones for the target-in run, and γ_o is the same ratio for the target-out run. The values of γ_o were larger than 0.95 in most cases. By taking the ratio γ_o/γ, uncertainties due to the counter inefficiencies and reactions occurred somewhere other than the target were eliminated.

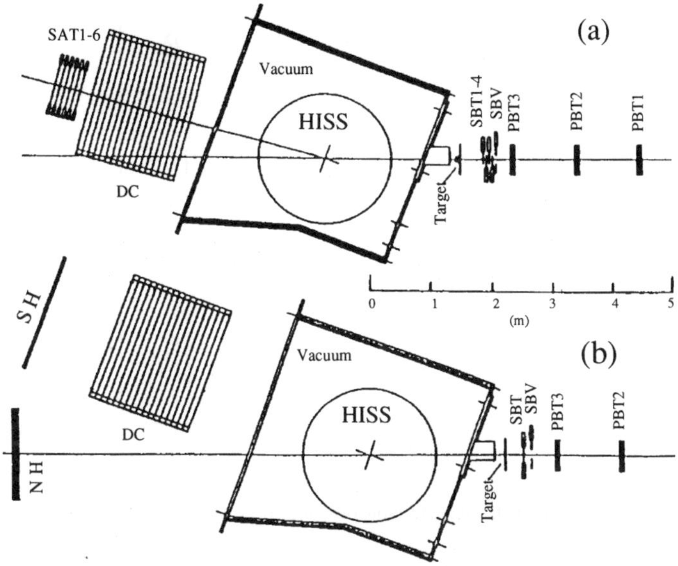

Fig. 5. Experimental setup for interaction cross section (a) and fragmentation cross section (b) measurements. PB's are MWPC's and SB's and SA's are scintillators.

For the studies of fragmentation processes, projectile fragments were measured at around zero degree with the HISS, equipped with a particle-detector system of a wide coverage that composed of multi-wire proportional chambers (MWPC), drift chambers (DC), and a scintillation hodoscope (SH). The range covered by the detector system was $\pm 20\%$ in rigidity and $\pm 2.0°$ in angle. Transverse-momentum distributions of fragments were measured as well as the partial fragmentation cross sections. In order to detect neutrons in coincidence with a specific fragment, a neutron hodoscope (NH) in addition was also used.

3.2 Radii of light nuclei

3.2.1 Interaction radii and rms radii

The nuclear sizes were studied from the interaction cross sections $\sigma_{\rm I}$. The measurements were made so far on most of p-shell and some sd-shell nuclei with targets of natural Be, C and Al. Since impacts given to participant nucleons at a nuclear collision of a few hundred A MeV are far beyond the internal motions, the cross section, $\sigma_{\rm I}$ reflects the nuclear sizes as represented by the formula,

$$\sigma_{\rm I} = \pi\{R_{\rm I}({\rm P}) + R_{\rm I}({\rm T})\}^2 \tag{7}$$

where $R_{\rm I}$(P) and $R_{\rm I}$(T) are the interaction radii of the projectile and the target, respectively. An assumption made here, to make the interaction radius meaningful, was that the interaction radii of the projectile and the target are separated in the equation (separability) [12]. The assumption of separability can examined by experimental values of

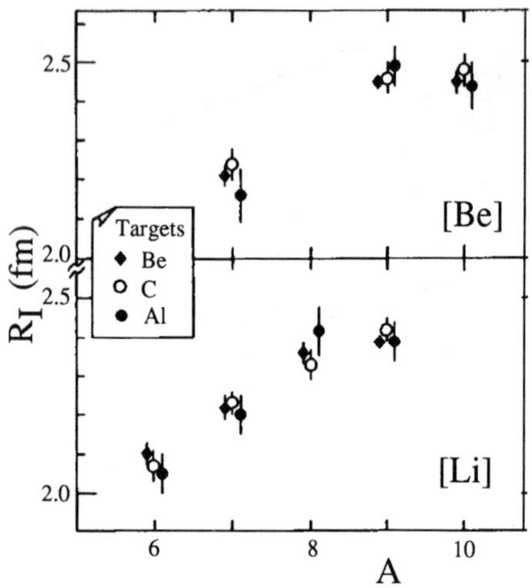

Fig. 6. Interaction radii $R_{\rm I}$ for Li and Be isotopes obtained with three different targets.

$\sigma_{\rm I}$'s for various projectile-target combinations. Figure 6 shows $R_{\rm I}$ of Li and Be isotopes obtained from Be, C and Al targets. Here the absolute scale of radius was determined from the least-square fit of $\sigma_{\rm I}$'s of ^{4}He +^{4}He, ^{9}Be +^{9}Be, ^{12}C +^{12}C, ^{4}He +^{12}C and ^{9}Be +^{12}C collisions. It is seen that a projectile interaction radius was in fact independent of the target selection. The separability of interaction radii was thus far demonstrated to hold within a few per cent for all combinations of He-, Li-, Be- or B-isotope projectiles with ^{9}Be, ^{12}C or ^{27}Al targets [12]. The separability was also demonstrated by a theoretical calculation based on the Glauber model [13]. It showed that the separability hold well if the surface diffuseness of a nucleus does not change drastically.

The interaction radii of individual nuclei so far determined are plotted in Fig. 7. Generally $R_{\rm I}$ follows an $A^{1/3}$ line as indicated in the figure. However large growth of radii is seen in neutron rich nuclei.

The measured interaction cross sections, $\sigma_{\rm I}$ were related to the root-mean-square (rms) radii, $R_{\rm rms}$ using the optical-limit calculationof the Glauber model [14], and $R_{\rm rms}$'s of light nucleihave been deduced [12]. The model relates the $\sigma_{\rm I}$ to the nucleon-nucleon cross sections, $\sigma_{\rm NN}$ and the nucleon density distributions, ρ(r) as

$$\sigma_{\rm I} = 2\pi \int_0^\infty \{1 - T(b)\} b\, db \tag{8}$$

with

$$T(b) = \exp\left\{ -\sigma_{NN} \int \rho_{T,z}(S)\rho_{P,z}\left(\left|\vec{b} - \vec{S}\right|\right) d^2s \right\} \tag{9}$$

where $\rho_{i,z} = \int \rho_i(\sqrt{s^2 + z^2})dz,\ i$=P(projectile) or T(target).

It was noticed firstly that the obtained $R_{\rm rms}$ was almost independent of the assumed density distributions, i.e., a Gaussian or a shell-model harmonic-oscillator distribution,. Secondly, the obtained $R_{\rm rms}$ for stable nuclei, ^{4}He, ^{6,7}Li, ^{9}Be and ^{12}C were compared

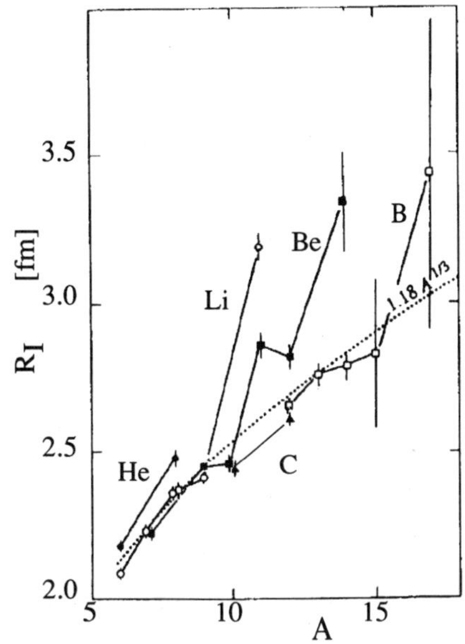

Fig. 7. Interaction radii of the p-shell nuclei.

with the rms radii R^e_{rms} of the electron-scattering data [10], and the agreements between the both radii for charge distribution were very good as shown in Fig. 8.

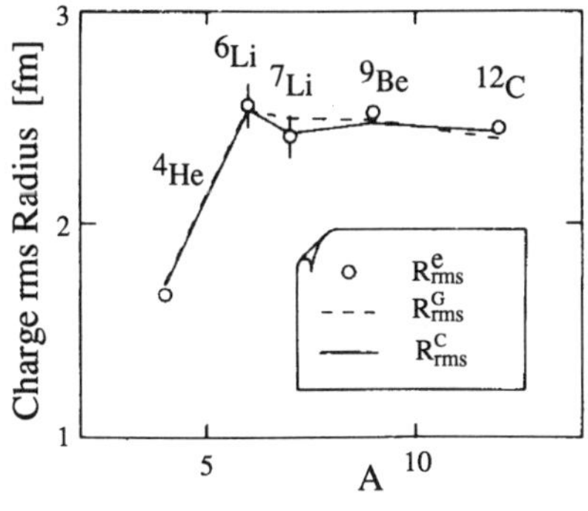

Fig. 8. Charge rms radii of light stable nuclei. The circles indicate radii determined by the electron scatterings. The dashed line shows the results obtained from fitting the interaction cross sections assuming the Gaussian density distribution, and the solid line shows the results assuming the harmonic-oscillator shell-model distribution.

3.2.2 *Isospin dependence of nuclear radii*

The isospin dependence of nuclear matter radii provides a new systematics which could not be studied without radioactive nuclear beams. The observed rms radii of nucleon distributions are plotted for isobars of mass number A from 6 to 12 in Fig. 9. Mirror pairs of isospin, ^{7}Li-^{7}Be,^{8}Li-^{8}B, and ^{10}Be-^{10}C, show equal radii within the experimental errors, suggesting that the Coulomb effect on the radii is small for these light nuclei. On the other hand, nuclei with larger isospins show larger radii except for ^{9}Be in $A = 9$ isobars.

The observed isospin dependence of isobar radii was compared with predictions of Hartree-Fock calculations using the Skyrme potential by Sato and Okuhara [13], as shown in Fig. 9. Here two kinds of potential parameter sets, SIII and SV, were employed; SIII includes a strong density-dependent interaction, while SV includes no such interaction. The both predictions gave fair agreements within 0.2 fm to the data in most cases. While the SV prediction only represented a much weaker isospin dependence than observed, the SIII prediction well reproduced the observed isospin dependence, except for $A = 9$ isobars. The comparison, therefore, indicated necessity of a strong density-dependent interaction to understand the observed behavior.

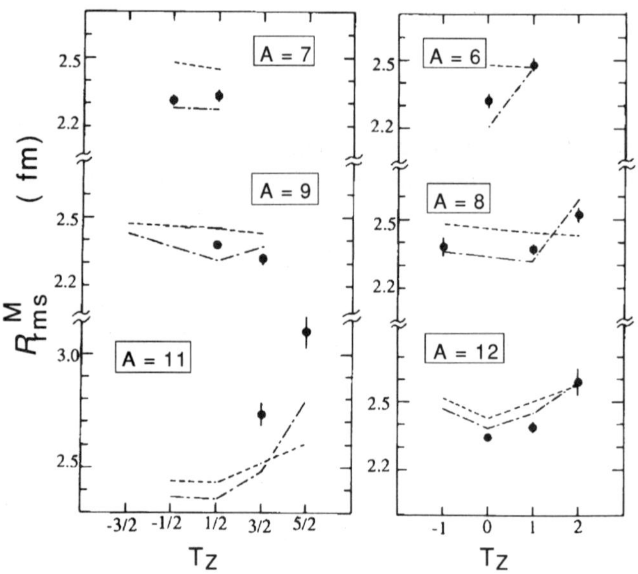

Fig. 9. Isospin dependence of rms matter radii. Experimental values are shown by solid circles. Hartree-Fock calculations with Skyrme potential are shown for SIII by dash-dotted lines, and for SV by dashed lines.

The irregularity seen in $A = 9$ isobars is considered to be due to a special cluster structure in ^{9}Be, i.e., ^{9}Be has the strong 2α+n configuration, and thus two α clusters are weakly bound by a neutron. Therefore the density distribution spreads more than usual nucleus.

4 Neutron Skin

How large difference can one see between the proton and the neutron density distributions in nuclei. It is a long standing question in nuclear physics. We discuss here the first empirical verification of a thick neutron skin ($\sim$ 0.9 fm) in ^{6}He and ^{8}He.

Matter radii of He isotopes derived from the interaction cross sections showed a drastic increase of the rms radius from ^{4}He (1.57 $\pm$ 0.05 fm) to ^{6}He (2.48 $\pm$ 0.03 fm) and ^{8}He (2.52 $\pm$ 0.03 fm). It was difficult, however, to draw any quantitative conclusion only from these data about the difference of the density distributions of protons and neutrons. Concerning this problem, an important relation between the interaction cross sections and the nucleon-removal cross sections was deduced by Ogawa, Yabana and Suzuki [15] by applying the Glauber model to a loosely bound system: Suppose that ^{6}He is well described by a ^{4}He core and two neutrons, then the interaction cross sections $\sigma_{\rm I}$ for ^{6}He and ^{4}He and the two-neutron removal cross section $\sigma_{\text{-2n}}$ of ^{6}He with the same target, are related as,

$$\sigma_{\text{-2n}}(^6\text{He}) = \sigma_{\rm I}(^6\text{He}) - \sigma_{\rm I}(^4\text{He}). \tag{10}$$

Here they used the fact that ^{5}He is particle unbound. This relation holds only if the ^{4}He core in ^{6}He is not modified from the free ^{4}He. Although multiple scatterings between the projectile and the target nucleons are included in the Glauber model, rescatterings, such as a collision between the ^{4}He core and a knockout neutron from ^{6}He itself, are not included. However its effect is presumably small for the present case because of a peripheral nature of the reaction and the tight binding of ^{4}He. This relation, in turn, provides a way to test experimentally how well does the core persist in the nucleus. In the case of ^{6}He, the data almost satisfied the equality of Eq. (10);

$$\sigma_{\text{-2n}}(^6\text{He}) = 189 \pm 14 \text{ mb, and } \sigma_{\rm I}(^6\text{He}) - \sigma_{\rm I}(^4\text{He}) = 219 \pm 8 \text{ mb,}$$

indicating that ^{4}He remains intact as a good core in ^{6}He. Where cross sections with C target at 800A MeV were used. In the case of ^{8}He, however, the relation did not hold at all;

$$\sigma_{\text{-2n}}(^8\text{He}) = 202 \pm 17 \text{ mb, and } \sigma_{\rm I}(^8\text{He}) - \sigma_{\rm I}(^6\text{He}) = 95 \pm 9 \text{ mb.}$$

The ^{6}He wave function is, therefore, strongly modified in ^{8}He and thus has no identity as a core. Instead, the relation,

$$\sigma_{\text{-2n}}(^8\text{He}) + \sigma_{\text{-4n}}(^8\text{He}) = \sigma_{\rm I}(^8\text{He}) - \sigma_{\rm I}(^4\text{He}), \tag{11}$$

an extension of Eq. (10) under the same assumption about the ^{4}He core in ^{8}He, hold well;

$$\sigma_{\text{-2n}}(^8\text{He}) + \sigma_{\text{-4n}}(^8\text{He}) = 297 \pm 19 \text{ mb and } \sigma_{\rm I}(^8\text{He}) - \sigma_{\rm I}(^4\text{He}) = 314 \pm 8 \text{ mb.}$$

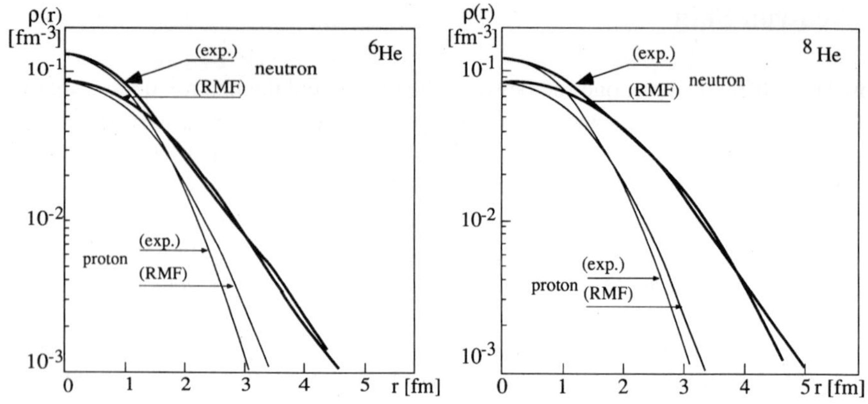

Fig. 10. Proton and Neutron density distributions in ^{6}He and ^{8}He nuclei. The solid curves show the distribution deduced from the interaction cross sections under the assumption that the ^{4}He core is not modified in ^{6}He nor in ^{8}He. Dotted curves show the predictions of the RMF model.

Noted here that both ^{5}He and ^{7}He are unbound. Therefore, ^{8}He is well described by a ^{4}He core and four neutrons orbiting around the core.

Under the assured assumption that ^{4}He forms a good core, the density distributions of nucleons in ^{6}He and ^{8}He have been deduced from $\sigma_{\rm I}$'s for ^{4}He, ^{6}He and ^{8}He on C target by the optical-limit calculation of the Glauber model. In the fitting, we used the density distribution of ^{12}C obtained from electron scattering data. For He isotopes we assumed the harmonic-oscillator type density distribution with different size parameters for $1s$ and $1p$ orbitals. Firstly the size parameter of $1s$ orbital was determined to fit $\sigma_{\rm I}(^4{\rm He})$. Then the size parameter of $1p$ orbital in ^{6}He and ^{8}He was determined to fit $\sigma_{\rm I}(^6{\rm He})$ and $\sigma(^8{\rm He})$, respectively. Here the center-of-mass motion of the core was taken into account assuming no strong correlation between the orbital neutrons. The resultant nucleon distributions of ^{6}He and ^{8}He are shown in Fig. 10. The distribution of neutrons extended far beyond that of protons, and the rms radius for neutrons, $R_{\rm n}^{\rm rms}$ was much larger than that for protons, $R_{\rm p}^{\rm rms}$; ($R_{\rm n}^{\rm rms} - R_{\rm p}^{\rm rms}$) $\sim$0.9 fm for both ^{6}He and ^{8}He. It was thus shown that these nuclei have thick neutron skins. It was not due to the halo effect of shallow binding neutrons as observed in ^{11}Li, (see next section) because the two neutron separation energy is 2.0 MeV for ^{8}He. An appreciable neutron halo only appeared in a nucleus with an extremely small separation energy of the last neutron(s), i. e., 0.3 MeV for ^{11}Li, and 0.5 MeV for ^{11}Be. The ^{6}He case was more or less marginal however because the separation energy is about 1 MeV.

The density distributions of ^{6}He and ^{8}He thus obtained were compared with predictions of the relativistic-mean-field (RMF) model [16]. As seen in Fig. 10, the empirical density distributions for protons and for neutrons were well reproduced by the RMF model, except at the central region where the present experimental data were not very sensitive. Stimulated by this success, the RMF-model calculations were extended to other nuclei. The results are shown in Fig. 11, in which the difference between the neutron and the proton rms radii ($\Delta R_{\rm rms} = R_{\rm n}^{\rm rms} - R_{\rm p}^{\rm rms}$) is plotted against the difference between the neutron and the proton Fermi energies ($\Delta E_{\rm F} = E_{\rm F}^{\rm n} - E_{\rm F}^{\rm p}$) for each nucleus.

The values of $\Delta R_{\rm rms}$ for a wide mass range lie in gross on a straight line and increase monotonously with $\Delta E_{\rm F}$, and the $\Delta R_{\rm rms}$ larger than 0.5 fm are seen for nuclei with $\Delta E_{\rm F}$ larger than 10 MeV. For stable nuclei ($\Delta E_{\rm F} \sim 0$), the $\Delta R_{\rm rms}$ are smaller than 0.3 fm even for neutron rich nuclei, such as ^{48}Ca and ^{208}Pb, in consistent with the observations [17].

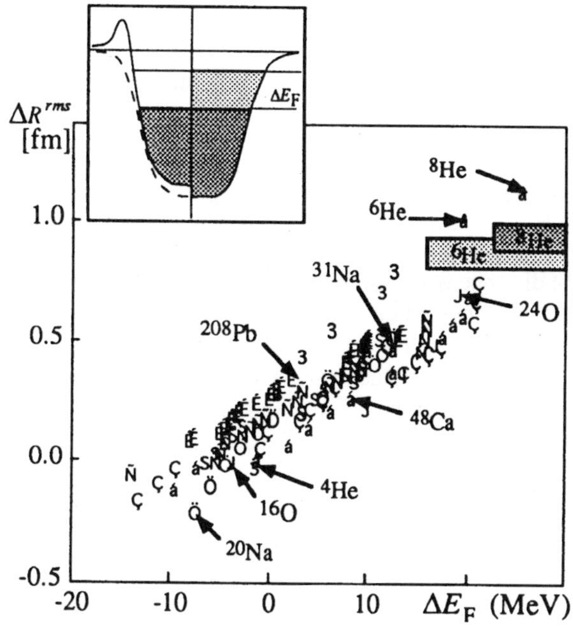

Fig. 11. The radius difference $\Delta R_{\rm rms} = R_{\rm n}^{\rm rms} - R_{\rm p}^{\rm rms}$ for protons and neutrons in a nucleus, as a function of the Fermi-energy difference, $\Delta E_{\rm F} = E_{\rm n} - E_{\rm p}$, obtained by the RMF model. The empirical values for ^{6}He and ^{8}He are shown by the shaded areas of which horizontal length have no meaning since the ΔE's are not known. The insert shows the nucleon occupation in a self-consistent potential. The dark area indicates the occupation in a stable nucleus. The lightly shaded area shows the extra neutrons which contribute to a formation of neutron skin in neutron rich nuclei.

The physical content behind the plot of Fig. 11 is as follows. In a stable nucleus, neutrons and protons occupy orbits up to almost the same Fermi energy, as illustrated by the dark shadow in the insert of Fig. 11. Then, if we assume that the nuclear potentials and the density distributions are self-consistently determined and temporarily drop the small isovector potential as a sake of discussion, the proton and the neutron potentials are the same except for the Coulomb potential. Only the existence of the Coulomb potential makes the radius for protons slightly smaller. No thick neutron skin is, therefore, expected for stable nuclei even for the case of a large neutron excess. In addition, the isovector potential hitherto neglected acts attractively for protons and repulsively for neutrons in neutron excess nuclei and tends to make the difference of the density distributions even smaller than discussed above. It is therefore not relevant to plot the skin thickness against the neutron excess. Instead, the Fermi energy difference, $\Delta E_{\rm F}$ is an appropriate quantity because it actually reflects the orbital sizes of the lastly filled proton(s) and neutron(s). In a beta-radioactive nucleus far from the stability line, $\Delta E_{\rm F}$ becomes large and thus a large difference is generally expected between the proton and the neutron radii.

5 Neutron Halo

Here we discuss a new aspect of very neutron rich nuclei, the neutron halo. A recent study has confirmed the formation of a neutron halo in ^{11}Li and ^{11}Be.

5.1 What is the neutron halo?

In a nucleus having a neutron halo, a low density tail of the neutron distribution is extends out to a large distance from the center of nucleus. Although configuration mixing of high lying orbitals may contribute, the main cause of such a long tail is considered to be the small separation energy of neutrons.

Let us assume a nucleus with a neutron loosely bound to an inert core. If we assume an interaction potential between the neutron and the core to be a square well, the size of the potential is that of the core nucleus itself. (Fig. 12) The wave function of the neutron outside the potential is written as,

$$\Psi(r) = \left(\frac{2\pi}{\kappa}\right)^{-1/2} \frac{e^{-\kappa r}}{r} \left[\frac{e^{\kappa R}}{(1+\kappa R)^{1/2}}\right], \tag{12}$$

where R is the width of the potential. Using this wave function the density distribution of the neutron is written as,

$$\rho(r) = |\Psi(r)|^2. \tag{13}$$

The parameter κ, which determines a steepness of the slope of the density tail is related to a separation energy of the neutron $E_{\rm s}$ as,

$$(O\kappa)^2 = 2\mu E_{\rm s}, \tag{14}$$

where μ is the effective mass of the system. As can be seen from these equations, as $E_{\rm s}$ decreases κ become smaller and thus the tail of the distribution gets longer.

Although the surface diffuseness is known to be constant for stable nuclei, that constancy can be understood simply as being due to the nearly constant nucleon separation energy (6-8 MeV) for stable nuclei. In general the surface diffuseness is expected to depend on the nucleon separation energy. The neutron halo is the most pronounced case for a small separation energy (<1 MeV).

On the other hand the momentum distribution $f(p)$ of the neutron is expressed by the Fourier transform of the wave function:

$$f(p) = C\,/\,(p^2+\kappa^2). \tag{15}$$

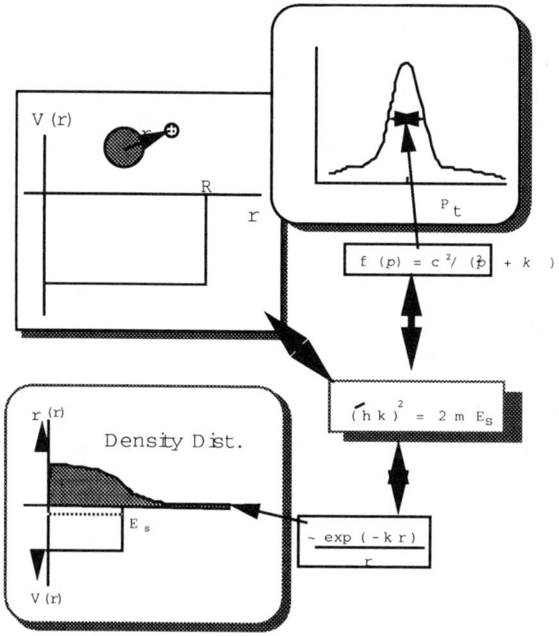

Fig. 12. Relation between density tail and momentum distribution.

The width of the momentum distribution is again related to the parameter κ. In contrast with the density distribution, the smaller the E_s, the smaller the width of the distribution. This is obviously a reflection of Heisenberg's uncertainty principle; when the distribution in coordinate space is wide, that in momentum space is narrow.

As shown above, the neutron halo is formed in a nucleus with a very small neutron separation energy. The formation of the neutron halo can be identified by a large extension of the density distribution or of an apparent large nuclear size and by a narrow momentum distribution.

5.2 Nuclear radii and density distribution

As shown in Fig. 7, sudden growth of radius for some of the neutron rich nucleus was seen. In particular the radii of ^{11}Li, ^{11}Be, and ^{14}Be are extremely large compared with those of neighboring nuclei. The separation energy of the last two neutrons in ^{11}Li is known to be 340 keV and that of the last neutron in ^{11}Be is 500 keV. Thus both of these nuclei have extremely small separation energy. The calculation based on the model shown in the previous section shows excellent consistency between the observed radius of ^{11}Li and the two neutron separation energies [18]. In this model, the radius of the core is assumed to be equal to that of ^{9}Li.

A strong deformation could, of course, produce the large apparent size of these nuclei. However the magnetic moment of ^{11}Li was determined recently and the value is found to be very close to the Schmidt value for $P_{3/2}$ proton. Therefore no large deformation is expected. It was also demonstrated by a deformed Hartree-Fock calculation that no deformations are expected for neutron rich Li isotopes.

The situation is much more complicated for ^{11}Be. The ground state of ^{11}Be is known to show an anomalous spin parity assignment ($1/2^+$) suggesting a strong mixing of *sd*-shell orbitals. The deformed Hartree-Fock calculation can reproduce the observed radii of Be isotopes reasonably well with a large deformation. Recently, a new data have become available for the reaction cross section of ^{11}Be at much lower energy (33A MeV). The measurement has been made at the RIPS of the RIKEN Ring Cyclotron Facility. Because the nucleon -nucleon cross sections at 33 MeV ($\sim$ 200 mb) are much larger than that at 800 MeV ($\sim$ 40 mb), a low density tail of the neutron halo can be seen more sensitively at lower energy.

The density distribution of ^{11}Be was investigated using energy dependence of the interaction cross sections by applying various trial function shapes [19]. The σ_R were calculated by a Glauber-type calculations with these distributions and the quality of the fit to the experimental data at 33A and 790A MeV for ^{12}C and ^{27}Al targets were examined. The density distributions for the stable nuclei, ^{12}C, ^{13}C, and ^{27}Al, were taken from the data of electron scattering [20].

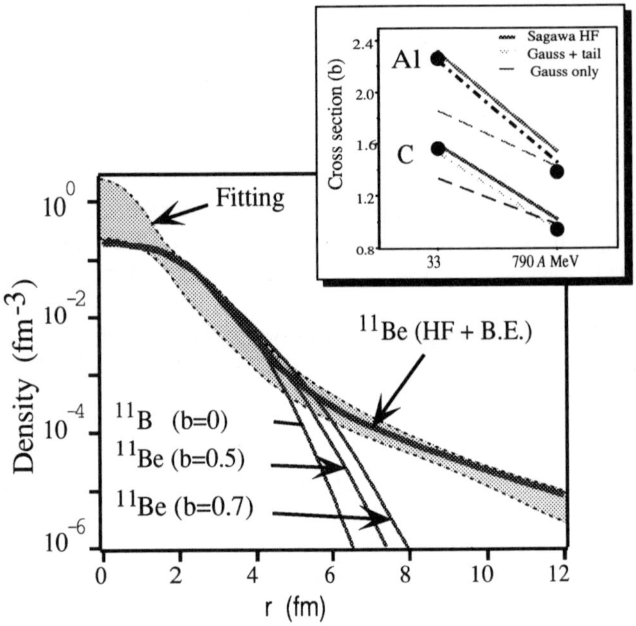

Fig. 13. Nuclear density distributions of neutron halo nuclei ^{11}Be determined by the energy dependence of the interaction cross sections. It shows the long tail associated with halos. Deformed density distributions for ^{11}Be with deformation parameter b=0.5 can reproduce the cross section at 800A MeV. However it fails to reproduce the energy dependence of the cross section.

We first examine the effect of deformation. The Glauber-type calculations with the deformation parameter β =0.5-0.7 reproduce the cross sections well at 790A MeV for both C and Al targets. In this calculations, density distribution of ^{11}Be was made by deforming the phenomenological ^{11}B density distribution. The angular average of the deformed density distributions with $\beta = 0.5$ and 0.7 are also shown in Fig. 13. However σ_R at $33A$ MeV are not explained by deformations of these β values. A broader spread of the nucleon density distribution is required to explain the large σ_R at $33A$ MeV.

We then test two other types of density distributions, a Gaussian and a Gaussian plus Yukawa-square tail. The former shape is known to be a good approximation for light nuclei. On the other hand, the latter may be a closer expression reflecting the effect of finite binding energy. The results were shown in Fig. 13. The best-fit by Gaussian distribution is shown by long dashed lines in the insert. In this fit the width parameter of the Gaussian was the only parameter. Clearly the fitting is very poor. In contrast, the σ_R are fitted well by the "Gaussian plus Yukawa-square tail" distribution as shown by the dash-dotted lines. Here, the Gaussian width, the tail amplitude, and the tail slope were taken as the free parameters. The density distribution yielding the best fit is shown by the shadowed area in Fig. 2. Although the density distribution in the central part is not well determined, the tail part of density distribution is restricted very much. The tail part of the distribution contains a significant part of the total nucleon number. The integrated number of nucleons outside the radius of 5 fm, for example, is 0.5.

In order to study the effect of surface diffuseness in ^{11}Li, the interaction cross sections of ^{11}Li were also measured with a proton and a deuteron target. It was noticed that in these collisions the separability of the interaction radii was strongly violated.

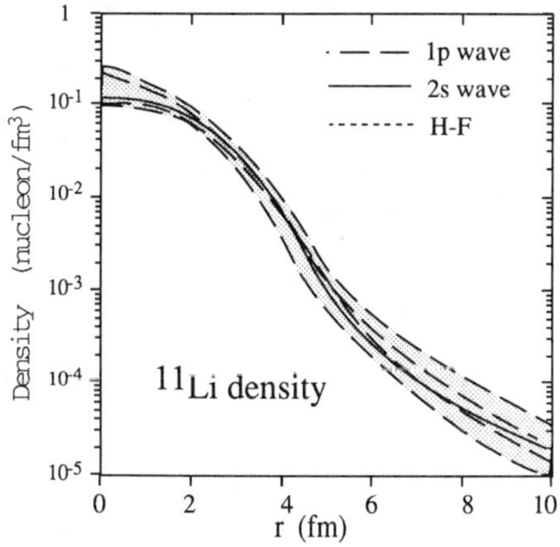

Fig. 14. Density distribution of ^{11}Li deduced from the interaction cross sections. The selection of orbitals, $1p$ or $1s$ for the halo neutrons, gave only a minor difference. The shaded area shows the fitting uncertainty. The dotted line shows the H-F model prediction constrained by the last neutron separation energy.

The σ_I's for ^{11}Li measured at $0.8A$ and $0.4A$ GeV with H, D, Be, and C targets were used altogether to derive the density distribution $\rho(r)$ of ^{11}Li by the same method as applied to ^{11}Be. The $\rho(r)$ of ^{11}Li was assumed to be composed of a core and a possible halo. The core density was given by a harmonic-oscillator type, assuming four nucleons in the $1s$ orbit and one proton and four neutrons in the $1p$ orbit. The remaining two neutrons bringing about the halo were assumed orbiting in a potential the shape of which was given by the core density normalized with a depth parameter. Altogether two fitting parameters were used to fit the observed σ_I's, i. e., the core width parameter of the harmonic oscillator, and the binding energy of the halo-neutron orbit as the depth parameter of the potential.

The best-fitted density distribution $\rho(r)$ for ^{11}Li is shown in Fig. 14. For the halo neutrons, $1p$ or $2s$ orbits were assumed, and the both fits gave essentially the same distribution as seen in the figure. A difference was seen in the orbit energy, which was ascribed to the centrifugal-barrier effect. It was noted here that the selection of orbital was more or less arbitrary and only the final density distribution had the practical meaning. The obtained density distribution gave the rms radii,

$$R_{\rm rms}(^{11}{\rm Li}) = 3.1 \pm 0.3 \text{ fm}, \text{ and } R_{\rm rms}(2{\rm n}) = 4.8 \pm 0.8 \text{ fm}$$

for the ^{11}Li nucleus and the halo neutrons, respectively. The size of the ^{9}Li core ($R_{\rm rms} \sim$ 2.6 fm) turned out to be slightly larger than that of the ^{9}Li nucleus ($R_{\rm rms}$ = 2.32 $\pm$ 0.02 fm), qualitatively consistent with the center-of-mass motion of ^{9}Li core in ^{11}Li .

Functional shapes other than the one described above were also investigated: a) Single Gaussian gave no good fit to the data and was definitely inappropriate. b) Gaussian core + Yukawa tail gave also a good fit and the obtained $\rho(r)$ showed essentially the same halo tail both with a slightly higher central density.

Intensive theoretical studies have been performed to understand the structure of these halos by various authors. Bertsch, Brown and Sagawa [21] made a Hartree-Fock calculation by constraining the separation energy of the last occupied orbit. The calculated density distributions are shown in corresponding figures (Fig. 13 and Fig. 14). They show good agreement with the present semi-empirical result. An important implication of the HF calculation is that the long tail of the density distribution mainly arises from the weakly-bound last neutron.

A microscopic model of ^{11}Li was developed by Ikeda, Suzuki and their collaborators [22] by a hybrid model combining a cluster-orbital-shell model with an extended cluster model. The model gave a plausible explanation of the binding mechanism of the three-body system of ^{9}Li+n+n, and both the rms radius and the last-neutron binding energy were consistently reproduced.

5.3 Momentum distributions

Studies of projectile fragmentation at high energy ($\geq$400 MeV/nucleon) shows that the momentum distribution of nucleons inside a projectile nucleus can be determined from the momentum distribution of the projectile fragments. Extending the method used for stripping reactions to many nucleon removal, the momentum width of the projectile fragment is expressed by the separation energy of last neutrons,

$$\sigma^2 = u\langle\varepsilon\rangle \frac{A_F(A_P - A_F)}{A_P} \tag{16}$$

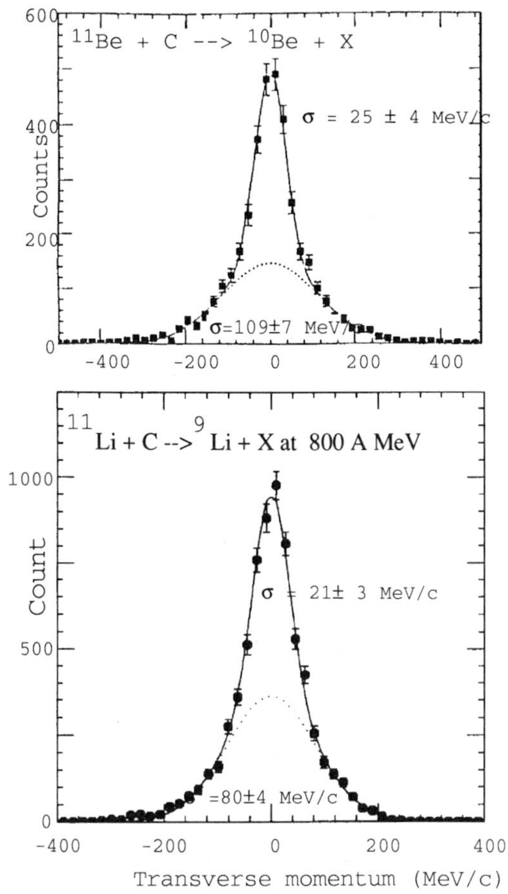

Fig. 15. Transverse momentum distribution of fragments from halo nuclei ^{11}Be and ^{11}Li

where u is the atomic mass unit and $\langle\varepsilon\rangle$ is an average separation energy of the removed nucleons. It is essentially the same conclusion with the discussion in Sec. 5.1. The transverse momentum distributions of fragments of ^{11}Be and ^{11}Li by carbon target were measured at 790 MeV/nucleon [12]. Figure 15 presents a transverse momentum distribution of (a) ^{10}Be fragment from ^{11}Be + C reaction and (b) ^{9}Li fragment from ^{11}Li + C reaction both at 800 MeV/nucleon. Both of the data show a very narrow peak on top of another wider peak. The fitting of the momentum distribution by two Gaussians gives the width to be σ_{narrow} = 25$\pm$ 4 MeV/c and σ_{wide} = 109$\pm$ 7 MeV/c for ^{10}Be spectrum. In ^{9}Li, σ_{narrow} = 21$\pm$ 3 MeV/c and σ_{wide} = 80$\pm$ 4 MeV/c. The narrow peaks indicates an existence of neutrons with extremely small momentum fluctuation as expected in the neutron halo. The σ_{wide} is consistent with the momentum fluctuations of usual nucleons.

The narrow width of ^{10}Be is consistent with the value 21 MeV which is estimated from the separation energy of a neutron (E_s(n) = 500 keV) using Eq. (6). It is a little more complicated for ^{11}Li because two neutrons are in same loosely bound orbitals.

However, the observed narrow width gives $\langle\varepsilon\rangle$ = 0.34±0.16 MeV which is consistent with separation energy of last neutrons; ΔE[(^{9}Li + 2n) – (^{11}Li)] = 0.34±0.10 MeV. Therefore both spectra are consistent with the expected narrow momentum distribution of weakly bound last neutrons and thus show formation of neutron halo.

6 Enhancement of the Electromagnetic Dissociation and Soft E1 Mode

Electromagnetic dissociation(EMD) provides important information on collective excitation of nuclei. For stable nuclear projectiles, observations of the EMD process at energies from 1 [23–26] to 200 GeV/nucleon [27, 28] have been reported and enhancements of fragmentation due to the EMD process have clearly been seen, in particular in the single-nucleon-removal channels on high-Z targets. This phenomenon was interpreted as resulting from the excitation and decay of an E1 giant dipole resonance (GDR) in a projectile by virtual photons associated with the rapidly changing Coulomb field.

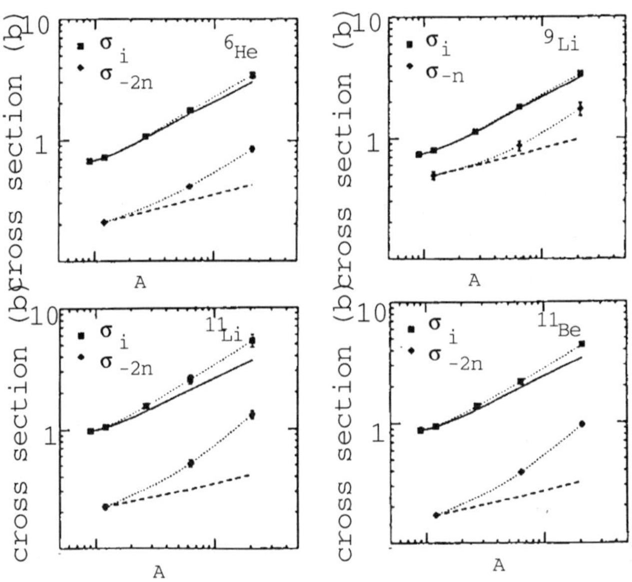

Fig. 16. Target mass dependence of $\sigma_{\rm I}$ and $\sigma_{-x\rm n}$ for ^{6}He, $^{9\text{-}11}$Li, and ^{11}Be projectiles at 0.8A GeV. Nuclear contribution to the cross sections are shown by the solid and the dashed curves.

Figure 16 shows the target mass dependence of the interaction cross sections, as well as the neutron (single or two-) removal cross sections, for selected neutron rich nuclei. The estimated nuclear part of the cross sections are also shown in the figure by solid curves for $\sigma_{\rm I}$ and dashed lines for $\sigma_{-x\rm n}$ reactions. In estimating $\sigma_{\rm I}$ the "separability" of the projectile and the target radii in the interaction cross section was assumed. However it is to be noted that the separability for neutron halo nuclei has not been tested. Experimentally the large EMD cross sections appear as a break of the separability for high Z target data. It is not possible, therefore, to estimate the nuclear part of the cross section with good accuracy unless we know the density profile of the neutron halo nuclei. We estimated the nuclear cross section by using several different types of density distributions, such as double gaussian, double harmonic oscillator, gaussian + exponential, and Sagawa's Hartree-Fock. An uncertainty of up to 50 % is expected in the resultant EMD cross sections. Further knowledge is necessary for more accurate estimation of the EMD cross sections. However the qualitative conclusion which can be drawn from the following discussion is not affected because the enhancement of the EMD cross section is so large.

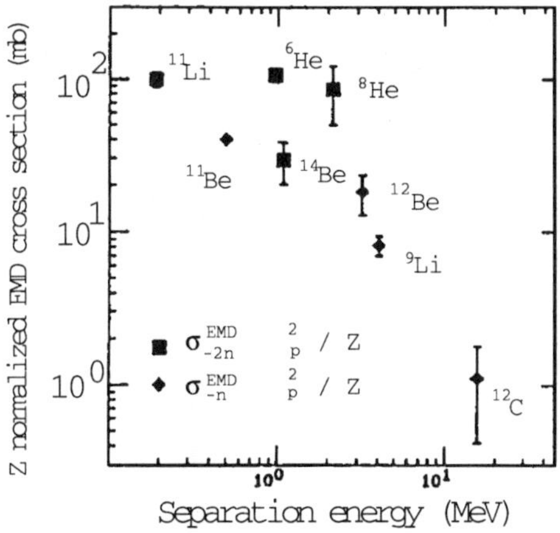

Fig. 17. Neutron separation energy dependence of the neutron removal cross sections after normalization by projectile Z.

The large EMD cross sections are seen in ^{11}Li and other neutron rich nuclei such as ^{8}He, ^{11}Be and ^{14}Be. Figure 17 shows the variation of the EMD cross section with the separation energy. It is clearly seen that a larger enhancement is observed for nuclei with smaller separation energy. Therefore, it is considered that the observed large EMD cross sections are related to a specific nuclear property of the extremely neutron-rich nuclei, probably to a neutron halo.

In order to compare the observed EMD cross section of neutron-rich nuclei with a typical cross section for stable nuclei, the total EMD cross section, $\sigma_I^{EMD}(^{12}C+Pb)$, was estimated to be 90±35mb. The value was obtained by adding the observed EMD parts of the single-nucleon-removal cross sections at 1.05 GeV/nucleon [24], with $\sigma_{-n}^{EMD}(^{12}C+Pb)$ and $\sigma_{-p}^{EMD}(^{12}C+Pb)$ being the major contributions. The value of σ^{EMD} for ^{11}Li is about 20-times larger than that for ^{12}C. If the projectile charge dependence of σ^{EMD} is assumed to be Z^2_{proj}, which is considered to be reasonable, the enhancement of $\sigma^{EMD}(^{11}Li)$ compared with $\sigma^{EMD}(^{12}C)$ is extremely large; by a factor of 80.

An EMD process is considered to occur by the excitation of a projectile nucleus by virtual photon field of a target nucleus. The EMD cross section is then calculated by ,

$$\sigma_{\mathrm{EMD}} = \int N_\gamma(E_\gamma)\sigma_\gamma(\gamma + \mathrm{B} \to \mathrm{X})\, dE_\gamma \,, \tag{17}$$

where N_γ is the virtual photon spectrum and σ_γ is the photo nuclear cross section. It is known that the electric giant dipole resonance (GDR) dominates σ_γ in high energy collisions. It has been shown [23–26] that the observed EMD cross section for stable nuclei can be reproduced well by this equation from the known photo-nuclear cross section $\sigma_\gamma(\gamma + \mathrm{B} \to \mathrm{X})$ and theoretically calculated N_γ.

Concerning the GDR of extremely neutron-rich nuclei, K. Ikeda [29] suggested the possibility that the GDR could split into two components for nuclei with a giant neutron halo. One component would correspond to the oscillation of core protons against core neutrons with a normal frequency, and the other to the oscillation of the core against the skin neutrons, as shown schematically in Fig. 18. Since the restoring force of the dipole oscillation is roughly proportional to the derivative of the density distribution, the frequency of the latter mode (we will denote it as the soft GDR) is expected to be very low.

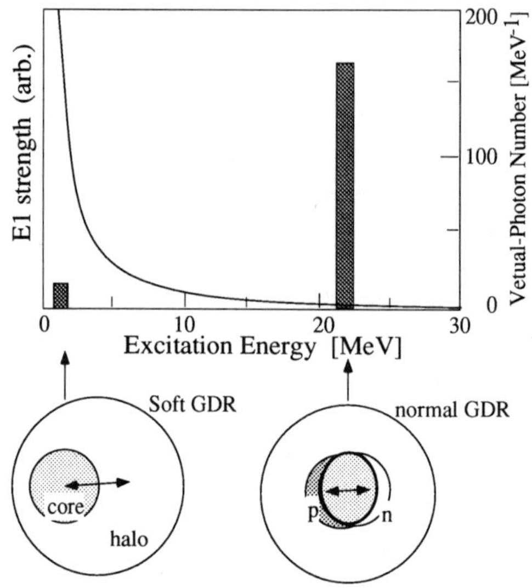

Fig. 18. Schematic pictures of the soft and the normal GDR in a neutron-rich nucleus with a neutron halo. The E1 oscillation between protons (shaded area) and neutrons (white area) is shown for each mode. The strength (black bar) of each E1 mode is shown together with the virtual-photon spectrum seen by a $0.8A$ GeV projectile on a Pb target.

Within the framework of the GDR model, the energy of the soft GDR can be estimated. The GDR at normal excitation energies ($E_{\mathrm{normal\ GDR}} \approx 22$ MeV estimated from the systematics) contributes to $\sigma_{\mathrm{I}}^{\mathrm{EMD}}$ only by about 80 mb which agrees well with EMD cross section of ^{12}C. Therefore, the main part of $\sigma_{\rm I}^{\rm EMD}$ is due to the excitation of the soft GDR. An integrated strength of the soft GDR was estimated to be 10% of the total E1 strength. (Since the E1 strength is roughly proportional to the square of the number of neutrons involved in each mode; the strength of the soft GDR : normal GDR = $2^2 : 6^2$ for the ^{11}Li case, as shown in Fig. 18.) In order to reproduce the observed $\sigma_{\rm I}^{\rm EMD}(^{11}{\rm Li+Pb})$, the energy of the soft GDR had to be $E_{\rm soft\ GDR} = 0.9^{+0.5}_{-0.3}$ MeV if $\Gamma_{\rm soft\ GDR}$ was assumed to be 0.7 MeV. The energy, $E_{\rm soft\ GDR}$, was rather insensitive to a choice of $\Gamma_{\rm soft\ GDR}$. Thus, the excitation energy of the soft GDR turned out to be remarkably low. Qualitatively, the most essential conclusion drawn here is that the observed large EMD cross section, together with the rapidly-decreasing virtual-photon spectrum, requires an appreciable strength of the GDR to be located at very low excitation energy irrespective of the detailed structure of the GDR.

Using the parameters of the soft GDR, $\sigma_{\rm -2n}^{\rm EMD}$ was estimated by integrating $N_\gamma \sigma_\gamma$ from $E_{\rm th}$(2n)= 0.19 MeV to $E_{\rm th}$(3n)= 4.2 MeV, since the photo-excitation up to 4.2 MeV produces only ^{9}Li. It gave 1.63 b for the ^{11}Li+Pb case, which was nearly a factor of two larger than the observed $\sigma_{\rm -2n}^{\rm EMD}$. In order to explain the observed difference between $\sigma_{\rm I}^{\rm EMD}$ and $\sigma_{\rm -2n}^{\rm EMD}$, more E1 strength has to be located above 4.2 MeV. This suggests other possibilities, such as: (1) the normal GDR is also located at lower excitation energy, or (2) the E1 strength is fragmented widely. Considering the lightness of the projectile, the latter case seems to be more realistic.

The location and strength of the soft E1 "mode" as estimated above have been confirmed by recent EMD measurements performed at low energies, 45 and 75 A GeV [30]. An excited state of ^{11}Li, as a possible candidate to the soft GDR, was found at the excitation energy, $E_{\rm x} = 1.2 \pm 0.1$ MeV, from an experiment on the double-charge-exchange (DCX) reaction induced by pions, $^{11}{\rm B}(\pi^-, \pi^+)^{11}{\rm Li}$, at T_π =164 MeV by Kobayashi et al. [31]. The spin and parity of the state was assigned as $1/2^+$, $3/2^+$ or $5/2^+$ from the measured angular distribution, so that the state can be excited by the E1 transition. However, it was not clear whether this state was the main component of the soft GDR or additional components were necessary, especially since the DCX reaction do not favor to excite a real collective state.

Bertsch and Foxwell [32] extended the Hartree-Fock model to calculate the E1 strength for ^{11}Li. (Remember that his model explain the neutron halo.) The Random Phase Approximation (RPA) was used for calculating the dipole strength. The calculation shows the softening of the dipole mode as shown in Fig. 19. However their model predict smaller strength than to explain the observed enhancement of the EMD cross section. The calculated cross section is only about the 30% of the observed value. Even including the uncertainty of the experimental determination of EMD cross section, more enhancement is needed. It is, therefore, suggested that a theory has to deal with correlation between halo neutrons.

Suzuki and Tosaka [33] recently extended the cluster orbital shell model calculation for soft dipole excitation in ^{9}Li + n + n system. They found that quite an amount of strength (10% of the energy-weighted sum rule or 30% of the non energy-weighted sum rule) is located at very low energy. As shown in Fig. 20, they also predict the EMD cross section to be 600 mb which agrees reasonably well with observed cross section.

Therefore a soft dipole model at low excitation energy of extremely neutron rich nuclei are supported by theoretical models. However quantitative study is awaited to

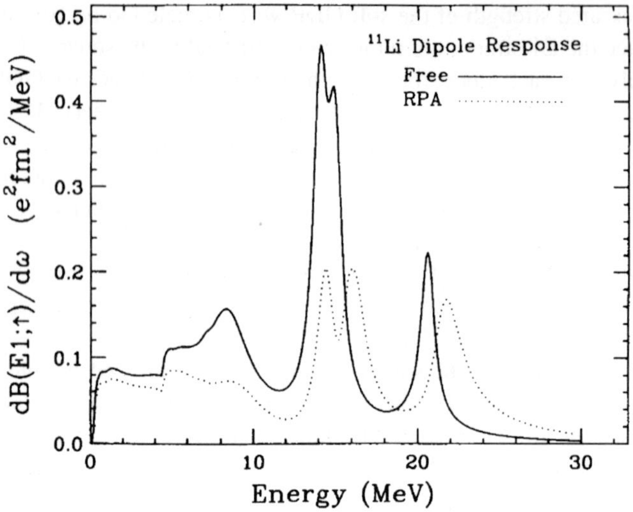

Fig. 19. The electric dipole strength distribution in ^{11}Li as calculated by Bertsch and Foxwell. The solid line shows the free response and the dotted line the RPA.

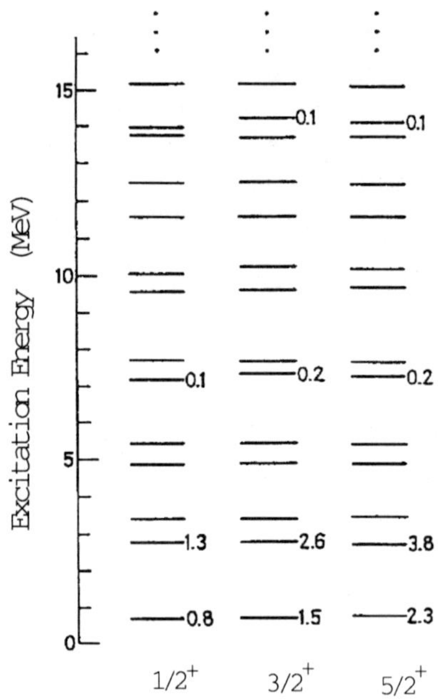

Fig. 20. Excited states correspond to the soft dipole mode of ^{11}Li . Numbers adjacent to the levels show the reduced electric dipole transition probabilities in unit of mb.

see the effect of correlation between halo neutrons.

Recently, the invariant mass of ^{11}Li fragments, i.e. ^{9}Li + n + n, were measured to see the excitation function by Coulomb interaction [30, 34]. Although enhancement of B(E1) were observed at very low excitation energy, it s not clear whether it is the soft E1 resonance or just the enhancement of B(E1) due to the small separation energy. Further detailed studies, both experimental and theoretical, are necessary to clear the situation.

7 Summary and Other Applications of Radioactive Nuclear Beams

We have seen the production and use of the radioactive nuclear beams for study of nuclear structure. In particular the nuclear radii and density distribution of neutron rich nuclei were discussed.

From the measurements of interaction cross sections, σ_{I}, interaction nuclear radii were determined, and the rms radii, R_{rms} were deduced with the aid of the Glauber model. The majority of R_{rms} of p-shell nuclei were nearly constant (2.3 $\sim$ 2.4 fm), except ones close to the neutron drip line where a large increase of radius was seen, e. g., from ^{9}Li to ^{11}Li, and ^{10}Be to ^{11}Be, etc. The isospin dependence of nuclear radii became also possible to study, and in fact it was observed that nuclei with different isospin have different radii. It then showed the importance of the density dependent nuclear interaction.

Neutron skin as thick as $\sim$0.9 fm was verified in ^{6}He and also in ^{8}He for the first time. The $\rho(r)$ of ^{6}He and ^{8}He were deduced from the measured σ_{I}'s for ^{4}He, ^{6}He and ^{8}He under the assumption that ^{4}He formed a good core in ^{6}He and ^{8}He. The core assumption was assured by the relation of the nucleon-removal cross section to the interaction cross sections that required from the Glauber model in the case of loosely-bound system. The obtained $\rho(r)$ of ^{6}He and ^{8}He were in reasonable agreement with predictions from the relativistic-mean-field (RMF) model. The RMF calculation were, then, extended to other nuclei in a wide mass range, and it was shown that the difference between the neutron and the proton rms radii, ΔR_{rms} was in gross in a linear relation to the difference between the neutron and the proton Fermi energies, ΔE_{F}. For nuclei far from the stability line a large ΔR_{rms} is generally expected, since ΔE_{F} becomes quite large, as in the cases of ^{6}He and ^{8}He.

Nuclei with extended neutron halos were discovered. In the measurements of transverse momentum spectra of projectile fragments, a very narrow peak was observed in the distribution of fragments clearly indicating the existence of a neutron halos The density distribution $\rho(r)$ of ^{11}Be and ^{11}Li were also deduced. The fitting requires definitely distributions with extended neutron halos in consistent with the fragmentation data.

From the study of EMD process of ^{11}Li, an extremely large EMD cross section, σ^{EMD} was observed. The large σ^{EMD} was understood as due to a soft-GDR excitation, a new mode specific to a nucleus with a giant neutron halo. The excitation energy of the soft GDR was estimated from the measured σ^{EMD} to be very low as $\sim$1.3 MeV in the framework of GDR model.

Before we gain these new studies using radioactive nuclear beams, three important character of the nuclear density distribution were determined from studies of stable nuclei. And following character was commonly believed.

1. The half density radius of a nucleus is proportional to $A^{1/3}$.
2. The proton and the neutron density distributions in a nucleus are proportional to each other even if the number of protons and the number of neutrons are different.
3. The surface diffuseness is almost same for all nuclei.

These characters were all broken in unstable nuclei. It is now obvious why such characters were seen. The first character is seen because one had not measured radii of isobars before. The second is just because a stable nucleus is balanced, namely the Fermi energy of proton and neutron are the same. Therefore this rule is broken when the two Fermi energies are different, which is the very case for unstable nuclei. The third is due to the fact that separation energy of valence nucleons are all 6-8 MeV in stable nuclei. When the separation energy is different, as seen in unstable nuclei, the surface diffuseness can be different.

Studies carried by radioactive nuclear beams thus shows the importance of extending the study of nuclear structure to wider range of instability.

Radioactive nuclear beams have been also applied to studies of astrophysical interest such as the stellar and primordial nucleosynthesis and solar neutrino problem. Although I have not discussed them in this lecture these are also extremely important and promising application. Interested students will gain a lot of information in references 1, 2, 3.

Low energy radioactive beams have been used for many studies of different discipline. The high-energy radioactive beams are new and have not been used yet in other field of research than nuclear physics, astrophysics, and cancer therapy [35]. However they are considered to be extremely powerful tool for studies in many discipline. Here, I show the important properties of radioactive nuclear beams for possible applications below;

1. *Elemental selection*: Any element may be delivered independent of their chemical characters and therefore the best element can be used for a specific problem. A wide variety of selections of elements is possible for appropriate uses. For example, interstitial sites and substitutional sites of crystals can be studied selectively by many available elements.
2. *High energy*: Because the energy is high, the beam control is easily done by magnetic system. Also implantation depth can be adjusted by changing the beam energy. Therefore it provides, for example, the first opportunity to start a tracer from any depth independent of an elemental composition of the sample material.
3. *Radioactivity*: By definition, a radioactive ion decays in a material where it stopped. Emissions of α, β, and/or γ rays can be used for probing the stopped position and the surrounding electric and magnetic fields with extremely high sensitivity.
4. $Wide range of lifetime and spin$: A selection of a lifetime enables the application to phenomena of various time range (1 μs - years). A selection of nuclear spin is also possible within an element.
5. *Polarized beam*: Polarized beams of radioactive nuclei can also be produced by several methods. The polarized implanted ion would also provide a great tool for studying the magnetic and the electric structure of a material.

Probes which have some of the advantages listed above existed before but none of them had them all. The RNB has all of these advantages so that it possesses large possibilities of applications.

References

1. Radioactive Beams, *Proc. of the First Int. Conf. on Radioactive Nuclear Beams,* Berkeley, Ca 1989, Myers, W. D., Nitschke, J. M., Norman, E. B. Eds., World Scientific, 1990.
   Proc. of the Second Int. Conf. on Radioactive Nuclear Beams, Louvain-la-Neuve 1991, Delbar, Th., Eds., Adam Hilger 1992.
   Proc. of the Workshop on the Science of Intense Radioactive Ion Beams, Los Alamos, 1990, LA-11964-C UC-413.
2. *Proc. of Int. Symp. on Structure and Reactions of Unstable Nuclei,* Niigata, Japan 1991, Ikeda, K. and Suzuki, Y. , Eds., World Scientific.
   Proc. of the Int. Workshop on Unstable Nuclei in Astrophysics, Tokyo, Japan 1991, Kubono, S. and Kajino, T. Eds., World Scientific.
   Nucleus-Nucleus Collision IV, Toki, H., Tanihata, I., and Kamitsubo, H., Eds. Special issue of Nuclear Physics, North-Holland, Nucl. Phys. A538 1992.
3. Isao Tanihata, On the Possible Use of Secondary Radioactive Beams, *TREATISE ON HEAVY-ION SCIENCE,* Vol. 8 , Bromley, D. A., Eds., (Plenum Publishing Corporation, 1989) p443. Detraz, C. and Vieira, D. J., Exotic Light Nuclei, *Ann. Rev. of Nuclear and Particle Science* (1989) p407.
4. Heckmann, H. H., Greiner, D. E., Lindstrom, P. J., and Shwe, H., *Phys. Rev. C* 17 (1978) 1735 . Greiner, D. E., Lindstrom, P. J., Heckmann, H. H., Cork, B., and Bieser, F. S., *Phys. Rev. Lett.* **35** (1975) 152.
5. Goldhaber, A. S. and Heckmann, H. H., High Energy Interactions of Nuclei, *Ann. Rev. Nucl. Sci.***28** (1978) 161.
6. Gelbke, C. K. et al., *Phys. Rev. Lett.* **37** (1976) 1191.
7. Van Bibber, K. et al., *Phys. Rev. Lett.* **43** (1979) 840.
8. Tanihata, I., *Hyperfine Interactions,* **21** (1985) 251. Tanihata, I. et al., *Phys. Letters* 160B (1985) 380.
9. Dufour, J. P. et al., *Nucl. Instrum. Methods in Phys. Res.* **A56** (1986) 34.
10. Kubo, T. et al., *Nucl. Instr. and Meth. in Phy. Res.* **B70** (1992) 309.
11. Asahi, K. et al., *Hyperfine Int.*, in press 1992.
12. I. Tanihata, T. Kobayashi, O. Yamakawa, S. Shimoura, K. Ekuni, K. Sugimoto, N. Takahashi, T. Shimoda and H. Sato,*Phys. Lett.* **B206** (1988) 592:
    I. Tanihata et al.,*Phys. Rev. Lett.* **55**(1985) 2676:
    I. Tanihata et al., *Phys. Lett.* **B160** (1985) 380.
13. H. Sato and Y. Okuhara, *Phys. Rev.* **C34** (1986) 2127.
14. P.J. Karol, *Phys. Rev.* **C11** (1975) 1203.
15. Y. Ogawa, K. Yabana and Y. Suzuki, *Nucl. Phys.* **A539** (1992) 295.
16. D. Hirata, H. Toki, T. Watabe, I. Tanihata and B.V. Carlson, *Phys. Rev.* **C44** (1991) 1467.
17. C.A. Whitten, Jr., *High energy Phys. and Nucl. Structure,* ed. D.F. Measday and A.W. Thomas (North-Holland, Amsterdam, 1980) p.419: G. Igo et al., *Phys. Lett.* **B81** (1979) 151:
    S. Shlomo and R. Schaffer, *Phys. Lett.* **B83** (1979) 5.
18. P. G. Hansen and B. Jonson, Europhysics Lett. **4** (1987) 409.
19. M. Fukuda, T. Ichihara, N. Inabe, T. Kubo, H. Kumagai, T. Nakagawa, Y. Yano, I. Tanihata, M. Adachi, K. Asahi, M. Kouguchi,M. Ishihara, H. Sagawa, and S. Shimoura; Physics Letters B **268** (1991) 339.
20. H. De Vries et al., Atomic Data and Nuclear Data Tables Vol.**36** (1987) 496.

21. G.F. Bertch, B.A. Brown, and H. Sagawa, *Phys. Rev.* **C39** (1989) 1154.
22. I. Tanihata, D. Hirata, T. Kobayashi, S. Shimoura, K. Sugimoto and H. Toki, Phys. Letters B **289** (1992) 261.
23. D. Olson et al., Phys. Rev. C **24** (1981) 1529.
24. H. Heckman and P.J. Lindstrom, Phys. Rev. Lett. **37** (1976) 56.
25. M. Mercier et al., Phys. Rev. **C33** (1986) 1655.
26. J. Hill et al., Phys. Rev. **C60** (1988) 1722.
27. P.B. Price et al., Phys. Rev. Lett.**61** (1988) 2193.
28. J. Hill et al., Phys. Rev. Lett.**60** (1988) 999.
29. K. Ikeda, private communication.
30. S. Shimoura, *Structure and Reactions of Unstable Nuclei,* ed. K. Ikeda and Y. Suzuki (World Scientific, Singapore, 1991) p.132.
31. T. Kobayashi, Structure and Reactions of Unstable Nuclei, ed. K. Ikada and Y. Suzuki (World Scientific, Singapore, 1991) p 187.
32. G. Bertsch and J. Foxwell, Phys. Rev. **C41** (1990) 1300.
33. Y. Suzuki and Y. Tosaka, Niclear Physics **517** (1990) 599.
34. K. Ieki et al., Phys. Rev. Letters. **70** (1993) 730.
35. Chatterjee, A. and Llacer, J., *Proc. of the First Int. Conf. on Radioactive Nuclear Beams,* Berkeley, Ca 1989, Myers, W. D., Nitschke, J. M., Norman, E. B. Eds., World Scientific, 1990. p 403.

Lecture given at the JSPS-INS International Spring School, March 1993 at Shimoda

Vacuum Fluctuation in Finite Space

Il-Tong Cheon

Department of Physics, Yonsei University, Seoul 120-749, Korea

Abstract Fundamental property of the physical vacuum has been revealed by analysing the nuclear energy levels in a finite space formed with two parallel metallic plates. Tiny energy level shifts, an order of magnitude of 1 neV, have been found for the nucleus placed between two parallel plates separated from each other by $b = 1$ mm.

1 Introduction

Vacuum is one of the most important and basic concepts. All physical phenomena are associated with what kind of vacuum is given. Therefore, it is extremely important to investigate properties of the vacuum. Energy of the particle moving in the one-dimensional harmonic oscillator pontential is well known to have the form

$$E_n = \hbar\omega(n + \frac{1}{2}).$$

The lowest energy is $\frac{1}{2}\hbar\omega$ which is usually called "zero-point energy". This ground-state actually denotes vacuum in the bound system. The zero-point energy is a general feature in quantum-mechanical bound states.

On the other hand, quantization of the Maxwell equations for the free radiation field gives

$$\epsilon = \sum_k (a_k^\dagger a_k + \frac{1}{2})\hbar\omega_k$$

where $n =< a_k^\dagger a_k >$ is the photon number. The second term remains even for $n = 0$. Since the radiation field energy should vanish for the case of no photon in the free space, this "zero-point energy" was disregarded as an unobservable quantity. In 1948, however, Casimir [1] pointed out that it could be observable by setting two parallel perfectly conducting plates. Namely, the difference between the zero-point energy inside the space of volume $bL^2(L \gg b$, where L is the size of rectangular plate and b is the separation distance) and the free value for this same volume is finite, i.e. ,

$$\sum_{ljn} \frac{1}{2}\hbar\omega_{ljn} - \frac{bL^2}{(2\pi c)^3}\int_{-\infty}^{\infty} \frac{1}{2}\hbar\omega d^3\omega = -\frac{\pi^2 L^2}{720}\frac{\hbar c}{b^3}.$$

The value on the right-hand side is well known as Casimir energy and was observed by the measurement of the force exerted on the two parallel plates [2].

This fact indicates that the vacuum state in the restricted space is different from the free vacuum state.

Another important property of the vacuum is "vacuum fluctuation". Let us consider a hydrogen atom. The proton is fluctuating in a tiny region governed by the uncertainty relation. Thereby, the Coulomb field produced by the proton is also fluctuating. The electron bound in that Coulomb field can feel this field fluctuation and, thus, a tiny electron energy-level shift takes place. This is a physical mechanism for the Lamb shift in the hydrogen atom. If the vacuum fluctuation could be modified by restricting the space with two parallel perfectly conducting plates, a tiny shift additional to the Lamb shift would be expected to take place. Possibility of such a shift was already discussed [3-6], but its observation is not yet attempted because of technical difficulties. Practical observation requires the plate separation distance b of an order of micrometers. There are numerous technical problems to be solved in order to measure the shift during the beam passing through such an extremely narrow gap.

However, for the case of nuclei, it can be observed around $b = 1$ mm [7].

2 Feynman integral in the finite space

Evaluations of all physical quantities in the quantum electrodynamics (QED) are generally based on the Feynman integrals. Therefore, the Feynman integrals should be calculated in the finite space $L \times L \times b$ formed by two parallel perfectly conducting plates. Since the photon spectrum in the finite space is discrete and given as

$$k^2 = k_0^2 - \boldsymbol{k}^2 = k_0^2 - \left[(\frac{\pi}{L} l)^2 + (\frac{\pi}{L} j)^2 + (\frac{\pi}{b} n)^2 \right], \tag{1}$$

the Feynman integral is expressed by summations [5]. For $L \gg b$, the summation over l and j can be replaced by integrals and, then, application of the Poisson's summation formula on the Fourier transformation to the summations over n gives the result [5]

$$\begin{aligned} &\frac{1}{(2\pi)^4} \int_{-\infty}^{\infty} dk_0 \frac{(2\pi)^3}{bL^2} \sum_{ljn=0}^{\infty} (k^2 - 2kp + i\epsilon - \Delta)^{-3} \\ &= -\frac{i}{32\pi^2} (\Delta + p^2)^{-1} - \frac{i}{64\pi b} (\Delta + p^2)^{-3/2}, \end{aligned} \tag{2}$$

where p is the electron four-momentum and Δ is an arbitrary constant. The first term on the right-hand side is equivalent to the result of Feynman integral in free space and the second term stands for effect of the boundary condition. Namely, eq. (2) can be expressed as

$$\begin{aligned} &\frac{1}{(2\pi)^4} \int_{-\infty}^{\infty} dk_0 \frac{(2\pi)^3}{bL^2} \sum (k^2 - 2kp + i\epsilon - \Delta)^{-3} \\ &- \frac{1}{(2\pi)^4} \int_{-\infty}^{\infty} d^4k (k^2 - 2kp + i\epsilon - \Delta)^{-3} = -\frac{i}{64\pi b} (\Delta + p^2)^{-3/2}. \end{aligned} \tag{2a}$$

This formula has a similar structure to third equation given in the section 1, a vacuum fluctuation in the finite space is different from that in free space.

3 Energy level shifts

The Lamb shift is usually obtained by evaluating the radiative corrections in the quantum electrodynamics(QED). All these calculations can be carried out with the Feynman integral given in eq. (2). Therefore, the result consists of the usual Lamb shift plus a contribution from existance of two parallel plates [3-6]. On the basis of analysis for the atomic system, it was also shown that tiny shifts of nuclear energy levels took place for the nucleus placed between two parallel plates [7]. This shift is expressed as [7,8]

$$\Delta E = \frac{\alpha}{16M^3b}\left[\frac{301}{120} + \frac{8M}{3}\frac{A}{\hbar\omega}\right]\langle\beta|\Delta V_N(r)|\beta\rangle + \frac{3\alpha}{64M^3b}\langle\beta|\boldsymbol{\sigma}\cdot\boldsymbol{l}\frac{1}{r}\frac{dV_N(r)}{dr}|\beta\rangle, \quad (3)$$

where

$$A = D^{-1}\sum_i |\langle\beta|\boldsymbol{r}|i\rangle|^2 \ln\{(E_i - E_\beta)/\hbar\omega\} \quad (4)$$

with

$$D = \sum_i |\langle\beta|\boldsymbol{r}|i\rangle|^2 \{(E_i - E_\beta)/\hbar\omega\}\,. \quad (5)$$

In order to evaluate the transition probability, let us take the simplest nuclear model described by a harmonic oscillator pontential with $\boldsymbol{l}\cdot\boldsymbol{s}$ and $\boldsymbol{l}^2$-coupling as

$$V_N(r) = \frac{1}{2}M\omega^2r^2 + V_0 + \xi\boldsymbol{l}\cdot\boldsymbol{s} + \eta\boldsymbol{l}^2, \quad (6)$$

where V_0, ξ and η are constants. This choice of model does not lose generality in description of the specific nuclei whose low-lying energy levels are determined by a single proton in the open shell, the energy level shift, (3), takes place only on the proton.

An example is ^{133}Cs for which we take the parameter values, $\hbar\omega = 8.8$ MeV, $\xi = 0.12112$ MeV, $\eta = -0.036066$ MeV, and $V_0 = -47.9815$ MeV for the low lying positive parity states, $\frac{7}{2}^+, \frac{5}{2}^+, \frac{3}{2}^+, \frac{1}{2}^+$ with $n = 5$ and $V_0 = -64.3304$ MeV for the state $\frac{9}{2}^+$ with $n = 7$. With these values, the five low lying positive states can well be reproduced as 0 MeV, 0.081 MeV, 0.3838 MeV, 0.4185 MeV and 0.7060 MeV for $\frac{7}{2}^+(g.s.)$, $\frac{5}{2}^+$(0.081 MeV), $\frac{3}{2}^+$(0.3838 MeV), $\frac{1}{2}^+$(0.437 MeV) and $\frac{9}{2}^+$(0.706 MeV), respectively.

In the harmonic oscillator potential model, we have only four possible states for nonvanishing matrix elements, $\langle i|p|\beta\rangle$, i.e. $|i\rangle = |n' = n+1, l' = l \pm 1, j' = l' \pm 1/2\rangle$ for $|\beta\rangle = |n, l, j = l \pm 1/2\rangle$. Therefore, $|i\rangle$ are the negative parity states when $|\beta\rangle$ has the positive parity. For the negative parity states, we can choose the value of potential parameter V_0 so as to reproduce the lowest negative parity state $\frac{11}{2}^-$(1.07146 MeV). The value $V_0 = -54.74$ MeV is found in this manner. Using the same values of ξ and η as those for the positive parity states, we obtain energies of the required negative parity states, $\frac{11}{2}^-, \frac{9}{2}^-, \frac{7}{2}^-, \frac{5}{2}^-, \frac{3}{2}^-$ and $\frac{1}{2}^-$, as 1.072 MeV, 1.738 MeV, 1.842 MeV, 2.266 MeV, 2.324 MeV and 2.505 MeV.

In this case, the principal quantum number $n = 6$. Thus, the transition probabilities for the first excited and ground states of ^{133}Cs are found as [9]

$$A(\frac{5}{2}^+; 1st.ex.) = -6.5600 \quad (7)$$

and

$$A(\frac{7}{2}^{+}; gr.st.) = -11.270. \tag{8}$$

Taking into account

$$\langle\beta|\Delta V_N|\beta\rangle = 3M\omega^2, \tag{9a}$$

$$\langle\beta|\boldsymbol{\sigma}\cdot\boldsymbol{l}\frac{1}{\boldsymbol{r}}\frac{dV_N(r)}{dr}|\beta\rangle = M\omega^2\left\{j(j+1) - l(l+1) - \frac{3}{4}\right\}, \tag{9b}$$

we find the energy level shifts as

$$\Delta E(\frac{5}{2}^{+}; 1st.ex.) = -4.423\times 10^{-8}\frac{1}{b(\mathrm{mm})}\ \mathrm{eV}, \tag{10a}$$

$$\Delta E(\frac{7}{2}^{+}; gr.st.) = -7.608\times 10^{-8}\frac{1}{b(\mathrm{mm})}\ \mathrm{eV}. \tag{10b}$$

Thus, variation of the difference energy between these two levels is

$$\Delta\Delta E = \Delta E(\frac{5}{2}^{+}; 1st.ex.) - \Delta E(\frac{7}{2}^{+}; gr.st.) = 3.175\times 10^{-8}\frac{1}{b(\mathrm{mm})}\ \mathrm{eV} \tag{11a}$$

which is equivalent to the velocity shift,

$$\Delta\Delta v = 2c\frac{\Delta\Delta E}{E_\gamma} = 0.235\frac{1}{b(\mathrm{mm})}\ \mathrm{mm/s}, \tag{11b}$$

where $E_\gamma = 81$ keV is the energy of gamma-ray emitted from the free $^{133}Cs^*$ nucleus in the first excited state.

Similarly, for ^{193}Ir, we obtain

$$\Delta\Delta E = -2.13\times 10^{-8}\frac{1}{b(\mathrm{mm})}\ \mathrm{eV} \tag{12a}$$

and equivalently

$$\Delta\Delta v = 0.175\frac{1}{b(\mathrm{mm})}\ \mathrm{mm/s} \tag{12b}$$

with $E_\gamma = 73.04$ keV.

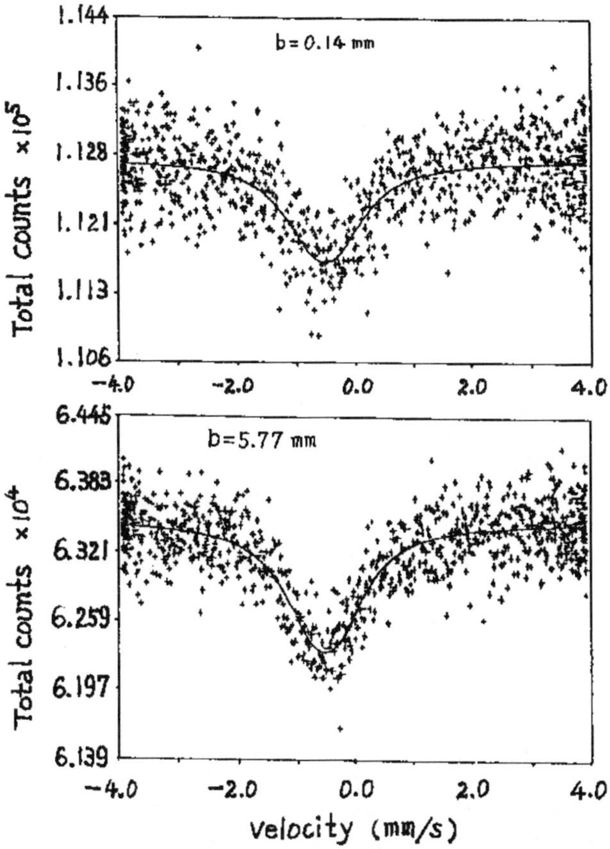

Fig. 1. Mössbauer resonance absorption spectra. They were taken from ref.10.

4 Experiment

Recently, an experiment was performed at low temperature 78 K with metallic $^{193}Os^*$ obtained by using the Research Reactor Facilities at Missouri University [10]. After 30.0 days of the half life, $^{193}Os^*$ yields a transition to $^{193}Ir^*$ in the first excited state. This γ-ray source, $^{193}Ir^*$, has a life-time of 6.2 ns. The absorber was prepared with a metallic plate of ^{193}Ir sandwitched between two thin parallel aluminum plates which were earthed during the experiment to prevent any surface-inhomogeneous electric potential.

The experimental results are shown in Fig. 1. They show the line positions at $\Delta v = -0.52 \pm 0.03$ mm/s and $\Delta v = -0.53 \pm 0.02$ mm/s for $b = 0.14$ mm and $b = 5.77$ mm, respectively.

These values contain the standard isomer shift which is assumed to be $v_0 = -0.51$ mm/s although it was, unfortunately, not measured. Therefore, the shifts of line positions due to existence of two parallel aluminum plates are $\Delta\Delta v = -0.01$ mm/s and $\Delta\Delta v = -0.02$ mm/s for $b = 0.14$ mm and $b = 5.77$ mm, respectively. They explicitly indicate that only when the energy of γ-ray emitted from the source is shifted by such amounts given above, this γ-ray can be absorbed by the absorber placed between two aluminum plates because the energy levels of the absorber nucleus shift in the finite space between two plates.

However, the data obviously do not obey the inverse law of the distance between two plates. This discrepancy may be due to neglect of plate opaqueness.

5 Effects of plate opaqueness

Equations (11a) and (12a) are valid only for the case of perfectly conducting plates. Therefore, the dependence of $1/b$ must be modified for the real material of plates. Namely, opaqueness should be taken into account because the vacuum fields can partially penetrate into the plates. In the case of perfectly transparent plates, such an energy shift completely disappears. In this paper, we investigate with conducting metallic plates.

Considering effects of the opaqueness of plates, let us try to modify the formulae (11a) and (12a). The transmission coefficient of the plane wave at the edge of the potential well is given as [11]

$$|T(k_z)|^2 = \left(1 + \frac{\gamma^2}{4k_z^2}\right)^{-1}, \tag{13}$$

where k_z is the momentum of incident wave coming along the z-axis which is perpendicular to the plates and γ is the magnitude of the delta-function type potential. Since the very low energy photon is mostly reflected by the potential well, it is sufficient to consider only the transmission of comparatively high energy photons. Then, for $4k_z^2 \gg \gamma^2$, eq. (13) can approximately be expressed as

$$|T|^2 \approx 1 - \left(\frac{\gamma}{2k_z}\right)^2. \tag{14}$$

Therefore, the field intensity is found by

$$\frac{dI}{I} = -\left(\frac{\gamma}{2k_z}\right)^2 \frac{dx}{\lambda} = -\frac{\gamma^2}{4k_z}dx. \tag{15}$$

Integrating this equation, we obtain

$$I(x) = N \exp\left(-\frac{\gamma^2}{4k_z}x\right). \tag{16}$$

Since the photon wave number is given as $k_z = \pi n/b$ in the finite space [5,7], eq. (16) can be rewritten as

$$I(x) = \sigma b \exp(-\sigma b x), \tag{17}$$

where $\sigma = \left\langle \frac{\gamma^2}{4\pi n} \right\rangle$ is called the index of opaqueness and the integration of $I(x)$ with respect to x is normalized to be unity. Thus, the energy shift (11a) and (12a) are found in the form

$$\begin{aligned} \Delta E(\sigma, d; b) &= \delta \int_0^{2d} \frac{I(x)}{b+x} dx \\ &= \delta \sigma b \int_0^{2d} \frac{\exp(-\sigma b x)}{b+x} dx \\ &= \frac{\delta}{b} B(\sigma, d; b), \end{aligned} \tag{18}$$

where d is the thickness of the plates, $\delta = 3.18 \times 10^{-8}$ eV · mm and -2.13×10^{-8} eV · mm for the cases of ^{133}Cs and ^{193}Ir, respectively and

$$B(\sigma, d; b) = D(\sigma b^2) - \left(1 + \frac{2d}{b}\right)^{-1} \exp(-2\sigma d b) D\left(\sigma b^2 \left[1 + \frac{2d}{b}\right]\right). \tag{19}$$

The function $D(\beta)$ is given by the exponential integral $E_1(\beta)$ as

$$D(\beta) = \beta \exp(\beta) E_1(\beta). \tag{20}$$

Since $D(0) = 0$ and $D(\infty) = 1$, we have

$$\begin{cases} \Delta E(0, d; b) = 0 & \text{(transparent)} \\ \Delta E(\infty, d; b) = \frac{\delta}{b} & \text{(perfect conductor).} \end{cases} \tag{21}$$

It is obvious that eqs. (11a) and (12a) are the case of $\sigma = \infty$ in the dielectric medium. Namely, $\sigma = \infty$ denotes perfect conductor while $\sigma = 0$ denotes transparent plates. For $\sigma = 0.15\ \text{mm}^{-2}$ and $d = 0.58$ mm, we obtain $\Delta\Delta v = -0.0081$ mm/s at $b = 0.14$ mm and $\Delta\Delta v = -0.0178$ mm/s at $b = 5.77$ mm. They are in good agreement with the experimental data, $\Delta\Delta v = -0.01$ mm/s and $\Delta\Delta v = -0.02$ mm/s.

6 Conclusion

When the space is restricted by two parallel plates, the photon spectrum becomes discrete and causes modification of vacuum field fluctuation. It has been shown that such physical phenomena could be observed as the energy level shifts of the nucleus placed between two parallel plates. The experimental results with ^{193}Ir show the tiny shifts of the nuclear energy levels. It is also very interesting that the effects of plate opaqueness are remarkable, particularly at short distance.

Another important point is that the Mössbauer resonance spectrum is independent of the position of nucleus between two plates. If it is position dependent, the experimental result should show a broad spectrum because the thickness of absorber is around 0.1 mm and therefore a certain nucleus is located at the central part of the space and another nucleus is located close to the plate surface. The experiment with ^{133}Cs is now going on at Leuven University, Belgium under collaboration with us and a preliminary result shows an extremely slim spectrum for the case of finite space formed by two plates. And it is very similar to that obtained in free space. This is also an indication that the shift is independent of the position of absorber. The ^{133}Cs case also shows substantially the energy level shifts.

It should be stressed here that the present investigation would be extremely important in a sense to have revealed one of the fundamental properties of physical vacuum, i.e. the spatial character of vacuum field fluctuation.

This work was supported by the Korean Ministry of Education.

References

1. H.B.G. Casimir, *Proc. Ned. Akad. Wet* **51** (1948) 793.
2. M.J. Sparnaay, *Physica (Utrecht)* **24** (1958) 751.
3. G. Barton, *Proc. Roy. Soc. London* **A410** (1987) 141.
4. Il-T. Cheon, *Abstracts of Proc. Xth Int. Conf. Atomic Phys. Tokyo, 1986,* p. 46.
5. Il-T. Cheon, *Phys. Rev.* **A37** (1988) 2785.
6. D. Meschede *et al.* , *Phys. Rev.* **A41** (1990) 1587.
7. Il-T. Cheon, *J. Phys. Soc. Japan* **60** (1991) 833.
8. Il-T. Cheon, to be published in the special issue of *"Laser Physics"*, *"Cooling and Trapping of Atoms, Molecules and Ions"*, Moscow.
9. The values given in ref.6 are divided by $\hbar\omega$ = 8.8 MeV, i.e. those values there represented $A/\hbar\omega$.
10. R. Wagoner, Ph.D. dissertation at Purdue Univ., 1992.
11. K. Gottfried, *"Quantum Mechanics"*, Benjamin/Cummings Pub. Co. Ins. 1966, **Vol. I**, Chap. II, Sec. 7.

Lecture given at the JSPS-INS International Spring School, March 1993 at Shimoda

JSPS-INS International Spring School

Sponsors

Japan Society for the Promotion of Science
Institute for Nuclear Study, University of Tokyo

President

T. Yamazaki

Organizing Committee

Y. Akaishi	INS, Univ. Tokyo
IL-T. Cheon	Yonsei Univ.
H. Homma *(Chairman)*	INS, Univ. Tokyo
M. Ishihara	RIKEN/Univ. Tokyo
I. Katayama	INS, Univ. Tokyo
K. Kikuchi	KEK(Nat. Inst. for High Energy Phys.)
S.M. Lee	Univ. Tsukuba
A. Masaike	Kyoto Univ.
K. Nakai	KEK(Nat. Inst. for High Energy Phys.)
S. Yamada	INS, Univ. Tokyo
K. Yazaki	Univ. Tokyo

International Advisory Committee

IL-T. Cheon	Korea	B.L. Reznik	Russia
S.P. Chia	Malaysia	H.M. Sen Gupta	Bangladesh
S.X. Fang	China	K.L. Shrestha	Nepal
J.C. Kim	Korea	B. Sinha	India
T. Methasiri	Thailand	Z.X. Sun	China
A.N. Mitra	India	A.W. Thomas	Australia
K. Namsrai	Mongolia	D.F. Walls	New Zealand
Parangtopo	Indonesia	F.J. Yang	China
T.H. Phat	Vietnam	L.M. Yang	China
K.K. Phua	Singapore	D.M. Yanga	Philippines

School Secretariat

K. Hata	T. Komatsubara	T. Nagae *(Chief)*
M. Sekimoto	M. Ueda	M. Wada

Host Institutes

Institute for Nuclear Study (INS), University of Tokyo
National Institute for High Energy Physics (KEK)
RIKEN (Institute of Physical and Chemical Research)

School Schedule

March 8 Shimoda

Welcome Address	S. Homma K. Kikuchi
School Orientation	T. Nagae
Lecture	
Trends and Perspectives of Heavy Ion Physics 1	P. Kienle
Welcome Party	

March 9 Shimoda

Lectures	
Non-Perturbative Aspects of Nucleon Structure 1	A.W. Thomas
Current Topics in Nuclear High-Spin Phenomena 1	I. Hamamoto
Nuclei with Strangeness	K. Imai

Talks by Participants

Byung Geel Yu(Korea), Jian-Wei Zhang(China),
Wally Melnitchouk(Australia), Chun Sil Yoon(Korea)

March 10 Shimoda

Lecture	
Trends and Perspectives of Heavy Ion Physics 2	P. Kienle

Talks by Participants

Il-Tong Cheon(Korea), Takumi Muto(Japan), Yoshitaka Itow(Japan)

Excursion

March 11 Shimoda

Lectures	
Astrophysics with Nuclei	J. Arafune
Non-Perturbative Aspects of Nucleon Structure 2	A.W. Thomas
Topics on Space-Time Symmetries	A. Masaike

Talks by Participants

Vo Van Thuan(Vietnam), MD Azizur Rahman(Bangladesh),
Anju Sharma(India), Darmadi Kusno(Indonesia)

March 12 Shimoda

Lectures

Current Topics in Nuclear High-Spin Phenomena 2	I. Hamamoto
Exotic Aspects of Hadronic Atoms	T. Yamazaki
Physics with Unstable Nuclear Beams	I. Tanihata

Talks by Participants

Arup Bandyopadhyay(India), Bernardine Renaldo Wong(Malaysia), Feng Jun(China), Yeongduk Kim(Korea), Yoshinori Nishino(Japan), Atsushi Kokubo(Japan)

March 13 Shimoda

Talks by Participants

Sachiko Takeuchi(Japan), Flordivino de Leon Basco(Philippines), Izumi Kumagai(Japan), Satoshi Nakamura(Japan)

March 15 INS Tour

Welcome Address	S. Homma
INS Tour Orientation	T. Nomura
Guided Tour to	
Cyclotron, TARN II and Electron Synchrotron	
Lectures	
Theoretical Activity	Y. Akaishi
INS Cyclotron General	S. Kubono
TARN II Experiment	T. Tanabe
ES General	K. Maruyama
Kamioka Experiment	T. Kajita (ICRR)
Get-Together Party	

March 16 RIKEN Tour

Welcome Address	M. Ishihara
Introduction and Tour Orientation of RIKEN	Y. Yano
Guided Tour to RIKEN Accelerator Research Facility	
Lectures: *Heavy-Ion Science at RIKEN*	
Nuclear Physics	T. Kobayashi
Atomic Physics	T. Kambara
Nuclear Chemistry	F. Ambe
Solid-State Physics and Other Research Activities	R. Kadono/ E. Yagi
Free Discussion and Get-Together Party	

March 17 KEK Tour

Welcome Address	H. Sugawara
Lectures	
Experiments at KEK-PS	K. Nakai
TRISTAN - Accelerator	K. Sato
TRISTAN - Experiments	M. Yamauchi
Guided Tour to	
TRISTAN - Accelerator + Detectors	
PS - Accelerator + E-Hall + N-Hall	
Lectures	
K-Decay Experiment at KEK-PS	J. Imazato
Nuclear Physics with High-Energy Light-Ion Beams	J. Chiba

March 18 KEK Tour

Lectures	
BSF - Meson + Neutron + Medical	K. Nagamine N. Watanabe
Photon Factory	H. Iwasaki
Guided Tour to	
BSF and Photon Factory	
Lectures	
B-Factory Project	K. Abe
JLC Project	K. Fujii
Mini-Symposium: *Future collaborations*	K. Nakai

March 19 University of Tokyo

Lectures: *INS Present and Future Project*	
JHP General	T. Yamazaki (INS)
Accelerator	Y. Yamazaki (KEK)
K Arena	T. Nagae (INS)
E Arena	M. Wada (INS)
M Arena	Y. Miyake (U.T.MSL)
BNL Heavy Ion Experiment	H. Hamagaki (INS)
HERA Experiment	T. Hasegawa (INS)
Farewell Party	

List of Participants

Ahn, J.K.	Kyoto Univ.	Korea
Akaishi, Y.	INS, Univ. Tokyo	Japan
Akiyoshi, H.	Kyushu Univ.	Japan
Arafune, J.	Univ. Tokyo	Japan
Bandyopadhyay, A.	Variable Energy Cyclotron Centre	India
Basco, F. de L.	Univ. of the Philippines	Philippine
Bazhanskij, I.	Far Eastern State Univ.	Russia
Bratkovskaya, E.	FarEastern State Univ.	Russia
Cheon, IL-T.	Yonsei Univ.	Korea
Choi, G.	Sung Kyun Kwan Univ.	Korea
Chung, M.S.	Kyoto Univ.	Korea
Davidson, P.M.	Australian National Univ.	New Zealand
Guimaraes, V.	INS, Univ. Tokyo	Brazil
Gurjav, G.	Inst. of Physics and Technology	Mongolia
Hamamoto, I.	Univ. of Lund	Sweden
Hata, K.	INS, Univ. Tokyo	Japan
Higashi, A.	INS, Univ. Tokyo	Japan
Higuchi, K.	Japan Soc. for Prom. of Science	Japan
Homma, S.	INS, Univ. Tokyo	Japan
Hu, W.	Fudan Univ.	China
Imai, K.	Kyoto Univ.	Japan
Ishihara, M.	Univ. of Tokyo	Japan
Itow, Y.	Kyoto Univ.	Japan
Itoh, S.	Tohoku Univ.	Japan
Jeong, M.T.	Dongshin Univ.	Korea
Jeong, S.	Soong Sil Univ.	Korea
Jun, F.	Inst. of Nuclear Research	China
Katayama, I.	INS, Univ. Tokyo	Japan
Kienle, P.	T.U. Munich	Germany
Kikuchi, K.	Japan Soc. for Prom. of Science	Japan
Kim, Y.D.	KEK(Nat. Inst. for High Energy Phys.)	Korea
Kim, E.S.	KEK(Nat. Inst. for High Energy Phys.)	Korea
Kokubo, A.	Tohoku Univ.	Japan
Komatsubara, K.	INS, Univ. Tokyo	Japan
Kumagai, I.	Hokkaido Univ.	Japan
Kurokawa, S.	KEK(Nat. Inst. for High Energy Phys.)	Japan
Kusno, D.	Univ. of Indonesia	Indonesia
Lee, S.M.	Univ. Tsukuba	Japan
Liu, G.	RIKEN	China
Mao, Y.	China Inst. of Atomic Energy	China
Masaike, A.	Kyoto Univ.	Japan
Melnitchouk, W.	Univ. Adelaide	Australia
Mizutori, S.	INS, Univ. Tokyo	Japan
Muto, T.	Kyoto Univ.	Japan

Nagae, T.	INS, Univ. Tokyo	Japan
Nagasaka, Y.	Univ. Tsukuba	Japan
Nakai, K.	KEK(Nat. Inst. for High Energy Phys.)	Japan
Nakamura, S.	Univ. Tokyo	Japan
Nishino, Y.	Osaka Univ.	Japan
Oda, M.	RIKEN	Japan
Ogawa, H.	INS, Univ. Tokyo	Japan
Ohkura, H.	Niigata Univ.	Japan
Outa, H.	INS, Univ. Tokyo	Japan
Phan, H.T.	HanoiState Univ.	Vietnam
Pu, Y.H.	RIKEN	China
Rahman, MD A.	Univ. Dhaka	Bangladesh
Ruber, R.	KEK(Nat. Inst. for High Energy Phys.)	Netherlands
Sekimoto, M.	INS, Univ. Tokyo	Japan
Sharma, A.	Univ. Delhi	India
Takayama, K.	Tokyo Inst. Tech.	Japan
Takeuchi, S.	Tokyo Inst. Tech.	Japan
Tanihata, I.	RIKEN	Japan
Teansomprasong, P.	Kine Mongkut's Inst. of Technology	Thailand
Thomas, A.W.	Univ. Adelaide	Australia
Thuan, V.V.	Inst. of Nuclear Sci. and Tech.	Vietnam
Tlusty, P.	INS, Univ. Tokyo	Czech Republic
Ueda, M.	INS, Univ. Tokyo	Japan
Wada, M.	INS, Univ. Tokyo	Japan
Widmann, E.	INS, Univ. Tokyo	Germany
Wong, B.R.	P.J. Community College	Malaysia
Yamada, S.	INS, Univ. Tokyo	Japan
Yamazaki, T.	INS, Univ. Tokyo	Japan
Yazaki, K.	Univ. Tokyo	Japan
Yoon, C.S.	GyeongsangNat. Univ.	Korea
Youn, M.	INS, Univ. Tokyo	Korea
Yu, B.G.	Yonsei Univ.	Korea
Zhang, J.W.	Peking Univ.	China

JSPS-INS International Spring School
March 8-19, 1993